Rethinking Medical Humanities

Medical Traditions

Edited by
Alain Touwaide

Volume 7

Rethinking Medical Humanities

Perspectives from the Arts and the Social Sciences

Edited by
Rinaldo F. Canalis, Massimo Ciavolella
and Valeria Finucci

Postface by Alain Touwaide

DE GRUYTER

ISBN 978-3-11-078800-6
e-ISBN (PDF) 978-3-11-078850-1
e-ISBN (EPUB) 978-3-11-078859-4
ISSN 2567-6938

Library of Congress Control Number: 2022938905

Bibliographic information published by the Deutsche Nationalbibliothek
The Deutsche Nationalbibliothek lists this publication in the Deutsche Nationalbibliografie; detailed bibliographic data are available on the internet at http://dnb.dnb.de.

Cover image: Collage of illustrations in medical treatises from the 1st to the 16th century CE, from Greece and Rome to the Arabic World and China (1. peaches from a Japanese edition from the Li Shi Zhen, Bencao Gangmu, A.D. 1714; 2. plant from the manuscript of Padova, Biblioteca Seminario, 194, Constantinople, ca. A.D. 1430; 3. scene of uroscopy from the manuscript of Paris, Bibliotheque nationale de France, grec, 2294 15th century; 4. a man climbing on a tree from a manuscript of Dioscorides, Arabic, Bologna, Biblioteca universitaria, 2954, A.D. 1244).
Typesetting: Integra Software Services Pvt. Ltd.
Printing and binding: CPI books GmbH, Leck

www.degruyter.com

This volume is dedicated to the memory of Dr. Rinaldo Fernando Canalis, a renowned surgeon, an enthusiastic educator, an impassioned scholar of the history of medicine, and a warm and generous good man. He will be missed.

Acknowledgments

Rethinking Medical Humanities would not have been brought into completion without the generous assistance of Professor Alain Touwaide, whose guidance and professional help permeates every structural element of the volume. Many thanks are also due to Lee Walcott and the Ahmanson Foundation for their continuous support of the academic activities in the field of Medical Humanities at UCLA, particularly those related to the Renaissance and Medieval periods. Lastly, but certainly not last, much gratitude is due to Nina Bjekcovic for her meticulous correlation of the book`s texts and references.

https://doi.org/10.1515/9783110788501-202

Contents

The Humanities in Medical Education

Postface

List of Figures

https://doi.org/10.1515/9783110788501-204

List of Contributors

Roberta Ballestriero teaches at the Academy of Fine Arts, Venice. She is also an Associated Lecturer, Central Saint Martins College, University of the Arts, London, as well as an Art Historian in Residence at the Gordon Museum of Pathology, King's College, London.

Donald Beecher is Chancellor's Professor of English (Carleton University, Ottawa), specializing in the literature and culture of the Renaissance. Among his many publications are those dealing with prose fiction, theatre, medicine, music, and cognitive science as it pertains to literature. He is currently investigating the psychology of revenge in relation to Chettle's *Tragedy of Hoffman* (ca. 1602).

Remo Bodei (†2019) was a renowned Italian philosopher who taught at the University of Pisa, ant Scuola Normale di Pisa, at the University of Pisa and was distinguished Professor at the University of California, Los Angeles.

Rinaldo D. Canalis (†2022) is an Emeritus Professor of Head and Neck Surgery-Otolaryngology in the Department of Head and Neck Surgery-Otolaryngology of the University of California, Los Angeles, and an Associate Faculty Center, of Medieval and Renaissance Studies at UCLA.

Massimo Ciavolella taught at Carleton University (Ottawa) and at the University of Toronto before coming to his present position as Franklin D. Murphy Professor of Italian Renaissance Studies and Comparative Literature at the University of California, Los Angeles.

Valeria Finucci is Professor Emerita of Romance Studies and Theater Studies at Duke University. She works on the early modern period and writes on prose fiction; drama; costume books, literary theory, gender studies, and medical culture.

Mariacarla Gadebusch Bondio is the Director of the Institute for Medical Humanities at UKB University Hospital in Bonn, Germany. She works on medical ethics, prediction and prognosis, blood in history, and esthetic beauty in the early modern period.

Francesco M. Galassi is an Associate Professor – Academic Level D at Flinders University (Adelaide, Australia), and the director of the FAPAB Research Center, Sicily. He also teaches human Anatomy at the University of Bologna, Ravenna Campus, Italy. His studies concentrate on the evolution and history of human diseases, often with a focus on historical figures.

Manuela Gallerani is Professor of Philosophy of Education and Ethics at the Alma Mater Studiorum-University of Bologna. Her studies and research activities focus on ethics, esthetik, gender studies, and medical humanities.

Jennifer Kosmin is an Assistant Professor of History at Bucknell University, PA. Focusing on childbirth, her work explores the history of medicine in early modern Italy by shedding light on how race, sexuality, health and illness inform our ideas about bodies and disease.

Jorge A. Lazareff is an Emeritus Professor of Neurosurgery at the Department of Neurosurgery of the David Geffen School of Medicine, University of California Los Angeles.

https://doi.org/10.1515/9783110788501-205

Pablo Maurette is an Assistant Professor of English at Florida State University. His research focuses on the intersections between literature, science, and the history of ideas in the period between 1400 and 1650 in England, Italy, France, Spain, and the New World.

Stephen Meredith is a Professor in the Departments of Pathology, Neurology, and Biochemistry and Molecular Biology at the University of Chicago, where he also teaches courses on literature, philosophy, and theology.

Ariella Minden is a doctoral candidate at the University of Toronto and is currently a Doctoral Fellow at the Kunsthistorisches Institut in Florenz- Max-Planck-Institut. Her current work focuses on how to evaluate Bologna's role as a center of printmaking in the early modern period.

Sowon Park is an Assistant Professor of English at the University of California, Santa Barbara. She works on British modernism, political fiction, world literature and cognitive neuroscience, and is the creator/convener of the Unconscious Memory Network.

Eduardo H. Rubinstein is an Emeritus Professor of Anesthesiology and Physiology. Departments of Anesthesiology and Physiology at The David Geffen School of Medicine, University of California, Los Angeles.

Paolo Savoia is a Senior Assistant Professor in the Department of Philosophy and Communication Studies at the University of Bologna. He has written on the history of sexuality, reconstructive surgery, pain management, and the relation between food and science in the early modern period.

Giovanni Spani is an Associate Professor of Italian at the College of the Holy Cross in Worcester, MA. His area of interest includes Italian literature of the Middle Ages, Medieval Italian history and historiography, and history of medicine.

Romy Sutherland is Assistant Adjunct Professor of Comparative Literature at UCLA where she teaches courses on film in comparative contexts. She has published several articles including "Commanding Waves: The Films of Peter Weir," an essay for the "Great Directors" series of the journal *Senses of Cinema*. She is currently working on a book on Weir's Australian films.

Elena Varotto is Research Fellow – Academic Level B at Flinders University (Adelaide, Australia), deputy director of the FAPAB Research Center, Sicily, and an occasional lecturer at the University of Catania. Her research focuses on biological and forensic anthropology, paleopathology and bioarchaeology.

Francis C. Wells is a cardiothoracic surgeon specializing in valve replacement and repair at the University of Cambridge, UK. He has been a pioneer in repairing leaking mitral valves and in the surgery of the aorta. His interest in beauty and harmony in the Renaissance has led him to organize exhibits on the heart and to publish *The Heart of Leonardo*.

Introduction: Setting up the Terms

Rinaldo F. Canalis, Massimo Ciavolella and Valeria Finucci

Rethinking Medical Humanities

Medical Humanities may be broadly conceptualized as a discipline wherein medicine and its specialties intersect with those of the humanities and social sciences. As such it is a hybrid area of study where the impact of disease and healing science on culture is assessed and expressed in the particular language of the disciplines concerned with the human experience. However, as much as at first sight this definition appears to be clear, it does not reflect how the interaction of medicine with disciplines in the humanities has evolved to become a separate field of study. In this publication we attempt to explore, through the analysis of a group of multidisciplinary essays, the dynamics of this process.

The idea for this book originated at an international conference held at the UCLA Center for Medieval and Renaissance Studies where the essays presented, unrestricted by historical periods, offered diverse topics reflective of the many ways medicine and its specialties interface with those of the humanities and the social sciences. This volume is based on those essays, but considerably expanded to include specifically solicited works by specialized scholars in disciplines not addressed at the conference. When structuring it, it became apparent that a cut and dry classification of the topics addressed was cumbersome to reach. We elected to avoid a rigid compartmentalization of the book based on individual fields of study and opted instead for a fluid division into five multidisciplinary sections, reflective of the complex interactions of the nineteen included works with medicine.

The Essays

Philosophical and Ethical Queries

The book opens with an ample overview of the field by **Donald Beecher**, "Pondering the Perimeters: Towards a Definition of Medical Humanism." Beecher addresses medical humanities as a component in the study of the intellectual history that presently incorporates broader intersections involving the arts and non-medical writing in which medical ideas and issues find representation. The combined terms also describe the branch of medical training concerned with patient-centered healing, medical ethics, and more compassionate practices. Finding a common definition for such diversity can only be an endeavor in perspectives and in setting up the abstract notion of margins regarding the inclusion and exclusion of materials and topics merely as a way of perceiving core concerns more critically and clearly. Beecher's study begins and ends with open approaches to the areas of investigation associated with the

https://doi.org/10.1515/9783110788501-001

medical humanities, approaches based largely on the cogency of the rationales of contributors concerning the analytical ties between the nature of humanity and the concerns of medical thought and practices. At the center, however, is a position statement regarding the study of medical ideas in the age of humanism, arguably the first and still valid meaning of the term "medical humanism." With the aid of historical hindsight this early modern period of medical humanism is easier to define, yet raises its own particular debates over core and peripheral issues. Much of relevance from that period remains to be discovered, being of pertinence not only to the history of medicine, but to the rise of enduring ideas, and to the birth of the modern clinic.

In that vein, what should doctors do when they become aware that they made mistakes in taking care of patients? Does medical fallibility – either caused by poor knowledge of the matter or by malpractice – require a correction on moral grounds? Do doctors need to be physician-philosophers to prevent mistakes along the lines suggested by Hippocrates and Galen? The question of moral responsibility is addressed by **Mariacarla Gadebush Bondio** in "Avoidable Mistakes: Premodern Medical Fallibility as an Ethical Problem with Epistemological Implications." In this essay, she shows that programs in logic that allow for self-criticism and the techniques to control passions can be useful guides for doctors in better diagnosing patients. A key text in her essay is *De cautelis medicorum* by Gabriele Zerbi, which argues for a program of *cautelae* as a device to avoid mistakes. This pattern is then echoed in the deontological works of Alessandro Benedetti, Giovanni Filippo Ingrassia, Giovanni Antonio Sicco, Leonardo Botallo and Roderigo de Castro. They all argue in different ways that no certitude can be attained in medicine and therefore to avoid mistakes a conscientious physician should study medical books, should be careful in prescribing medications, should develop practical skills which are especially important in surgery, and should avoid being deceived by the patient's self-analyses or by easy diagnoses. In short, the physician should be neither ignorant nor arrogant, but rather a *bonus medicus* or a *medicus prudens*, a recommendation that applies just as well to today's medical professionals as it did in the early modern period.

Examining the cultural intersections that inform medical practice, ethics and teaching in her essay, "When the Fetus Becomes a Child: Reflections from the Long Eighteenth Century," **Jennifer Kosmin** brings into focus the moral, social and political debates that define fetal life from the early modern period to the Eighteenth century. The concepts of fetal life and viability changed through history and these changes had often less to do with medical or theological insights into what makes a fetus a living child than with the social and cultural anxieties that the various periods reflected. A case in point is that of a baby girl born by cesarean section to Angiola Chiozzi of Casalmaggiore, Italy, in 1812. Angela died during childbirth and the baby too appeared to be born dead. In the legal case brought to court by Angiola's husband (she died intestate) the lawyers argued that the baby girl had been born alive because she had moved, ever so slightly. This made her a legitimate successor and the husband therefore had the legal right to Angiola's property. Not so, argued Angiola's relatives: for them the baby was dead at birth, therefore having Angela died childless, her inheritance was to be returned to her family. The legal issue was clear: what signs make a fetus vital? Does life begin at conception, as preformationists believed, in which case everything should be tried to save it even if this means that the mother's life acquires a secondary importance, or does a fetus need to be viable too, a future member of the body politic, as it was argued in the Eighteenth century? The questions were riveting then as much as they are today, i.e., current debates over abortion rights, assisted reproductive technologies, and end of life care.

Health and illness are key human preoccupations and therefore a culture of cure and prevention should be foundational in our society, **Manuela Gallerani** argues in her essay, "Ethical Aspects of Responsibility in the Curing/Caring Relationship." Such a culture involves active listening to the patients as well as establishing a relationship that understands their needs and emotions to be effective. Cure/care, she writes, require both *cura sui* (take care of yourself in its Latin meaning) and taking care of a relationship. The process of cure/care can only start when there is mutual trust between patient and caregiver, one established through both verbal and paraverbal communication; diagnoses should be personalized, she adds, along cultural and emotional lines so the patients understand them. The move from "a predominantly *biomedical* model to a *biopsychosocial* approach, where the *person-patient* and that person's pathology/illness are studied" is paramount in pursuing appropriate therapies and in avoiding misunderstandings and anxieties. Only when patients are considered subjects, and their personal experiences and cultural conditioning are understood as bearing on the mental and social depictions of their illness can an effective cure be provided. This shift from a disease-centered medicine to a patient-centered medicine involves gentle listening, that is, understanding the patient's inner world, and also active listening, that is, non-judgmental understanding of a patient's reality. In the second part of her essay Gallerani illustrates the curing and caring relationship through examples from contemporary writers and poets whose words deliver cogent, deeply-felt, often painful universal messages.

Ailing or Hale: The Body in the Arts and Literature

Disease is usually understood as suffering, but what if instead it is spiritualizing, or desired, or even natural in the sense that we are all destined to die? This is the view of disease that **Stephen Meredith** examines in his essay, "Disease and the Problem of Evil in the Novels of Thomas Mann." Mann was obsessed with infirmities and had a morbid imagination. His characters are mentally ill, physically obese, and neurologically challenged; they have phobias and are entrapped in environments where people die of cholera, tuberculosis, rickets, or typhoid fever; they have horrible migraines, cancer, syphilis, septicemia and meningitis. In short, their bodies are "unnatural," grotesque, for their physical disability is the visual manifestation of a profound ailment of the soul. Mann describes how the progression into the abyss unfolds and how the unconscious comes to the fore as man falls away from God's love, so that at the end the reader is confronted with the inner rottenness, the moral irresponsibility, the childish protestations and the inevitable deterioration from grace of all the main characters. Meredith cogently examines this mode of dissolution, this near Manichean rush to death through an impressive tour-de-force review of some of Mann's short stories and of his major novels, *The Buddenbrooks*, *The Magic Mountain,* and *Dr. Faustus*.

Choosing a seemingly unconventional approach in thinking through medical humanities, **Pablo Maurette** has Georges Canguilhem's classic "Monstrosity and the Monstrous" (1952) bear on the pioneer work of Fortunio Liceti's *On the Causes and Nature of Monsters* (1616a). In his essay, "Monstrosity and the Monstrous Revisited: Fortunio Liceti's Medical Imagination," Maurette argues that for the philosopher and physician Canguilhem the difference between normalcy and pathology does not exist at the biological level; for him disease is not a divergence from the normal, because an anomaly does not mean an abnormality as there is no stasis in the flow of life. The monstrous therefore is just a juridical term, what violates religion or morals, or is an often-terrifying figure of the imagination; monstrosity, on the other hand, refers to individuals with genetic deviations. Turning to Fortunio Liceti, we know that his work appropriately reflects a common early modern view of monstrosity: a monstrous newborn can visualize parental sin, it was thought, can function as a sign from above, can be the horrid fruit of maternal hyper-imagination or a wonder that nature creates (*lusus naturae*). But Maurette contends that for Liceti monstrosity also happens in the natural world, thus coupling a pre-modern understanding of the body as influenced by imagination with a new attention to proto-scientific hypotheses. In short, Liceti can believe in secrets and myths but can also diagnose like a doctor. Thus, in Maurette's reading he is "a Janus-like creature with two faces, one looking at the past, the other glancing into the future," a scientist interested in serious pathologies and a philosopher/artist who believes in deformities coming from dreams, from myth, from book learning, from visual illustrations and from poetry – scientific investigation approached in a ludic, culture-enhanced way.

In her essay "The Flesh of Wax: The Use of Scientific Collection in Medical Humanities," **Roberta Ballestriero,** who works on the history of wax modelling in normal and pathological anatomy, discusses the history of this fascinating art form, from its inception as a technique to make jewelry in ancient times to the Italian Renaissance, when it was first used in the votive industry for the creation of body parts, ill or miraculously saved by holy intervention. Later wax modeling was used to teach anatomy with tridimensional models in order to overcome the chronic scarcity of unpreserved and rapidly decaying cadavers available for dissection. Beyond its usefulness in the study of normal structures, Ballestriero points out that, in the eighteen hundreds, wax models and moulages became an excellent method for the reproduction of the pathological changes found in many illnesses. The value of these models continues in our own days as a vehicle to show medical students the clinical findings of diseases now nearly extinguished and in some instances, particularly in dermatology, they give them the opportunity to leisurely examine the results of disease, especially advanced disease, without causing undue distress to a patient. The essay contains eight illustrations that highlight the interaction of medicine, wax modelling and teaching.

The renowned pre-Vesalian surgeon Berengario da Carpi (1460?–1530), the son of a barber-surgeon who became a famous anatomist, stands "at the crossroads of the history of medicine, the history of art, and a visual and material history of death and religious ritual." In their essay, "The Body Between Life and Death: Berengario da Carpi and the Anatomical Imagery of the Sixteenth Century," **Paolo Savoia** and **Ariella Minden** trace first the unusual path that led Berengario to become a professor at the University of Bologna and one of the most sought-after medical experts of his time, physician to the pope and to the Medici, expert in cranial fractures as well as syphilis. Savoia and Minden then concentrate on how, as a pioneer of the new anatomy, Berengario not only disseminated his knowledge through commentaries and manuals, especially through the very successful *Isagoge Breves,* but also visualized his discoveries through a most innovative set of woodcuts of life-like figures. In the *écorchés* revealing their own internal parts, for example, we can see how his representation of corpses for strictly medical purposes was influenced by the ways in which the cadaver – be it that of a saint or a criminal, the revered or the reviled – was handled at that time in Italy. Anatomical illustrations for Berengario – and all the anatomists after him that made use of the new medium of print to teach students about body parts – can "speak" to us even today because they show that the human body should be understood as fluid, that is, a body informed by legal, cultural, artistic and religious norms.

In "The Anatomy Lesson of Dr. Jan Deijman and the Social History of the Brain," **Jorge Lazareff** presents a historical analysis of Rembrandt's second and less known "Anatomy Lesson," painted in 1656. Originally, the painting portrayed Jan Deijman, Amsterdam's *Praelector Chirugie et Anatomie,* dissecting the brain. The scene is incomplete as a fire in 1723 eliminated the likeness of six attendees as well as Deijman's. Doctor Gijsbert Calkoen's image survived incineration, however, and he is shown as about to begin dissection of the brain. This structure is, probably for the first time, depicted in color, albeit the wrong one: red instead of light gray. It is far from perfect, but it does show the two hemispheres, a suggestion of sulcus and gyrus and the dura split, and reflected laterally in two equal halves as customarily done at the time, as shown by van Calcar in Vesalius' *Fabrica.* Lazareff goes on to note and describe the dissected body focusing on the feet to show the erosion and damage caused by years of habitually walking without shoes. The individual whose life history of crime destined him to the dissecting table was a poor man and Lazareff gives a concise, but essentially complete history of the deleterious effects of poverty and childhood malnutrition on the structural elements of the brain and the resulting functional deficiencies.

Secrets of the Dead

In her essay: "The Bio-turn in History Writing: Death, Last Wishes and Lasting Wishes," **Valeria Finucci** enters the debate between cultural studies and the medical arena through narrative histories of death and burial practices. The obsession with the remains of famous people can teach us something about the way we respond to death, Finucci surmises, since one's biological demise does not necessarily mean the end of one's story. In fact, the afterlife of the corpse often unduly engages the imagination of the living, who may choose to discard for personal reasons a dying person's directive on how to care and dispose of his/her own body. Taking as examples, among others, the cases of Gabriele Falloppio, Francesco Petrarca, Leonardo da Vinci, Galileo Galilei, Gaspare Tagliacozzi, Grazia Deledda, and Gaspare Pacchierotti whose bodies were, for a variety of reasons, exhumed years, decades or even centuries after their passing, the author shows how the living, time and again, "re-wrote" upon the dead their own fears, desires, religious aspirations, political and gender preferences, moral struggles, philosophical beliefs, artistic desires, and academic conflicts. Moving to the present, health care agents too, she argues, would benefit from understanding how much deliberations on end-of-life choices are inflected by cultural, ethical, even environmental drives.

An anthropological and historical approach, based on texts selected from the chroniclers of the Conquest of the Incas and modern knowledge of the diseases afflicting the peoples of the Andean regions in ancient times, was used by **Rinaldo Canalis** in "The Fatal Disease of the Last Reigning Inca" to develop a differential diagnosis of the probable causes of the death of Huayna Capac Inca. These causes narrowed to three infectious processes characterized by severe cutaneous manifestations: epidemic typhus, Peruvian wart, or Carrion's disease and syphilis. Although many previous authors had attributed Huayna Capac's demise to smallpox, a readily available parallel with the collapse of the Aztec empire, no evidence was found of an epidemic disease and widespread mortality among the empire's inhabitants, a fact that also excluded typhus. The study predominantly focused on Juan de Betanzos' chronicle, the one providing the more detailed clinical history of the Inca's death and probably the more reliable among the better documented accounts. Notably, most previous authors looking into the mysterious death had no access to de Betanzos' complete account discovered only in 1985, but only to versions that did not include an account of the Inca's life. Of the two other infections discussed in the essay, Carrion's disease was considered the less likely because Huayna Capac was born and lived in an endemic area and probably had a grade of acquired immunity that would have prevented the lethal, hemolytic phase of this disease. Furthermore, a closer analysis of the record strongly suggested that his malady was less acute than previously considered. Thus, syphilis probably in the third stage (neurosyphilis) was thought to fit better as a diagnosis, strenghtened by the fact that, according to de Betanzos' and several other accounts, his newborn son had died of the same disease as his father. Syphilis is the only congenital disease among the ones discussed and it remains lethal, even in our own days. As a corollary, the study negates the often-accepted theory that an epidemic, rather that the brutality of the early years of the conquest, was a leading factor in the massive depopulation of the empire.

Over the past sixty to seventy years our knowledge of diseases afflicting humans in ancient times has been greatly enhanced by numerous paleopathological studies. Earlier focused on the more readily available osseous remains, more recently these studies and our knowledge of diseases in the past have been enhanced by the study of naturally and artificially mummified specimens and by pathologies described in classical literature and incidentally discovered in paintings and other art works. In "Paleopathology and Anthropology of the Renaissance: From the *Morbus Dominorum* to the Alleged 'Michelangelo's Shoes,'" **Francesco Galassi, Giovanni Spani** and **Elena Varotto** discuss the state of the art of these approaches with emphasis on pathologies found on historical figures of the Italian Renaissance. Several examples are presented to illustrate the use of modern medical imaging and photographical techniques along with laboratory results to discover, confirm or disprove traditionally accepted historical information. Of particular interest in the essay are the life histories of two Renaissance personalities as they relate to their clinically analyzed data and a detailed anthropological study of a pair of shoes and a slipper alleged to be Michelangelo Buonarroti's.

Reason, Affects and Madness

Remo Bodei's essay, "The New World Opened by Madness: Philosophy and Psychiatry," as Manuela Gallerani explains in an eloquent Afterword, "offers a lucid reflection on the complex and unresolved interweaving of reason, affects, and madness." Psychic distress is understood as an expression of the disconnection between internal and external worlds (which gives rise to so-called deviant behaviour in relation to *norm* and *rationality*); it is therefore, re-contextualised, historicised, and interpreted with subtlety and depth by Bodei. In fact, Bodei's inquiry focuses on the relationship between rationality and truth and the link between reality and temporality, seen as interpretative and foundational categories of our society that may nonetheless be perceived and experienced in *another* and *different* way by schizophrenic or psychotic individuals. Bodei proposes that we accept these 'apparently absurd experiences' and 'abnormal logics of madness' or 'alternative interpretations,' with a greater willingness to listen, as they reveal *other* possible worlds, but only to those able to surpass the rigid dictates of standardised rationality, which tends for align us to a common reality rather than respecting diversity. Nevertheless, every society creates and reproduces an orthodoxy of reality through the power of institutions (political, economic, or religious), culture, language, and traditions. A *hospitable reason* must recognise that there are thresholds that distinguish one world (one experience) from another and every identity among different forms of identity. In the final analysis, to better understand phenomena linked to madness it is necessary to learn to understand, and attribute new meaning to the term, rather than labeling it – as western cultures often hasten to – as an irrational world. In adopting a new, authentic, and original *vie sauvage*, one recognises that madness reveals the "latent fragility of everyone's experience."

Sowon Park's essay "Why Listen to the Mad? What *Schizophrenic Girl* offers to Narrative Medicine," stems from the premise that traditionally most psychiatrists would consider an error to attribute any meaning to the speech of the mad, thus placing an unsurmountable wedge between doctor and patient. The aim of Park's essay is to disprove this engrained belief, to make patient narratives more relevant to bio-medical discourses, as part of an emerging interdisciplinary movement called "Narrative Medicine," that brings powerful life stories from the humanities and the arts to those who seek and deliver healthcare. The focal point of Park's discussion is the analysis of the *Autobiography of a Schizophrenic Girl*, the chronic of a young woman's descent into schizophrenia and her difficult return to reality. In analyzing Renee's self-description–that is the pseudonym used in the memoirs – the paper discusses 1) why self-presentation is important for the psychotic; 2) how patient narrative can be integral part of the wider design of clinical and socio-political assessment, and 3) what the implications of this case-study are for translational medicine and narrative medicine." In the final analysis, just as Remo Bodei tells us in this volume's essay, to better comprehend madness-linked phenomena we must learn listen to the patient's story, and try to understand why certain behaviors are enacted.

Between the years 1275 and 1280, the renowned Catalan physician and natural philosopher Arnaldus of Villanova wrote what can be considered the first treatise on the subject of love as a disease, *Tractatus de amore heroico*. Although the doctrine of erotic melancholy was already part and parcel of the culture of the Latin West, Arnaldus' treatise frames the analysis of unrequited love in a scientific discourse that moves from the etiology of the disease to its possible treatment and cures. **Massimo Ciavolella's** essay, "The Malady of Love in Early Modern Medical Thought," discusses the pathogenesis of the disease and its impact on the culture of the age, pointing out that early modern medicine considered every form of love as a disruption of the humors of the body, as a shift in the material condition of the individual. In other words, love, whether happy or unrequited, was thought to determine a change in the body, both at the psychological and physiological levels, considered inseparable. The essay discusses the physiological mechanism that causes the image of a perceived person to move from the external senses to those located in the ventricles of the brain and why this *phantasma* (its technical term in the language of natural philosophy and medicine) can become an obsessive image imprinted in the lover's mind thereby causing a breakdown of psychic and physical functions that can lead to severe disease and even death. Ciavolella's essay points out that as lovesickness became part of the European Universities' medical *cursus studiorum* and the subject of medical dissertations, it begun emphasizing the pathological side of excessive erotic desire, linking it to madness and erotomania.

The Humanities in Medical Education

The value of the plastic arts, particularly painting, in the training of physicians in a modern setting is addressed in "Art Images and Medical Teaching" by **Eduardo Rubinstein.** The author reviews the current results of the experimental approach undertaken by several American and European medical schools aiming at enhancing the students' compassionate attitude in their relationship with patients through a self-analysis of their emotional response when exposed to art appreciation classes along with an attempt to improve their ability to analyze medical images. Rubinstein further explores the latter by advancing a hypothesis that art images simultaneously presented along medical figures enhance the students' ability to learn how to "see" and how to differentiate normal from abnormal findings. The hypothesis is supported by a detailed review of the physiology of vision at the eye and cortical levels and the presentation of numerous examples of pairing artistic and medical images with common features.

The importance of the arts and of the humanities in general in the professional formation of physicians is further explored by **Francis Wells** in a very personal communication. In his essay "The Humanities and Medicine: A Tautology or a Necessity?" he contrasts his longer than four decades experience as a heart surgeon with today's challenges to practice a humanistic profession, amid an increasingly detached patient-physician relationship, an ever more litigious society and the health systems' financial difficulties to pay for care. Wells gives special attention to the challenges of teaching students to be compassionate physicians and expect them to achieve top academic success while burdened with an ever-growing technical curriculum. He asks whether exposure to the humanities may be a positive force in the education of present-day medical students and whether altruism and caring can be taught and ultimately be of benefit to the society at large. These questions, contrasted with current experiences, are answered through an insightful analysis of the historical role that the humanities have played in education, particularly during the introduction of the classics in the early modern era and the Enlightenment. The author concludes that the humanities should have a significant place in contemporary medical education through the exposure to works of great artists, musicians and thinkers guided by dedicated teachers, conscious of the limited time that learning the science leaves for these additional endeavors.

In her essay, "Teaching PTSD with Film: The Case of Peter Weir's *Fearless*," **Romy Sutherland** focuses on how to incorporate the study of a feature film that deals with Post Traumatic Stress Disorder into "Mind over Matter: The History, Science and Philosophy of the Brain," a year-long, team taught, multidisciplinary Health Humanities course at UCLA. In the course, she offers a series of lectures on filmic portrayals of mental illness, and includes *Fearless* after her students have studied PTSD from neurological, psychiatric, and historical perspectives. The film portrays two plane crash survivors who appear to have little in common and suffer dramatically different forms of the disorder, and yet play a pivotal role in each other's healing. She outlines here her interpretation of the film intended to show students that their training in film analysis combined with their medical understanding of PTSD can lead them to uncover crucial keys to the narrative, keep deliberately concealed beneath the surface of the film, but which nevertheless can be rewarding for an attentive viewer to discover. In this article she also outlines the content of a seminar she offers for twenty students in the final term of the year, which explores current research into what takes place in the brain while we watch film. The seminar includes a Film Lab in which the students produce a short film about a neuroscience topic in order to deepen their understanding of the topic and to sharpen their appreciation of the techniques directors use to elicit specific responses from an audience.

The book last essay is **Jorge Lazareff's** "The Humanities and Global Health: Travels with Philippa Foot and Karl Popper." The author presents his experience developing a program, based on the principles of analytical philosophy, aimed at improving the clinical thinking of medical students in Central American schools. It is part of the UCLA-Global Health courses, currently offered as a minor in International Studies and constituted by free lectures at the UNAN-Managua, Nicaragua auditorium delivered in Spanish from the Lockheed room at UCLA Medical Center. The course, titled "Critical Thinking," includes students at different career levels. Lazareff discusses its structure as a branch of analytical philosophy and as such oriented to the development of logical techniques with the goal of attaining conceptual clarity. To this effect the students are introduced to Karl Popper's and Philippa Foot's writings. Throughout the course the students are invited to develop their own medical ethics booklet, keeping at the forefront the special circumstances of a region lacking many state-of-the-art techniques and legal conditions quite different than those encountered in the developed countries. Lazareff concludes that introducing the concept that logic and epistemology applied to healthcare ethics provides the young practitioner with the methods for searching evidence and determining treatment as a whole and not as an ethical alternative. As a measure of the program's success, when the UNAN Universidad Nacional Autonoma de Nicaragua created a three-year long Master/Doctor degree in Science in 2018, the critical thinking/analytical philosophy lecture series was included in the program.

Concluding Comments

An analysis of the essays that make up this volume reveals a large number of complex interdisciplinary contacts between the humanities, the social sciences and medicine. Although in any given essay the representative discipline can be readily identified, it is nearly always associated with others of equal or near equal importance, such as art, history, and education (Ballestriero, Lazareff, Rubinstein, Sutherland); history, art and anthropology (Finucci, Canalis, Galassi, Savoia/Minden); ethics, psychology and philosophy (Bodei, Kosmin, Gallerani, Gadbush-Bondio); literature, history and psychology (Maurette, Meredith, Park, Ciavolella); education, philosophy and psychology (Lazareff, Bodei, Park), to highlight the more evident. These complex interactions are at the core of medical humanities and the principal reason why an all-inclusive definition of the field escapes a simple approach. Furthermore, this plurality is complicated by a diverging view in academic medicine that limits the field to traditional courses in ethical practice and by the different outlook of the disciplines that comprise the humanities that, *by themselves*, are often unyielding to a simple classification. Of lesser importance, but still playing a complicating role in defining the field, common usage has enfranchised a confusing terminology. Among the more common examples, medical art more frequently refers to illustrations in anatomical textbooks than to dramatic depictions of, say, the plague, and medical literature is generally used to define publications in scientific journals and specialty books.

Regardless of the outlined difficulties to find a clear understanding of what the field entails, the nexus between medicine and the humanities is natural and has been so from their very inception as human endeavors, although evolving separately. In present days several factors have contributed to their re-approaching under the Medical Humanities umbrella, but two appear as the more important. The first is a response to the realization that academic disciplines are a reflection of the society from which they stem and encompass the knowledge of their age, including medical knowledge. The second is the growing challenge for medical schools to graduate physicians with a humanistic sense in a practice teaching environment overtaken by mechanization and economic pressure. This conjecture finds footing on the fact that most academic activities in this evolving field are generated, as in the essays here included, by centers and professionals in the humanities and by a growing interest to make courses in literature, the plastic arts and philosophy available to medical students. The latter is a demanding experiment, given the heavily packed medical school curriculum, but it appears to consistently bear fruit when used to enhance the participants' ability to recognize pathologies in medical imaging and, when properly guided, to direct their emotional responses towards a compassionate doctor-patient relationship.[1] It bears remarking that four of the essays presented directly address these issues (Rubinstein, Wells, Sutherland, Lazareff), while several others (Ballestriero, Galassi, Gallerani, Lazareff, Gadebush-Bondio), include them as an important component of their discussions.

In closing, it can be confidently stated that Medical Humanities is the product of a relatively recent movement aimed at promoting and strengthening the natural ties between the humanities and the healing sciences. The result is a unified field of study characterized not by the interface of a given discipline in the humanities with the healing sciences, but by the *simultaneous intertwining* of several with medicine and its specialties.

1 Naghshineh et al. 2008. See also Mann 2017.

Philosophy and Ethical Queries

Donald Beecher

Pondering the Perimeters: Towards a Definition of Medical Humanism

Defining the term "medical humanism" is largely a matter of compartmentalizing the diversity of materials and disciplinary orientations already actively pursued under this generic designation. It is rather late, in any case, to be prescriptive in the matter, although I might pretend to be so for the sake of debate. Beginnings are always possible from a semantic perspective, as in thinking critically about the combinatorial relationships implicit in the terms 'medical' and 'humanism.' For the historian of ideas, the twinned terms designate the medical philosophies developed and debated between 1400 and 1650. But these terms also pertain to medicine and the arts in all ages, as well as to studies in the ethical practice of medicine, with each domain posing questions around its respective criteria of inclusion and exclusion where topics and materials are concerned. The exercise brings to mind the sometimes-exasperating debates over the working term 'Renaissance,' which, after enjoying a long century as the name given to the historical and cultural period following the Middle Ages, was found to be connotatively wanting and thereafter largely replaced by the pedestrian but more comprehensive term, 'early modern.' The debate arose because too many aspects of that chronologically defined period were not part of a 'rebirth' based essentially on the same humanist values which are implicit in the term under investigation here. Once again, the problem is that 'humanism' may refer both to an historical period associated with the 'Renaissance,' and with the human-centered values first prioritized under that sobriquet which pertain to the practice of medicine in all times and places. The discussion to follow is bound to retrace some of those discussions, given the diversity of distinctions and characteristics attributed to the term. Essentially, meanings have moved on from the specific moral and educational program called 'humanist' which arose in fourteenth-century Italy. Yet that humanist era deserves our continuing investigation for the values and practices it foregrounded, values which remain of interest in our own times.

Despite these historical considerations, we might still hope from the outset that a sensible and cautious definition might garner some degree of consensus. But that haste must be made slowly. Of the twin terms 'medical' and 'humanism,' medical appears the less challenging, perhaps, insofar as it pertains to diseases, including their origins, modes of transmission, manifestations in the body, and their cures. But what of psychological well-being, patient care, disease prevention, the ethics of the profession, the financing of medical research, hospital administration, the salaries of ambulance drivers? There we can stop, because clearly, we are heading toward the disputably marginal, which is where exercises in definition invariably end

https://doi.org/10.1515/9783110788501-002

up. As for humanism, the term pertains to an educational program during the Renaissance centered in rhetoric, ancient texts, the virtues, the pre-eminence of humanity in all things, and trained civility. Already we might be in for debate, well before the term is extended to encompass all productions of the human spirit, all of the arts arising from creative human endeavors, or all philosophical speculations about the condition of man, as well as the virtues of inclusiveness, compassion, and respect for human dignity in the practice of medicine. Narrowly conceived, as stated above, medical humanism implies the study of early modern medical philosophy, with an emphasis on the representation of medical ideas in the arts or public fora. More broadly, it may incorporate all historical periods and concentrate more particularly on medical ethics and compassionate, patient-oriented care, driven rather more by pragmatic application than by knowledge largely for its own sake (although our plastic and pragmatically oriented minds have a remarkable capacity to draw wisdom and perspective from the most esoteric of investigations. It is a large part of how the learning brain works. Moreover, the juxtaposing of these terms means that each is qualified in the spirit or substance of the other, thereby creating a combinatory challenge of increasing complexity: humanism which is medically informed; medicine which is humanist oriented. Ultimate definitions will settle in the minds of individuals largely in accordance with their respective inclinations to think historically and philosophically, or to think broadly, ethically, and pragmatically. Either way, there will be peripheral cases and issues which will challenge all core definitions.

To begin, the term 'medical humanism' has already been claimed by medical schools across North America. It pertains to training programs in medical ethics, empathy for patients, and related 'biopsychosocial' aspects of patient care. In the words of Steven Lange, 'medical humanism and narrative medicine are the popular phrases *du jour* which buzz in the modern medical classroom.'[1] Narrative medicine has to do with patients and their stories and the means whereby they might be involved, or at least made to feel involved, in their own diagnostics and cures. This emerging discipline looks for philosophical grounding in the humanism that persevered in the nineteenth century as an outgrowth of the German *Aufklärung*, as in the writings of Johann Gottfried Herder. Humanism, for him, was not only achieving the best of human potential through the perfection of reason, but an attitude of fairness and of respect, bringing all right-thinking persons in many walks of life to collaborate and share their reflections, whether teachers, poets, philosophers or legislators. Lange also turns to those writings of Jean-Paul Sartre in which he tilts his thinking in the direction of care in a humano-centric world. Scaled down to a practicum, it is about compassion in the doctor-patient relationship, an attitude which is both implicit and expressed in the modernized versions of the Hippocratic Oath. As early as 1948, the concern for ethical

1 Lange 2016: 2.

guidelines for the world's physicians was addressed by the World Medical Association, resulting in the Declaration of Geneva which urges the 'utmost respect for human life.'[2] One modern version of the oath in the tradition of Hippocrates goes so far as to say that 'warmth, sympathy, and understanding may outweigh the surgeon's knife or the chemist's drug.'[3] This may strike some as pointlessly sentimental where the science of healing is concerned, and there has been resistance from the perspectives of those invested in the hard science of medical technologies and skills. Their views constitute the principal opposition to the prioritizing of the term as the more progressive orientation for the study of modern medicine. Patients are not the most reliable authorities in diagnostics, making the accommodation of their views more a matter of diplomacy than collaborative healing. This is but an internal tiff, however, regarding the mission of medical humanism in the modern world in relation to those practitioners invested in the scientific technologies of medicine addressing the body as a complex machine.

Nevertheless, ethical issues remain a principal common denominator linking the modern clinic to the *studia humanitatis* of yore, through which the intrinsic values of our humanity were bright to a higher consciousness. The particulars differ considerably from age to age, but always the practice of medicine has subsumed humanitarian values as well as codes of honorable and professional conduct. Anecdotally to this point, Jacques Ferrand, in his treatise on the diagnostics and cures of lovesickness (first published in 1623), arrived at the difficult matter of cures by coitus, raising for him a crux in early modern clinical practices. The act, according to the Arabic physicians, was a sovereign remedy for erotomania, and Ferrand, according to his scientific understanding, was in complete agreement, so long as it was performed 'in ways permitted by religion and law.' To underscore his point as a Christian physician, he states that it is 'totally absurd and immoral to prescribe, as Avicenna and Haly Abbas do, that lovers be purchased and regularly exchanged by way of a cure.'[4] But to pursue the question a bit further, does the same crux still constitute a point of medical ethics when Shakespeare and Fletcher take up the coitus cure in *The Two Noble Kinsmen*? In that context, the patient is likewise destined to die for want of a marriage to the person inciting the infatuation – one made mortal by the degree of the patient's pathological desire and the surfeit of seed.[5] Ferrand would have replaced coitus with an elaborate regimen of cures, mostly pharmaceutical,

2 WMA Declaration of Geneva.

3 Marks 2021.

4 Ferrand 1990: Ch. 33, p. 334.

5 Shakespeare, *The Two Noble Kinsmen*, passim (ed. G.R. Proudfoot, 1970). Be it recalled that Shakespeare was taken by the idea of solving social problems by resorting to the coupling arrangements constituting 'the bed trick,' the employment of which brought both *All's Well the Ends Well* and *Measure for Measure* to dubiously happy conclusions. In both instances, however, it was betrothed husbands who were tricked, after both had refused to consummate their unwonted marriages. Shakespeare, for the sake of the women, brought them into binding matrimony under duress of the law.

to eliminate the offending humoral imbalance. The dramatist's solution, however, is pure comic theatre, yet most effective according to the principles of Avicenna. Send in another boy who is in love with the girl after first disguising him as the man she adores. As the doctor in the play states, the act 'cures her, *ipso facto* / The melancholy humor that infects her.'[6] Thus, when the girl's father demurs in the name of honesty, he is told by the doctor, 'That's but a niceness; / Ne'er cast your child away for honesty.'[7] In the play, the physician puts life ahead of ethical niceties and thereby gains a medical cure through social manipulation, even as he serves as an agent of comic plotting. We smile, but the anecdote maintains the dilemma, even as it passes to the theatre where literary representations and medical practices join forces to explore the experiential side of mental diseases. Similarly unstable relationships between the urgency of the medical and the social contexts of patients will reappear in many of the more recent stories appearing at the end of this study. In a word, imaginative literature has its value in illustrating the liminal questions concerning medical practices.

As for our own times, ethical questions will only become more acute in theory and in practice, both in the clinics and in speculative fiction, as self-improvement and self-empowerment, longevity, and mental acuity are sought increasingly through elective medical procedures – procedures made possible by the often-controversial advancements already under development in such areas as surgical implants, organ transplants, edited genetics, and faculty-enhancing pharmacology. The definition of medical humanism will require constant tweaking and adjustment to keep up with the philosophical implications of human 'deification' through a medically oriented, post-religious quest for the next best forms of immortality. It has been said of our present era that 'our greatest medical achievements were the provision of mass-hygiene facilities, the campaigns of mass vaccinations and the eradication of mass epidemics.' But if Yuval Harari's supposition is right, 'the age of the masses may be over, and with it the age of mass medicine.'[8] As the elite seek their expensive quests for perfected states of health in a new economy based, not on the labour of the masses, but on technology and artificial intelligence, health care for the millions may lose all relevance except in purely ethical terms. Should that prospect prove true to any degree, medical humanism is destined to become one of the salient intellectual battlegrounds of the future as the rights to patient care are marginalized by a commodification of humanity through a supply and demand approach to human worth and health. In such a new world, the subtleties of the doctor-patient relationship will become a sentimental byway as universal healthcare systems slip into irrelevance and decay.

---

6 Shakespeare, *The Two Noble Kinsmen*, V.ii.36–37 (ed. G.R. Proudfoot, 1970: 113).
7 Shakespeare, *The Two Noble Kinsmen*, V.ii.30–31 (ed. G.R. Proudfoot, 1970: 113).
8 Harari 2017: 354.

From the perspective of a Renaissance scholar, explicitly and historically, medical humanism takes us back to the early modern centuries when humanism was still in vogue as a literary, philosophical, and educational system. As this intellectual and pedagogical context defines the term, it applies specifically to the medical treatises in that period, their modes of exposition and investigation, and the dissemination of their ideas throughout the general culture – modes which were gradually and systematically displaced throughout the seventeenth century by a gradual revolution in scientific goals and methodologies.[9] To name the period is not to suppose a monolithic approach to learning, for their treatises abounded in debates about propriety, relevance, and inclusion. They too had their many controversies concerning issues at the margins; whether, for example, demons were active in pathological capacities and could use diseases to imperil the soul. Those debates increased exponentially as medical ideas and practices impinged upon the interests of the era's theologians, philosophers, dramatists, botanists, historians, and even composers, or inversely, their ideas held implications for the practice of medicine. So once again to the margins. If the German *Empfindsamkeit* movement in music is a theory-based exploration of the relationship between musical styles and the moving of the passions, it is an implicit extension of the investigation of the ancient Greek modes so central to musical humanism, as in the work of Pietro d'Abano (1250–1315) and Giorgio Valla (ca. 1480).[10] For them, not only is music a motor behind the forms of creation, but in a fully medical sense, it is an occult power that can raise and quell all the passions of the mind. As represented by arguments in the *Medicus-politicus* (1614) of Rodrigo de Castro, music might also be adapted and employed as a therapeutic tool in the treatment of disordered affections through its mood-altering powers. Hence, the *Affektenlehre*, grounded in Greek antiquity and Descartes' *Passions de l'ame* (1649), and defined by such musical theorists as Kircher, Mattheson, and Marpurg, might claim a rightful place in the humanist clinic.[11] The point was under debate. But the fact remained that through analysis based on a theory of correspondences linking the natural and spiritual worlds, music becomes causally linked to the temperaments which in turn control the human passions. Just such analogical associations allowed Robert Burton, in *The Anatomy of Melancholy* (the first of its seven editions published in 1621), to include nearly every quality and memorable exploit in the entirety of the human experience under the diverse headings of melancholy, thereby medicalizing all of

---

9 Hiroshi Hirai 2011 evokes medical humanism as the context for several specific studies in early modern medical philosophers such as those by Fernel and Sennert. Others include Shirley and Hoeniger (eds) 1985; Wear et al. (eds) 1985; Nutton 1988; Maclean 2002.

10 See Palisca 1985. Again, Ferrand 1990: 348, will serve to make the link when, in Chap. 36 of his *Treatise on Lovesickness*, he tells the story of Pathagoras who averted the house invasion and rape intended by revellers by putting them to sleep with "grave, solemn and spondaical music" after ordering the minstrels to change the musical mode.

11 See Athanasius Kircher 1650; Johann Mattheson 1739; Friedrich Wilhelm Marpurg 1755.

human history. Abetting those connections was his choice of an expository form, for his treatise was designed according to the rhetorical partitions of the standard humanist medical treatise. In this remarkable literary exploit, we have what may constitute the epitome of medical humanism, including the encyclopedic reading habits that enabled the creation of a medically organized compendium. But in this same omnium gatherum, there lurks a death-dealing parody of the humanist scholarly enterprise in a maze of pseudo-analysis signifying nothing but morbid self-indulgence – a prophylactic exercise to stave off the meaninglessness of life. Burton is the periphery at the core of medical humanism with his run-away encyclopedism nevertheless based exclusively on the formal principles of humanist medical investigation.

In the narrowest sense, medical humanism pertains historically to the work of the medical philosophers active in the fifteenth and sixteenth centuries, and more particularly to those who worked according to the methods of humanist scholarship, beginning with the recovery, translation, and annotation of ancient medical texts. This led, in turn, to synoptic citations and commentaries, and an application of the doctrine of correspondences by which world orders were symbolically and causally interconnected in comprehensive ways. As historians, they looked to Pliny and Dioscorides as founding investigators of the natural world, and contributed to the growing pharmacopeia through the classification of newly discovered plants, animals, and minerals, particularly from Asia and the Americas. To be sure, the human body and its diseases remained the focal point around which they organized their investigations, yet their work led them to a substantial number of contributing topics including signs, astral systems, theories of causation, the nature of seed and the principles of generation, the occult features of nature, modes of contagion, natural magic, ideas of the soul, the mysteries of sensation, faculty psychology, dreams and omens, the constitution of matter, and many more such phenomena both visible and invisible. In all of these pursuits, they manifested a number of attitudes concerning the primacy of authority, methodical and polemical procedures, and the criteria for the demonstration of truth, as guided by values of the profession, a scientific secularism, and a qualified degree of skepticism.

Such studies in the history of medicine would seem to find justification only as knowledge for its own sake, and yet medical philosophy was one of the channels whereby certain *idées forces*, or prevailing and defining ideas, of the ancient and early modern worlds survived down to modern times. It is well-known that ancient medicine was based on a view of the body as a system of contending or opposite parts, and that health consisted in maintaining those parts in a state of perfect complementarity by permitting no humour to achieve an ascendency over the others. Hence, the body was viewed as an isonomic system which, in the equal representation and balancing of the parts, carries overtones not only of equilibrium and health, but of fairness, impartiality, and parity. Allopathic pursuits arose to the degree physicians recognized the means for rebalancing those elements deemed to be under- or over-represented within the organism. It is fascinating to think that this approach to health may be correlated to the rise of democracy in ancient Greece, for it is first recorded in the writings of the sixth-century B.C. physician Alcmaeon of Croton, who points out that "'monarchy' or the domination of any one power over the others was the cause of disease and destruction . . . ."[12] Democratic thinking assumes similar values concerning the body politic and the equality of its representative members in the agora, acting as their own physicians to maintain, through debate, the health and stability of the state. Thus, medical *isonomia* and the logic of democracy became synonymous, each sphere of regulation carrying the proviso that without cooperation, the entire organism would perish. That point was simply but elegantly demonstrated in the celebrated Aesopic fable about the 'Rebellion of the Hands and Feet against the Belly (or Head).'[13] The metaphor of the body politic underlying the fable seems merely convenient, but it ran deeply and long as an underlying principle of inclusive fitness for both organisms, body and state. John Adams confirms the point in the *Novanglus* where he states: 'some physicians have thought, that if it were practicable to keep the several humors of the body in exact balance of each with its opposite, it might be immortal; and so perhaps would a political body . . . '.[14] Only through the purging of 'ill humors' in body and state, there is hope for long life. The force of the metaphor is located in its representation of the harmonization of equal parts, which was to become a salient idea in Western thought with its singular potential for modifying the hierarchies of power represented by patriarchs and potentates.[15]

As for the players, the list of early modern physicians innovatively and polemically engaged in these matters is already long and is destined to grow as research endeavors expand. Those coming most readily to mind include Julius Caesar Scaliger, Pietro Pomponazzi, Pietro d'Abano, Justus Lipsius, Rodrigo de Castro (Amatus Lusitanus), Nicolò Leoniceno, Johann Wier, Girolamo Cardano, Levinus Lemnius, André du Laurens, Ambroise Paré, and Johann Winter (Andernacus). That is a sampling, but there are many more: Daniel Sennert for his work on living atoms and the nature of matter, Girolamo Fracastoro, for his germ theory of diseases, Andreas Vesalius for his work on human anatomy, Paracelsus for his opposition to superstition and dogma in the practice of medicine, Jean Fernel for his Platonic interpretation of Galen, and Leonard Fuchs for his work in medicinal botany, to whom we might add a number of non-physicians who nevertheless contributed to theories

12 Sahlins 2008: 217.

13 Aesopus, *Fabulae*. See Ogilby 2021: *Fable 47*, 262–266.

14 Adams, 2000: 134.

15 This is a further example of the emergent psychological states which permitted the rise of the cultural configurations conducive to the unique historical emergence of the commercial, religious, political, and social achievements of the Western World so comprehensively set out by Henrich 2020.

and methodologies in the study of natural history, such as Jacopo Zaberella, Bernardino Telesio, Nicholas Monardes, and Charles L'Ecluse.

Just how the scientific and philosophical perspectives of the physicians coincided with the overarching values of the *studia humanitatis*, with its initial interest in grammar, poetry, history, rhetoric, and moral philosophy, is a point for pondering if the medical writers are to share in the term 'humanism.'[16] Much thought has gone into the quest for a universally satisfactory definition of this movement initially epitomized by Cicero's secular approach to the perfectibility of man through education leading to an emphasis on virtue, morality, and dignity. Humanism comes closest to the discipline of medicine in its concern for what makes us human, the value of life itself, combined with a love for learning, and a credo of honorable conduct. Medical humanism signals, as well, the degree to which physicians, as philosophers and rhetoricians, reflected the training they received from the disciplines and methods espoused in the schools, and in the many ways that medical philosophy drew from a common store of problems and received ideas.

The field of medical humanism in the Renaissance may be defined further by the many problematic points of philosophical speculation scattered throughout their treatises. Because these matters pertained to traditions and authorities, as well as to clinical observation, they gave rise to considerable citation and debate in the pursuit of a reasoned synthesis, including case studies and wide speculation upon causes and the occult. Modes of contagion were of primary concern, whether by miasmas, by touch, or through poisons, including contaminated vapors, congenital inheritance, psychosomatic disturbances, or the unbalancing of the bodily humors.[17] Considerations also turned toward the patients, their constitutions, diets, obsessions, and the external signs of pathological corruption. After diagnostics came prognostics, the course of the malady, and the chances for recovery through time or treatment, while closing chapters dealt with the methodical, surgical, and pharmaceutical cures and their respective merits. Within these sequential diagnostic structures, authors were at liberty to return to first principles, as when Jacques Ferrand first seeks to define love as the force of attraction which unites all the elements of creation into their respective forms in the fullest humanist sense before addressing its forms of pathological expression. In fact, Ferrand engages in speculations on many topics drawn from a wonder cabinet of medical lore and anecdotes graced with quotations and exempla. Those topics ranged from the physiology of sight to the relationship between procatarctic, remote, and efficient causes, or from the medicinal properties of artesian waters to the degrees of coldness in lettuce and mint and their effects upon the production of sperm. Also to be found are debates with theologians over the healing roles of the clergy, or with botanists over the

16 Further to these matters, see Grafton 1996: 203–23.
17 Carlin (ed.) 2005: passim.

properties of medical simples, or with past practitioners over the ethical use of folk magic in diagnostic matters. How corresponding and semiotic forces in the universe altered the body remained occult, but not for want of highly sophisticated theories about the causations arising in the properties of food, geographical environments, temperaments, life styles, curses and spells, astral forces, and demonic agency. All that might impinge upon the health of the body from poisons in the form of noxious vapors to despair over the uncertainty of divine election belonged to the realm of medical speculation. To understand the ways to health, the physician had to know a great deal about the world and all its mysterious forces.

Up to this point, the domain of medical humanism has been assigned to the medically trained philosophers. Not only are they practitioners, but through their treatises, they became the principal framers of the medical ideas and procedures we are calling humanist. But investigations expand rapidly with the acknowledgment and validation of medical debates in a wide variety of discursive and imaginative modes from theology, history, and philosophy to the visual, narrative, and musical arts. Medical thought found representation in ways as diverse as rhetorical structures and poetic imagery to points of academic debate or imaginary case studies. Just as the physicians might recite the ancient poets to their purposes, dramatists like Ben Jonson might establish the tilted eccentricities of his characters according principles derived from Galen by urging that every man is either in or out of his humor.[18] Reason would hold that medical humanism must now be extended to the artists, philosophers, dramatists, and prose fiction writers who incorporated into their creations a high degree of medical literacy. Hylomorphism might serve as a case in point. It is the theory that all natural things consist of both matter and form, that biological materials are inert and neutral until a defining pattern is settled upon them. There were difficult sticking points, however, concerning how such forms were transmitted, designed, and preserved outside of matter. How indeed could souls be fitted upon amorphous and inert matter?[19] Perhaps the ultimate challenge was posed by the doctrine of spontaneous generation (abiogenesis), as in the production of maggots, suggesting that matter sometimes or always had the capacity to generate its own forms such as the fire contained within the flintstone.[20] Despite these controversies, the theory prevailed as a leading received idea in the Renaissance, making it available in both structural and thematic terms to creative artists, such as Edmund Spenser, who, in canto vi of the third book of the *Faerie Queene*, employs this challenging doctrine as the defining principle of an allegorical

18 These matters are most explicitly laid out in the preface to Jonson 2001, *Every Man Out of his Humour*, which became his working credo in the representation of human eccentricity and hyperbolical behavior in his worlds of satiric excess.

19 The question is closely related to seed, or the 'semina formarum' discussed in the works of Marsilio Ficino. See Hirai 2002: 257–352.

20 See Lüthy and Newman 1997: 215–352.

garden devoted to the pleasures of Venus with her seasonally resurrected lover, Adonis.[21] All around their secret bower was a grand seminary, a place of matter to which Old Genius assigned new forms before releasing them into the world. Spenser's poetic vision is a grand paean to the creative forces of nature causally and symbolically aligned with human love and its place in the eternal cycles of life, decay, and death. For after the creations of Old Genius had spent their lives, they returned to the back door of the garden to be disassembled, their materials stored for limitless numbers of reincarnations. What is Old Genius but a homunculus in a system that otherwise lacks clear mechanisms for the genesis of forms. He becomes the missing cause, which was otherwise vaguely attributed to levels of heat or astral influences.[22] Thus, through allusion and reference, the scientific concerns of the medical humanists imprint themselves upon culture at large.

Inversely, the narrative arts of poets and prose fiction writers are a reminder that the early physicians as clinicians were likewise practitioners of narrative. They profiled diseases according to the order of story insofar as maladies have their beginnings, crises, reversals, and endings and are traditionally described in that order. That sequence found literary representation when Boccaccio, at the outset of the *Decameron*, famously anatomized the Black Death in this manner, beginning with its arrival in the city, the initial signs of contagion, the emerging lumps and buboes, followed by the encircling rings that signaled the tragic end.[23] One of the salient legacies of the Hippocratic physicians was their careful narrative descriptions of diseases, the durations of fevers, the conditions of the eyes, the color of the skin, the bouts of sweating, and all such related signs of origin and progression by which diseases might be discriminated one from another. Diseases became their own stories and bodies became their own semiotic witnesses. Even poets, on occasion, might become interested in such narratives. Sappho was celebrated for her introspective profiling of the symptoms of unrequited love with a clinical precision, many features of which were confirmed centuries later by physicians.[24]

---

21 For an overview of Adonis worship and the early beliefs concerning the dependence of the entire plant kingdom on the sexual activities of the goddess, see Frazer 1967 (1922): 427–429.

22 The putative role of heat led Aristotle, in an inattentive moment, to attribute to it the differentiation of the sexes, leading to a complex Renaissance debate over the capacity for sex changes during the course of a lifetime. What had been lost in the debate and its alleged applications during the Renaissance to explain sexual transformation is that heat was initially deemed to be a quality invested in the seed, the pneuma that imprinted traits at the time of conception, and not an element apt to transform sexes by driving the genitals outward (or sucking them inward?) later on in life.

23 Boccaccio, *Decameron*, "Introduction to the First Day" (ed. Beecher and Ciavolella 2017: 29–36).

24 Ferrand 1990: 272, says of her, "Does it not appear that Sappho was as wise and as experienced in this art as our Greek, Latin, and Arab physicians in light of the fact that they mentioned no indisputable signs that this lady did not already know?".

When diseases are essentially social or psychological in nature, the medical narratives become even more story-like, taking on the profile of case studies drawn from clinical encounters with specific patients and their experiences. Jean de Veyries builds his entire medical treatise and all its scientific observations around the romance of a young marquis legitimately betrothed to the daughter of a duchess, but who must undergo treatment for frenzied love before the marriage would be allowed.[25] Such another narrative, later to be medicalized, was created by Julius Capitolinus about Faustina, the wife of Marcus Aurelius. To reverse her amorous fixation upon a Roman gladiator, the warrior was slain and she was given his blood to drink in a potion. Presumably it was to have been a cure by disgust. But instead, the blood became a residual component of her body which polluted her imagination at the moment her next child was conceived. The result was the birth of Commodus who would later become one of the bloodiest emperors in the history of the empire. An intended cure for love frenzy became an example of the transmission of parental traits to the offspring and a lesson in the equivocality of certain kinds of medical analysis. This tale, with its medical implications, was repeated by Pietro Crinito in Bk. 25 of *De honesta disciplina*,[26] and again, straight out of Capitolinus, by André du Laurens some three-quarters of a century later. Narrative is also involved in the prescription of regimens, for they are, in the case of diseases of mental disorders, elaborate scenarios of prescribed activity involving the administration of drugs and blood-letting séances, but also diet, exercise, baths, recommended entertainment and travel, along with proscribed reading and other counter-productive indulgences. Once more, the physician is the humanist in the presence of a partially uncooperative patient who must be induced with care and consideration, as in the administration of a complex regimen to a rich merchant living in Arles, the entire procedure closely and strategically described by François Valleriole in his *Observationum medicinalium libri sex* (1588). Arguably, in all of these instances, whichever direction the influence flows, we are investigating informed expressions of medical values through a broad range of cultural media under the aegis of medical humanism, circumscribing the domain by example.

That we are attempting to define an academic discipline simultaneously is largely a naming game, for precision in such things is merely a matter of debate around margins. Valleriole's extensive case study of the lovesick merchant is a magisterial examination of the medical mechanisms whereby an excess of erotic desire is philosophically and systemically bound to pathological effects which are proper only to medical assessment and treatment. Yet by definition, his patient operates in a world of desires which are culturally conditioned and subject to resolution

25 Veyries 1609. *La genealogie de l'amour divisée en deux livres.*
26 Pietro Crinito, *De honesta disciplina libri XXV*. Basileae: Henricus Petrus, 1532: 24–25; André du Laurens, *Des maladies melancholiques et du moyen de les guarir* in *Toutes les oeuvres*, Théophile Gelée (trans). Paris: Chez P. Mettayer, 1613.

in cultural terms. Even the physicians had to acknowledge that erotic frenzy or melancholy was partially a poetic disease. And yet, the bias of Renaissance medicine was to fortify the definition of lovesickness as a matter of clinical urgency. In that vein, the opening of *Romeo and Juliet* fully qualifies as a medical anecdote insofar as Romeo manifests many of the symptoms of the solitary and melancholy lover. Even Benvolio's tactic of driving the elusive Rosalind from his mind by sneaking him in to see Juliet is an approved measure among the medical practitioners to free the patient from an amorous obsession. Such diversionary techniques go back as far as the Arabic physicians for curing erotic depression or mania. But in the process, we have come to another ambiguous crossroad where the imaginative writer borrows from current professional ideas to profile an imaginative character in love, even as that character provides a case study through which such medical ideas might be exemplified. Hence, the binary relationship between literature and medical philosophy which goes all the way back to the ancients.

Even where minimalist definitions are allowed, what constitutes proper medical humanism, and when is the term bankrupted by merely marginal or irrelevant concerns? Is the trained physician, Sir Thomas Browne, still on topic when he declares that 'there is a Phytognomy, or Physiognomy, not only of Men but of Plants and Vegetables; and in every one of them, some outward figures which hang as signs or bushes of their inward forms,' insofar as he concludes that all their sundry constituent parts may be reduced to a single 'word that doth express their natures,' providing the means whereby Adam named all living things?[27] Should a dramatic character be assessed in medical terms who feigns madness in order to publicly discredit his intellectual accountability in ways that render him below suspicion, while equivocating in code to a confidant that he is mad only 'north-north-west' and can distinguish between a hawk and a heron when the wind blows just right? Is Hamlet's condition sufficiently pathological by cause or simulation to evoke medical consideration? Is King Lear truly mad or merely raging in the rain at the ingratitude of his daughters? Does clinical analysis ever apply? Or consider the speculations on philocaption by Kramer and Sprenger in the *Malleus Maleficarum* and their nod toward the medical treatments of love, only to marginalize them rhetorically in favour of demonic causation. Or how do we categorize *A Treatise of Melancholy* (1586) by the clergyman and physician Timothy Bright, written specifically for a friend in the throes of sadness, but more generally for the comfort of all those radical Protestants suffering from despair over the uncertainty of the signs of election and the future of the soul? Are they not truly depressed and mentally suffering? Or how should we consider the artistic treatments of the story of Antiochus and his secret and unrelenting love for his young step-mother, as told by Valerius Maximus and Plutarch? Through the diagnostics of the physician Erasistratus whereby he detects the secret cause of the patient's

---

27 Browne 1904: I.86.

malaise, he simultaneously produces a foundational anecdote in which clinical practitioners, three quarters of a millennium later, were still finding authority for the foundation of their clinical practices. But would that fact suffice to designate all subsequent representations of the story as expressions of medical humanism, whether by Camoëns in his *El Rei Seleuco*, or Ingres in his painting of the bedroom scene, or Etienne Nicholas Méhul in his late eighteenth-century opera, *Stratonice*?[28] It is a nice point, for the artists were not physicians and their purposes were not medical. Yet, each work, of itself, illustrates the medical aspects incorporated into Renaissance medical diagnostics. How medical is Boccaccio's quite wonderful tale of Girolamo and Salvestra (IV.8) in which both lovers die of spontaneous grief (a topic treated by Ferrand in the second chapter of his *Treatise on Lovesickness*), or Straparola's equally compelling and detailed tale 'Of Philomena the Hermaphrodite Nun' (XIII.9) as a case study in sexual ambiguity? For each example there are grounds for reasoned accommodation, although those grounds become increasingly difficult to define.

We are reminded again of the medical literacy among artists when, in Milton's *Paradise Lost*, Eve awakens Adam to report her frightening dream about meeting Satan as 'the tall, dark stranger' in the garden, inciting Adam to comfort her through reassurances based on faculty psychology which prove that she has not yet crossed the fatal line that constitutes sin.[29] Faculty psychology was a free-floating *idée force*, a generalized model of sensation and cognition available for citation and synthesis. In essence, medical thought, through all its many scientific registers, had produced a vast referential cupboard of analytical procedures, medical anecdotes, case studies, points of regimen, biological lore, botanical insights, orders of treatment, and definitions of disease apt for imitation or paraphrase in a variety of non-medical contexts. When is an allusion to any of them, however transformed in expression and function, not a little piece of medical humanism? At the same time, physicians were being drawn into public debate over such matters as hermaphroditism, witchcraft, demon possession and exorcism, divinatory magic, folk medicine, and institutions for the treatment of incurables. Increasingly, they were becoming public intellectuals, writing for wider readerships in their respective vernaculars, engaging in debates which escaped the clinic in ways that often aroused the antipathy of the Church or the ruling order.

In the final analysis, however, why should the term 'humanism' confine the study of the intersections between medicine, philosophy, and the arts to the early modern period? With a further broadening of the term 'humanism,' the entire discipline may be liberated from such restrictions to include all cogent representations of medical ideas in any mode of expression or enquiry. To that end, humanism need only be redefined as pertaining to any form of writing or artistic expression

28 Méhul 1996.
29 Milton, *Paradise Lost*, Bk. V. ll. 95–128 (ed. David Scott Kastan 2005: 147–149).

concerned vitally and centrally with the human experience. Such an allowance now prioritizes the medical dimension of medical humanism as the defining condition, one that must be established discursively by each researcher. The effect of such a broadening is to draw the matter of marginality to the very center of the discipline. The question of relevance and inclusion becomes a matter of criteria, judgement, and persuasion. Is it a genuinely medical perspective that souls may be sin-sick, and that Christ is the balm of Gilead that cures them? Or looking into the future, which is here now, is it still medical humanism to assert that human nature is the by-product of a designed human brain through eons of evolutionary selection and adaptation – a design that circumscribes our cognitive abilities and emotional ranges, and determines the thresholds that constitute social functionality and dysfunctionality? Does the term medical humanism, in either of its parts, still apply if this new science materializes the soul as a mere emergent property of blind neural circuitry? Where is the humanism in these cold scientific calculations? Is it enough that bodies subject to disease are considered the natural objects of medical analysis, given that the human psyche, as the seat of computation and the emotions, is an embodied instrument? How can our potential for mental or physical dysfunctionality ever escape the fused literal-metaphorical application of the concept of disease and the interests of the medical establishment? And yet, as humankind increasingly becomes a set of evolutionary designed algorithms emerging through a community of uncomprehending modules contributing blindly to the production of behaviors adequate to sustain reproduction, we can imagine that both the medical and the humane will find their places only in much qualified and inferentially extended definitions of the nature of homo sapiens.

As for the discipline, perhaps it is now more useful to forego formal limits altogether, as long as some minimal connection between medical substance and the representative arts can be demonstrated. With that, as stated above, the indeterminacy of the discipline becomes an integral part of every speculative contribution, the validity of which is a matter of rhetorical efficacy. Hence, the pursuit of the margins increases as an intellectual goal, making the search for a definition the foremost feature of the discipline. More vital than ever are the cases to be made for the dubious, insofar as perpetually redefining the margins serves to confirm or modify the core of a now generic discipline which, with the volatility of the present, will remain in an emergent state. When fictional writers become concerned with medical issues through social narrative, equivocality is bound to arise, as when medical questions and non-medical motivations contend with each other. In William Carlos Williams' 'The Use of Force,' a young patient violently refuses a throat examination, compelling the doctor to resort to equally violent force for the patient's welfare. The episode is clearly medical in substance, for it takes place in a doctor's office and involves a contest of minds between a terrified or willful child and an adult whose professional authority and efficiency is called into question by resistance. As such, it is a clinical issue dealing with medical procedures versus patient-centered care.

Yet as a story, it is a contest in which the urgency of medical analysis is a mere catalyst to a more primal struggle, one that fundamentally has nothing to do with medical issues. The contest between patient and professional is a confrontation between primal fear and impatient authority – a non-clinical issue, except that it is central to the medical humanism of respect for the rights of the patient, and how those rights are to be defined.

Or consider D.H. Lawrence's protagonist in 'The Horse Dealer's Daughter' who attempts to drown herself in the local pond in a fit of despondency. She is rescued by the village doctor's assistant, who wades into the pond, resuscitates her, then responds positively to her desperate erotic advances. Does the tale qualify merely by dint of the young man's profession and the heavy overtones of insanity blended artistically into their amorous negotiations? It all depends on whether the attempted suicide is pathological or rhetorical. And when are such attempts not inherently both? Or consider A.J. Cronin's story, 'Doctor, I can't . . . I won't have a child,' in which a callow society woman is refused an abortion by the family doctor and resorts to an underground butcher clinic, while convincing her husband that her worsening condition, including hemorrhaging, is due to influenza. Disgusted, the family doctor, nevertheless, does his Hippocratic duty to save her life, only to be told at the end that she, her lover, and her husband are going off for some relaxation together, her husband none the wiser, and the doctor cursing the duplicity of the sex. The context is medical, but the concerns are more patently social. There are many more to choose from in kind in Joseph Ceccio's anthology, *Medicine in Literature*.[30] The sum of the contents defines the book, the humanism in the literary, and the medical in the sometimes nearly incidental allusions to the profession. The anthologist does not attempt to define the occasional minimalism of the medical references and associations. Yet the point is well taken, that medicine has always been practiced within complex social environments.

Medical humanism can hardly be more broadly defined than it is in this anthology, although the editors of the new review, *Synapsis: A Journal of Health Humanities*, have come close, for they describe their domain as 'anything that connects medicine with the humanities – critical reading, looking, listening. [Their] interests are wide ranging: historical précis, new takes on books, investigation into cognition and imagination, and, of course, medical practice.'[31] They especially encourage articles ranging from 500–1000 words drawn from history and literature, as well as neuroscience, in the belief that fiction and memoires can influence the ways of thinking about health and disease as readily as the practices of medicine can find reflection in imaginative representations, either way potentially inspiring new genres in the arts. In this spirit, the present study concludes not far from where it began in the medical school

30 Ceccio 1978.
31 Hegele (founding editor) 2017: https://medicalhealthhumanities.com/about/.

courses in ethics, combined with medicine as seen through the representational arts.[32] Why then consider at all restricting that same intersection to the early modern period defined as humanist? That question answers itself through the historical precision that can be imposed upon the term according to the wishes of those who might so desire to define it. It is a choice, pure and simple. The extreme alternative is an open signifier which comes to its validity through mere whiffs of the medical from any part of its range of primary or secondary concerns as alluded to in any form of representation or communication, according to intersections assumed or defined by the reporter.

The analysis has now come full circle, taking us back, after a long detour, to stories which are humanist by dint of their highlighting of the human experience in relation to the practices and purposes of modern medicine. We are back to patients and doctors, and to the ethical issues that arise as a result of the ever-changing codes of professional conduct, the ideological conflicts over the rights of patients, the principles and practices of medicine, or the interpersonal and extra-professional relationships that emerge between patients and practitioners. These issues run in parallel to the ethical concerns related to genetic engineering, abortion, assisted suicide, questionable interventions and calculated risks, malpractice, and medical errors. Imaginative literature will dwell increasingly on these dilemmas as medical technologies become more invasive and transformative, and whether they are deemed to be for the better or for the worse. The imagination, as an emergent mental property designed to create the provisional drafts of future options before making volitional choices, is a faculty that projects the self into multiple states of possibility. In the present age, the imaginative writer as medical humanist may well concentrate, increasingly, on human stories associated with these advanced medical technologies. Above all, they may dwell on our ethical anxieties associated with artificially fortified bodies and minds, and particularly those faculties through which we formulate and experience both the external world and our own selves. What is our moral obligation to observe the limits imposed upon us by the long, slow adaptive legacy of evolution now that we can imagine self-perfection through our own technologies? That world is already with us through psychopharmacology, the pain-killer spectrum, as well as lifestyle and performance enhancing drugs from anabolic steroids to Ritalin as a mere beginning. In this new economy, there is potential for material salvation as well as injury to both mental and physical health. Meanwhile, only time will confirm the

---

32 Literature and medicine, as a component of medical studies, has grown significantly throughout the last fifty years. There were pioneers in the area, such as Joanne Trautmann, a professor of English at Pennsylvania University (Hershey), who was invited as early as 1972 to teach such a course as part of the 'medical humanities' component in the College of Medicine. Literary accounts of patients and their experiences have proven value in broadening the perspectives of students in training as doctors. Among the benefits is a greater measure of empathy from seeing the healing process through patients' eyes.

progress currently sought under the rubric of medical humanism toward a more humane and patient-centred medical profession. As a general research discipline, the pertinence and relevance of contributions will, of necessity, require their own respective rationales in relation to the generic terms and emerging themes of the discipline – contributions which might range from a biography of Albert Schweitzer or Clara Barton to sensationalizing stories on the abuses of medical marijuana, or cautionary reports on the risks of epidurals as an analgesic in routine child births. In such ways, the margins will continue to define the ever-expanding coherence of a discipline centered in the abnormal states of body and mind and the systems and philosophies of care brought to them through medical regimens, procedures, and therapies.

Mariacarla Gadebusch Bondio

Avoidable Mistakes – Premodern Medical Fallibility as an Ethical Problem with Epistemological Implications

Fallibility, or rather the awareness that mistakes can be made, is part of being human. This applies in particular to the medical capacity for self-understanding. For Hippocrates, a physician who could confine themselves to making only minor errors was to be lauded (. . . *And the physician who makes only small mistakes would win my hearty praise*).[1] The smaller the mistake, the better the physician. Galen believed that a philosophically educated physician, committed to the prevention of mistakes, could effectively protect himself against them.[2] He divided mistakes into two categories: on the one hand, there were the false opinions which arose from ignorance and a lack of reasoning power, and on the other, there was malpractice, which had its roots in irrational drives and passions. Accordingly, he advocated sound training in logic and dialectics as strategies for avoiding medical errors, accompanied by a technique for the control of passions. For their development and reinforcement Galen recommended that one should wilfully embrace the criticism of an honest friend, whose duty it would be to continually call attention to mistakes and urge their correction.[3] The ideal of a physician-philosopher was to prove enduring.[4] An echo of this approach to the prevention of mistakes resonates in the early modern treatise of Roderigo de Castro, the *Medicus Politicus* (1614), perhaps more than from any other contemporary author of deontological texts.

1 Hippocrates, *Ancient medicine*, ed. and transl. Jones, *Hippocrates, Works*, vol. 1, 2004: IX, 27.
2 Galen, *Quod optimus medicus sit quoque philosophus*, ed. Kühn, *Galeni Opera omnia*, vol. 1, 1821: 53–63. See also Galen, *Quod Optimus Medicus*, transl. Erasmus, ed. Waszink, *Desiderius Erasmus, Opera omnia*, 1969: 1.665–669.
3 Galen, *De cognoscendis curandisque animi morbis* I.5, ed. Kühn, *Galeni Opera omnia*, vol. 5, 1823: 21–26; Galien, *L'âme et ses passions, les passions et les erreurs de l'âme. Les facultés de l'âme suivent les tempéraments du corps*, eds. Barras, Birchler and Morand 1995: 3–74, Galen, *De cuiuslibet animi peccatorum dignotione atque medela libellus*, ed. Kühn Galen, *Opera omnia*, vol. 5, 1823: 58–103, chapter 4: 77–80.
4 See Galen 1969: 665–669.

Note: I thank Christian Kaiser for the valuable examination of the Zerbi section and for the inspiring exchange.

This article was written within the framework of two research projects supported by the German Research Foundation. The research on medical fallibility was started but not published during the bilateral Project "Fallibility and Error Culture in Medicine. Historical, Epistemological and Ethical Dimensions (1500–1650)." (GA 1086/6–1). The interrupted work could be carried out in the fruitful context of the Collaborative Research Center 1369 "Vigilance Cultures. Transformations - Spaces - Techniques", in which I lead a subproject on "Vigilance as Ideal, Strategy and Method in Pre-modern Medical Culture.

https://doi.org/10.1515/9783110788501-003

Two questions present themselves with regard to medical fallibility: how do authors of texts on the *medicus optimus*, *cautus*, *politicus* deal with medical fallibility? Is it perhaps precisely this darker side of human fallibility, predisposing us to deception, error, and mischance, which itself is the motivation for physicians to concern themselves with their moral profile and medical ethics?

These questions have accompanied my examination of those texts which we today describe as deontological, and which enjoy a tradition as old as medicine itself. In the Early Modern period such texts flourished. I will begin with Gabriele Zerbi's programme of cautelae as a device against mistakes, and then focus on the "right knowledge question"; Finally, I will discuss how some physicians attempted to develop a learning programme with the goal of facilitating the avoidance of mistakes and errors through the practice of criticism and self-criticism. Such programmes are presented in the deontological works of Alessandro Benedetti, Giovanni Filippo Ingrassia, Giovanni Antonio Sicco, Leonardo Botallo and Roderigo de Castro.

Gabriele Zerbi's Programme of 'Cautelae' as a Device against Mistakes

Gabriele Zerbi's (1445–1505) *De cautelis medicorum* (ca. 1495) is the first text on the topic of rules of caution for physicians, and was reprinted several times after 1495.[5] The anatomist and professor of medicine Gabriele Zerbi had first held a position at the University of Bologna (1475–1483), and then as pontifical physician at the Curia (1483–1494) before he was again appointed to the well-known medical faculty in Padua (1494–1505), where he held a chair for theoretical medicine. He published his deontological compendium at the beginning of the last phase of his academic career at the University of Padua.

In *De cautelis medicorum*, Zerbi lists a large number of medical and philosophical authorities from antiquity and the Middle Ages, ranging from Hippocrates and Aristotle, through Galen and Boethius, to the Arabic authors on medicine Haly Rodoan, Johannes Mesuë, Haly Abbas, Al-Farabi, Avicenna, Isaac Judaeus (Israeli), Averroes, and many more. All these names were at that time firmly established figures in the scientific canon. However, as concerns most of the cited passages, Zerbi most probably does not quote directly from first-hand sources. Rather, he includes sentences, and often entire sections, from other works, whose authors are much closer to him in time and culture. Zerbi himself openly names the two most important, Pietro d'Abano (ca. 1250–ca. 1315), who was

5 See the new bilingual edition which for the first time collates all previous text versions: see Gadebusch Bondio, Förg and Kaiser (eds), *Gabriele Zerbi, De cautelis medicorum*, 2019. This edition will be referred to in the following quotations. On the edition history, see French 1993: 72–97; Linden 1999: 19–37.

became known as the *Conciliator* after the title of his main work, and Niccolò Falcucci (Nicolaus de Florentia; † ca. 1412). A third source, unmentioned by Zerbi, is Alberto de' Zancari.[6] Since he not only lifts large passages from these medieval writings, but also the source references and citations mentioned in them, Zerbi gets by with relatively few sources in *De cautelis medicorum*: in addition to Pietro d'Abano's *Conciliator*, Falcucci's *Sermones medicinales*, and Alberto's *De cautelis medicorum*, he also draws on the Hippocratic texts *Lex* and *Iusiurandum*, and the *Regalis dispositio* by Haly Abbas (ʿAli ibn al-ʿAbbās al-Madschūsi; † 994), in Stephen of Antioch's 1127 Latin translation.[7] In addition, scattered passages from Galen and medical textbooks, together with a handful of quotations from poets and philosophers, are woven into the argumentation.

The significance of Zerbi's *De cautelis medicorum*, then, is not primarily in its content, which on the whole adheres to approved teachings from antiquity and the Middle Ages; rather, it lies in the unification and arrangment of existing material.[8] Zerbi is the first to assemble the diverse deontological texts circulating among learned physicians toward the end of the 15th-century into a coherent system.

In the present context of medical errors, Zerbi's reception of Niccolò Falcucci's work is especially important. Falcucci, who was a professor of medicine at the *Studium Florentinum*, wrote his *Sermones medicinales* toward the end of the 14th-century. It is a comprehensive manual that summarises medical knowledge and was reprinted several times in the 15th- and 16th-centuries.[9] Zerbi subsumes much of the medical ethical teaching discussed in Falcucci's *Sermones*, very often *verbatim*. The medieval tradition of the medical *cautelae* thus finds its way into Zerbi's book since Falcucci had also incorporated the precautions of older authors from around the medical school of Salerno. Examples include the admonition that the patient must have confessed his sins to the priest before the doctor begins treatment, or the prohibition against leering at women in the sick person's house. These traditional provisions can be found in the treatise *De adventu medici ad aegrotum*, written between the end of the 11th and the beginning of the 12th century,[10] from where they migrated to *De cautelis medicorum*, which – probably erroneously – was attributed to the famous Catalan physician

6 See Ziegler 2014: 117–129, 118.

7 In the apparatus, the new edition of *De cautelis medicorum* compares in detail in the textual material of the corresponding editions: Niccolò Falcucci, *Sermones medicinale*, 1491; Pietro d'Abano, *Conciliator differentiarum philosophorum etmedicorum*, 1472; Alberto de' Zancari, *De cautelis medicorum habendis*, ed. Morris 1914; Haly Abbas, *Liber medicinae sive Regalis dispositivo*, ed. Vitalis 1492; *Articella*, ed. Argilagnes 1483.

8 For the arguments supporting this observation, see the introduction of our edition: Gabriele Zerbi 2019, 24–30.

9 Park 1985: 220; Mucillo 1994: 401–404.

10 See the editions in *Collectio Salernitana*, ed. de Renzi 1853: 2.74–80 and Stroppiana 1956: 85–96.

Arnaldus de Villanova (ca. 1235–1311).[11] Falcucci elaborated on this material from older medical literature, and Zerbi rewrote it in his subsumption of Falcucci's text.

Fallibility runs like a thread through the six chapters of Zerbi's dense work. Errors are defined typologically: *error* (error), *falsum* (false), *fallacia* (swindle / intrigue), *fraus* (fraud), *mendacium* (lie), *peccatum* (oversight, mistake), *deceptio* (deception), *simulatio* (simulation), *sophisma* (sophism), *vitium* (mistake). These termini recur throughout Zerbi's book.[12] In the prologue, we can already see from the definition of the core concept of "cautela" that the author's primary objective is the avoidance of mistakes. *Cautelae* as rules of caution are the instrument by which this objective can be achieved. They should help the physician in all practical situations to make as few mistakes as possible (*peccare*), to err as seldom as possible (*errare*), and accordingly not, or hardly ever, be liable for prosecution (*delinquere*).

> In order that physicians who in the performance of their duties live with the challenge of making as few errors as possible and not making themselves liable to prosecution, so that they may dedicate themselves intensively to the art of medicine without any distractions, and directly achieve their ultimate objective, we wish with the help of God, the guardian of all life, to turn our attention to the work on the physicians' cautelae. We will deal with the cautelae summarily, as a detailed knowledge of them all is impossible. They are as infinite as human understanding, particularly when this is based upon sensory perception. More particularly, an error does not occur less randomly in human actions than errors or monsters in nature, produced by different material conditions.[13]

Zerbi took a realistic view of himself and his colleagues when he observed that mistakes (*peccata*) among people occur at least as frequently as they do in nature.

Investigation of the text shows four thematic areas of fallibility:

1. Medical knowledge: here Zerbi discusses a) what a physician should know in order to be able to recognise false teachings and doctrines which they should take care to avoid, and b) what they should know in order not to make mistakes themselves. The study, understanding and memorising of good medical texts, and the transfer of correct knowledge, contain a moral dimension and are the essential duty and area of responsibility of a good physician. Zerbi is of the opinion that a good

---

11 See Arnaldus de Villanova, *De cautelis medicorum*, ed. de Gabiano 1504: ff. 256vb–257vb. For some serious doubts concerning Arnaldus' authorship, at least with regard to the medical precautions, which correspond word for word with those of the Salernitan treatise, see Sigerist 1946: 136–143, 138.

12 See s. v. "decipio," in Lehmann and Stroux 2003: 3.95–96; s. v. "erro," also 3.1370–1396; s. v. "error," :3.1371–1373; s. v. "fallacia": 4.49–50.

13 Zerbi Gadebusch Bondio, Kaiser, eds., 2019: 44: "Ut igitur in difficultate non peccandi, sive non errandi in medicando minus delinquant medici, ut ad extimationem artificialem non impediti, neque distracti propinquius accedant, ut recte consequi possint finem intentum tractatum de cautelis medicorum aggrediamur cohoperante illo qui omnium viventium est tutela. Cautelas autem per capita tangemus quia omnium earum in singulari impossibilis est cognitio, quia in infinitum procedunt, sicut in infinitum contingit intellectum humanum prout reflectitur ad sensum errare precipue in actionibus humanis non minus forte quam peccata, seu monstra in natura eveniant propter varias materie dispositiones."

physician should never deny the truth and should be a "friend of the truth." Thus, he should not allow himself to be confused or deceived by "false prophets."[14]

2. Medical practice: Ward rounds, urine analysis, prognosis, bedside consultation with colleagues and the prescription of medication are all situations in which physicians should act with great care and attention.
3. Dealing with truth and deception: The bedside lie is legitimate if it is helpful in encouraging the sick person not to give in. In conversation with the patient, for example during anamnesis, there may be occasions when the patient lies and attempts to simulate. The physician should therefore do his best to pay close attention and not let himself be deceived or confused, as this could lead to a false diagnosis and treatment. After treatment, when concluding his business, the physician should also avoid being deceived or confused.
4. Mistakes by others: These include: a) mistakes made by patients (*peccata*) who as a result of incorrect behaviour (disorderly lifestyle) and a lack of compliance become ill. If these mistakes are of a sinful nature, such as intemperance (*intemperantia*), then the physician must motivate the patient to confession; b) relatives and assistants (*astantes*) may also make mistakes when caring for the patient by taking incorrect action. The physician must remain vigilant and govern and correct them; c) colleagues may also make mistakes. In these instances, the physician has to remain tactful and collegial. He may not draw attention to these mistakes in front of either the patient or other colleagues.

At the very beginning of his compendium,[15] Zerbi discusses the knowledge which goes towards improving a physician. A physician should respect the rules of medicine and avoid false teachings.[16] One basic precept is that the physician should constantly defend himself against deceptions and disgrace (*utque evitet cum diligenti attentione deceptiones et infamiam*).[17] After this more general warning, Zerbi arrives at the question, like Falcucci before him, of correct knowledge, which also prevents

14 See footnote 15.

15 The question of what a physician has to know was discussed by Pietro d'Abano (1250–1316) in the first *differentia* of his above-mentioned *Conciliator*, which is now available in a critical edition: Kaiser and Schenkel, eds., 2019: 191–247. See Jacquart 2012: 129–146.

16 Zerbi Gadebusch Bondio, Förg and Kaiser, eds., 2019: 40: "Tertio adimpletur observatione Canonum medicinalium diligenti evitatione false doctrine, nec non deceptionis, et infamie nonnunquam" [Thirdly, he respected the medical rules conscientiously avoiding false doctrines, and most certainly deception and sometimes disgrace]; "Tertio evitatione false doctrine et deceptionum quum ibidem dicatur: attendite id est diligenter cavete a falsis prophetis" [Finally, false doctrines and deceptions should be avoided: 'Pay heed, and beware of false prophets]'.

17 Zerbi Gadebusch Bondio, Förg and Kaiser, eds., 2019: 46. See also Id., III, 64: "Inter sapientes convenit medicum nonnullis presidiis se tueri oportere, ut fallacias et eminentes sibi deceptiones evitet" [Amongst scholars it is unanimously agreed that it is necessary for a physician to arm himself with several means of protection in order that he may avoid fraud and deception].

errors. For Falcucci, logic – following Galen – and astrology – following Hippocrates – were the effective disciplines for the avoidance of mistakes. Zerbi wished to see the physician educated in the *artes liberales*, first and foremost in "dialectics" (citing Pietro d'Abano or Falcucci respectively) "without which, as Galen says, the truth will not be recognized and nobody will be protected against mistakes."[18] Rhetoric and grammar, as well as arithmetic, music, geometry, and astrology (*Quadrivium*), and in addition natural philosophy – and philosophy in general – are all mentioned by Zerbi as fields of knowledge in which the physician should be knowledgeable.[19] Haly Abbas, given here by Zerbi as a source, and Pietro d'Abano stress how productive it is to learn these disciplines during youth.[20] One further aspect which Zerbi and Falcucci discuss is misstatements by authorities: how should a physician respond to this? The answer is diplomatically, the recommended approach is to be gracious and tolerant. The writer or translator should be held responsible for the mistake and not the respected author! (*Quandocumque in libris sapientum invenitur aliquod falsum non debet extimari: quod illud dictum sit sapientis sed potius debet imputari scriptori vel alteri: ut translator*).[21]

---

18 Zerbi Gadebusch Bondio, Förg and Kaiser, eds., 2019: 52–53: ". . . dyaletica, videlicet sine qua ut ait Galienus non scitur veritas, neque salvatur quis ab errore, ..." See Galen, *De constitutione artis medicative*, VII, in *Galeni Opera omnia*, vol. I, fol. 126v (ed. de Brescia Venice, 1490); Pietro d'Abano, *Conciliator differentiarum philosophorum et medicorum* 1472, I, propter tertium, 216; Niccolò Falcucci, *Sermones medicinales* 1491, Sermo I tract. I cap. 6, fol. 4r.

19 Zerbi Gadebusch Bondio, Förg and Kaiser, eds., 2019: 52: "Incumbet igitur pro viribus futurus medicus a primis annis erudiri, non solum in his artibus quas de trivio dicunt dyaletica videlicet sine qua ut ait Galienus non scitur veritas, neque salvatur quis ab errore, rhetorica pariter et grammatical, et eis quas de quatrivio aritmetica videlicet musica, geometria et astrologia, quibus ille de trivio adminiculantur in eis saltem tantum proficiendo, quantum medicinalis scientia poscit. Philosophia tamen naturali ut principali non neglecta, primo et ab adolescentia medicine scientie insudandum est, auctore Halyabate" [Consequently, the future doctor should from an early age dedicate himself to education. Not only in those arts that are part of the trivium – that is to say dialectics, without which, as Galen says, truth will not be recognised and no one will be saved from error, as well as rhetoric and grammar, and those of the quadrivium – that is arithmetic, music, geometry and astrology – supported by those of the trivium. He may continue in these disciplines at least as far as medical science requires. In this context natural philosophy as the most important should not be neglected. Firstly and 'from youth onwards,' according to Haly Abbas, 'one should sweat [with effort] for the sake of the art of medicine'].

20 Haly Abbas, *Regalis disposition* I.2, ed. Vitalis 1492: f. 2v–a: "Oportet autem hec ab pueritia observare et adoloscentia" [These [the *artes liberales* are meant] must be pursued from childhood and adolescence on]. See Pietro d'Abano 1472: 2, f. 8v.

21 Zerbi Gadebusch Bondio, Förg and Kaiser, eds., 2019: 80–81.

Right Knowledge as Moral Capital

The conviction that thorough and appropriate education is a protection against false teachings and mistakes can also be found in later deontological writings, above all in the 1540s. Whilst in his short *De medici et aegri officio* (1505/1506) the anatomist Alessandro Benedetti (*c.* 1450–1525) confined himself to only a few references, such as astrology[22] and simple substances,[23] to the issue of the appropriate knowledge of a *bonus medicus* or a *medicus prudens*, for the physician and philosopher Giovanni Filippo Ingrassia (1510–1580) from Sicily, the same subject was the basis for very detailed presentations.[24] Each chapter of his work *Iatrapologia liber quo multa adversus barbaros medicos disputantur . . .* (Venezia: G. Grifio, 1544) is devoted to one mistake. The main thrust of this polemical work points to the need to unite not only medicine and surgery, but also theory and practice (with the not insignificant example of autopsy), as well as subordinating medicine to philosophy. His objectives are: 1. not to leave surgery to uneducated barbers; 2. to establish autopsy as an essential mainstay of medical education, and 3. to define precisely both the methods and the aims of medicine as a worthy and noble science. Within this programme, Ingrassia dedicates the seventh chapter of the second book to the errors in defining what a doctor (*medicus*), physician (*physicus*), or surgeon (*chirurgus*) should know in order to be perfect.

Ingrassia's *medicus optimus* is steeped in knowledge of the natural (*cognitio rerum naturae*)[25] and supernatural (*nam praeter illorum naturam morbi consistunt*)[26] causes of disease. Ingrassia adds that only the doctor whose actions are based upon

---

22 Alessandro Benedetti, *De medici et aegri officio aphorismi Lib. I*, in Alessandro Benedetti, *De re medica opus insigne . . .*, 1549: 546: "Dubiae valetudini assidere medicus debet, et venarum pulsu vires saepe perquirere, caeterasque notas inspicere, qui et sideralis scientiae haud ignarus sit necesse est" [it is necessary that he should not be ignorant of astronomy! In cases of uncertain health the physician must support the patient, regularly check his physical condition by taking his pulse, and keep an eye on all other symptoms; furthermore he must be versed in astrology]. Translation from the original by Mariacarla Gadebusch Bondio. See also: Alessandro Benedetti, *De medici egri officio collectionum caput primum*, in Symphorien Champier, *De medicinae claris scriptoribus* [. . .] *tractatus* 1506: ff. 15–17.

23 Benedetti 1549: 546: "Simplicius effectus tenere summa cura debet . . ." [Knowledge of] the effects of simple substances should be taken careful note by the physician]. Translation from the original by Mariacarla Gadebusch Bondio.

24 Ingrassia is said to have studied medicine in Padua, Ferrara and with certainty in Bologna. In Palermo, Sicily, he was the personal physician of Isabella Di Capua, wife of Ferrante Gonzaga, the "viceré" of Sicily; he followed his patron to Naples where he lectured in theoretical medicine from 1545–1546 at the medical faculty. Between 1546–1547 he became lecturer of practical medicine and anatomy at the same institution. In 1563 he was appointed 'protomedicus' in Sicily, which allowed him to develop health policy reforms. See Preti 2004.

25 Ingrassia, *Iatrapologia, liber, quo multa adversus barbaros medicos disputantur . . .* 1544: 302.

26 Ingrassia 1544: 303.

this knowledge will not err (*non errabit*) – in as far as human power allows (*quantum humanis viribus permittatur*).[27] This physician-philosopher will possess the logic not to err beyond his own limits.[28] Like an avalanche, even small mistakes at the start of a logical argument would gradually become bigger and eventually be devastating. Ingrassia's language is impressive and rich in metaphor: owing to a lack of knowledge of philosophy, the mind would tire in the darkness of eternal wanderings and grow old (*Quicunque in medicina se exercent, eius autem philosophiae ignari sunt, . . . animus in tenebris perpetuo aberrans fatigatus consenescit*).[29] For Ingrassia, the significance of the rational method in medicine cannot be emphasised enough. As well as knowledge in disciplines such as geometry, arithmetic, and astronomy, he underlines how physiognomic knowledge can help physicians to guard against mistakes (*At enim eorum animum errore confici, ac defatigari qui usum physiognomiae ignorant*).[30] With reference to Pietro d'Abano's first *Differentia*, he also names all the areas in which a physician should be skilled. In the last part of this long chapter, Ingrassia states precisely what a physician should know about the symptoms, the causes, the localisation of illnesses, and the power of the bodily fluids (*vires humorum*)[31] in order to be able to make a correct prognosis and to determine an effective treatment.[32]

A few short years after Ingrassia had published his polemic, Joannes Antonius Siccus's *De optimo medico* (1551) appeared in Venice. We do not know the dates of Siccus's life; we do know, however, that he worked between Venice and Padua and was a valued teacher of the physician Luigi Luisini (born 1526).[33] In this work, Siccus intended to present prospective physicians with a model in the shape of the outstanding physician from Crema (or Venice), Vittore Trincavella (1496–1568), to whom his book was dedicated and whom Siccus praised as a true *medicus optimus*. For Siccus, the aim of good medical education is also to instil the future physician –

27 Ingrassia 1544: 303–04.

28 Ingrassia 1544: 305: "Logicas enim methodos ignorantes medici non solum in hoc maxime peccant (alibi narrat) quod finibus propriis excedant, ac foede in rationibus hallucinentur: sed etiam impossibile est exacte aliquod dogma citra logicam speculationem constituere . . . parvus autem error in principio (philosopho etiam teste de caelo tex. 33) in fine maximus efficitur". [Those physicians who shut themselves off to the principles of logic not only commit an error in doing so, that they thus exceed their own limits and make speculations on the causes; rather it is impossible to establish any sort of theorem without logical thought. A small mistake at the beginning, however, will become exceedingly great at the end]. Translation from the original by Mariacarla Gadebusch Bondio.

29 Ingrassia 1544: 306.

30 Ingrassia 1544: 306–307. Ingrassia assigned physiognomy, as *philosophia iudicatoria*, to astronomy.

31 Ingrassia 1544: 310.

32 Ingrassia 1544: 310–311.

33 See Poma 2014: 161–174.

who should have a certain inherent, intellectual aptitude (*ingenius*) – with the ability to distinguish between true and false.[34] In addition to the *artes liberales*, Siccus, like Galen, held the conviction that a physician should also be a philosopher.[35] At this point, he draws attention to the necessity for a good physician to be familiar not only with the *artes liberales*, but also with gymnastics and its effects, in order to be able to recommend the correct physical exercises to the patient. In the spirit of the 16th-century, Siccus decreed that the *medicus optimus* should carry out many dissections and be well trained in anatomy, and thus be able to determine precisely parts of the body and their functions, their diseases, and corresponding prognoses.[36] Pharmaceutical knowledge,[37] in other words, detailed knowledge of medications and their composition and effects, was necessary for the restoration of health. In the endeavour to acquire this knowledge, the physician had to be able to distinguish good books from bad, and truth from lies.[38]

The physician Leonardo Botallo (1519–1588), born in Turin and active in France, ensured a continuation of this tradition. His deontological text, *De medici et de aegri munere / On the Duties of Physician and Patient* (1565), expands the point of view to include the patient. The text is dedicated to Jacob of Savoy (1531–1585). Botallo introduces the topic of the required knowledge for a physician with the question: "What have the writings of Cicero, Livius, Virgil, and Lucian in common with those of Hippocrates, Galen or Aristotle?"[39] "Very little" is the answer, for the writings of those authors merely serve the vain display of learning. A certain pragmatism, along with the tradition of criticism of the medical profession, which since Petrarca had emphasised the dangers of useless knowledge for both doctor and patient, are clearly

---

34 Siccus, *De optimo medico*, 1551: f. 5v: "Ut ergo in omni quaestione, quid verum, quid falsum sit, diiudicare possit, acri primum sit ingegno opus est. " [In order for him to be able to distinguish between right and wrong in every question, it is first and foremost necessary that he be of keen mind]. Translation from the original by Mariacarla Gadebusch Bondio.

35 Siccus 1551: f. 6r.

36 Siccus 1551: ff. 7r–v.

37 Siccus 1551: ff. 7v–8r.

38 Siccus 1551: f. 8v. "Hinc factum est, ut multi suos libros variis mendaciis, et nugis asperserint" [Thus it happened that many have besmirched his books with all sorts of lies and fooleries]. Translation from the original by Mariacarla Gadebusch Bondio.

39 Botallo, *Tractatus de medici et aegri munere*, in Botallo, *Opera Omnia*, 1565: 4–5: "Quid commune habet Ciceronis, Livii, Virgilii, Lucanique lectio cum Hippocrate vel Galeno, vel Aristotele? Juvat quidem ad vanam ostentationem, verum detrahit haec non pauca ab artis opere ad quod omnem collimare ingenii aciem debet opifex . . ." [What has the reading of [the works] of Cicero, Livus, Vergil, and Lucan in common with Hippocrates, Galen or Aristotle? It promotes only inane boasting and in truth detracts considerably from the proper exercise of (medical) art, upon which those who practice it should focus all their sharpness of mind . . .]. Translation from the original by Mariacarla Gadebusch Bondio. See also the Italian translation: Leonardo Botallo, *I doveri del medico e del malato*, eds. Caretri and Bogetti Fassone 1981; Mariacarla Gadebusch Bondio 2005: 663–694.

recognisable here. Along with Galen, Botallo expresses the opinion that in order to avoid mistakes which might cause harm to the patient, emphasis should be placed upon a knowledge of things themselves rather than on their definition.[40] With his criticism of superfluous and useless knowledge which has no real relation to fact, Botallo established the question of what constitutes indispensable knowledge for a physician. He comes to the pragmatic conclusion that the time an individual has at his disposal in order to acquire well-founded knowledge is limited. As a result, no certainty can be achieved in medicine. The complexity of the subject and the conjectural nature of medicine lead to frequent false conclusions (*Medicina est, in conjecturis tota reposita*):[41] a physician thinks before he acts; should he err in his judgement, however, he often makes mistakes (*non raro fallatur*),[42] which can have disastrous and even fatal consequences for the patient.[43] In order not to harm the patient, therefore, the physician must dedicate himself to study, perfect his knowledge and skills, and be effective and quick without losing himself in useless details.[44] To logic, philosophy, astronomy, arithmetic, and geometry, Botallo adds optics (*perspectiva*), which is necessary for an understanding of the function of the eye.[45] For Botallo, as for Siccus (and almost the entire medieval medical tradition), anatomy, pharmacy (effect and application of simple substances), and experience in the so-called *sex res naturales* all belong to the fundamentals of medical education.[46] The focus on those areas of knowledge which are necessary for the successful practice of medicine becomes a moral obligation for the conscientious physician, who will no longer tolerate any distraction.

---

40 Galen, *Methodo medendi* II.2, VI.1 and VII.2, ed. Kühn, *Galeni Opera* omnia, vol. 10, 1825: 81–82, 385, and 459, respectively.

41 Botallo 1565: 7.

42 Botallo 1565: 7.

43 Botallo 1565: 8. ". . . Si fallax judicium, cur non et plerunque fallax erit opus ab eodem judicio institutum? Verum exigui momenti tale erratum non est, cum hominis perniciem quandoque inferat" [. . . If the assessment is incorrect, why then should not the work which is based upon this assessment be mostly flawed?]. See also 16: "Sed et assiduum hominem exigit medica ars, cuius errores in patientium jacturam vertuntur" [But even medicine requires an industrious man whose mistakes are at the expense of the patient]. Translation from the original by Mariacarla Gadebusch Bondio.

44 Botallo 1565: 15.

45 Botallo 1565: 8.

46 Botallo 1565: 11.

The Avoidance of Mistakes through Criticism and Self-Criticism

Around the middle of the 16th-century, observations on the medical branches of knowledge become an integral part of the *optimus-medicus* texts, although the argument for the avoidance of mistakes through right knowledge is differently evaluated. Roderigo de Castro (*c.* 1546–1627),[47] who addresses himself in his *Medicus politicus* to the virtues and faults of physicians with the aim of helping good physicians protect themselves against errors and mistakes, supported by a girdle of virtue, deals at several points with issues relating to knowledge.[48] The philosophy which "frees the soul from the passions," just as medicine frees the body from illness, should be twinned with traditional medicine.[49] De Castro criticises those who regard astrology, cosmography, mathematics, music, and geometry as disciplines a physician should learn, and states more precisely that "all this lies beyond that which we regard as the foundation of medicine" (*Haec omnia citra id, quod artium*

47 Also known as David Namias de Castro. See Sarraga and Sarraga 1997: 661–720 (especially 710). Rodrigo de Castro was born in Lisbon and studied medicine and philosophy in Salamanca. In Hamburg in 1596 he battled against the plague and wrote his *Tractatus brevis de natura et causis pestis quae hoc anno MDXCVI Hamburgensem civitatem afflixit* (Hamburg, 1596). As a result, an exception was made for him as a *marrano* and he was allowed to purchase a house in the Old Town. Benedict de Castro (born in Hamburg in around, died on 31st January 1684 ebd.), one of Rodrigo's sons, was like his father much in demand as a doctor at the princely courts of Europe. His many patients included, for example, Christine of Sweden. See Kellenbenz 1958: 323–331 (327); and more detailed: Arrizabalaga 2009: 107–124 (especially113–114); Gadebusch Bondio and Förg 2020: 83–114.

48 Roderigo de Castro, *Medicus Politicus, sive de officiis medico-politici tractatus*, 1614, *Preface*: f. a2v: "Itaque orsus ab eiusdem studiis, et genere disciplinae, ad medico virtutes amplexandas, fugiendaque vita me confero: nec tamen in omnes virtutum, aut vitiorum locos egregio, sed in eos duntaxat, qui maxime videntur medicis peculiares: tunc deinde mores candidiores moneo, et statuta medicis servanda propono, ac fraudes detego quondarum, quibus haud dubie odiosus noster hic labor erit". [Thus I have deferred my previous studies and am dedicating myself to the question of which virtues the physician should embrace, and which errors he should flee from. I do not address all virtues and errors, but only those which I find of particular importance for the physician. I exhort purer morals and establish rules which a physician should adhere to; I uncover some deceptions which undoubtedly cause our work to become an object of hate. But it is nothing new that the malicious hate those who give good advice]; see also de Castro 1614: 53–57: "De Disciplinis medico necessariis." Translation from the original by Mariacarla Gadebusch Bondio.

49 de Castro 1614: f. 53: "Democrates sapientiae cognitionem dicebat medicinae sororem esse, qui aut sapientia animam ab affectibus liberat, ita medicina morbos a corporis aufert" [Democrates states that the knowledge of philosophy is the sister of medicine. Just as philosophy frees the soul from the affects, so too does medicine remove away illness from the body]. Translation from the original by Mariacarla Gadebusch Bondio.

harum fondamenta teneamus).[50] A knowledge of the languages in which medical texts are written (Latin, Greek, and Arabic) is of far more value for a physician, he stresses, and so finds himself in the tradition of the criticism of the neglect of languages shaped by Roger Bacon (1214–1292 or 1294).[51] However, for the Portuguese Sephardim working in Hamburg, a physician should also have knowledge of the "vulgar" languages, allowing him to converse with sick people from many nations. Dialectics will be of use to him as an instrument for rational consideration, as it sharpens the mind and makes it possible to distinguish between what is applicable to medicine and what is not.[52] Arriving at anatomy, de Castro emphasises the necessity for autopsy (*Tum deinde Anatomes et oculorum inspectio medico pernecessaria est*).[53] Practical skills can guard against treatment errors: surgical skills are mistakenly regarded as superfluous for a physician, de Castro, on the other hand, is convinced that a perfect physician should be skilled in the art of surgery, even if he does not practice it with his own hands (*tametsi eam manibus non exerceat*).[54] His comment that "a master who has enough to command cannot simply give a slave/servant a command if he understands nothing of what he is commanding" makes eminent sense.[55]

De Castro has a similar attitude towards pharmacy. Knowledge of the characteristics of everything necessary for medicines, as well as of procedures (selection, date of gathering, storage and processing of the *Simplicia*) are, in his opinion,

---

50 de Castro 1614: f. 54: Nevertheless, de Castro dedicates a whole chapter of his book to the excellence of music: *Musicae excellentia, atque praestantia, rationibus, auctorum suffragiis et experimentis comprobatur* Roderigo de Castro, *Medicus Politicus, sive de officiis medico-politici tractatus*, IV, 270–77.

51 Bacon, *De erroribus Medicorum*, in Little and Withington (eds), Bacon, *Opera hactenus inedita*, 1928: 9.154–163; see also Gadebusch Bondio 2012: 291–311.

52 de Castro 1614, II.1 f. 55: "Est itaque legitimo futuro medico necessaria dialectica, ut quaecunque dixerit, scienter dicat, non contentus opinione quadam, quaquam recta, qua opinari quidam potest, rerum tamen medicarum sciens esse non potest" [Thus dialectics is necessary for the legitimate future doctor in order that everything he states, he states knowingly, and does not content himself with any opinion which may be correct, as some might believe, but who nevertheless does not need to be an expert in medical things]. Translation from the original by Mariacarla Gadebusch Bondio.

53 de Castro 1614, II.1 f. 55.

54 de Castro 1614, II.3 f. 67. De Castro criticises the situation in the German Empire and in Portugal where surgery is entrusted to a barber's shears, whilst in Spain, Italy and France medical surgeons, well-trained in both dietetics and in other rules of the art of medicine, practice honourably.

55 de Castro 1614, II. 3 f. 68: "Non enim simpliciter, ut servo dominus, qui satis habeat imperare, ipse eorum, quae jusserit, inexpertus, sed ut imperator militi qui quaecunque praeceperit, ea exacte calleat, et si opus sit, exequenda aggrediatur: omnia tamen proprie fecisse dicatur . . ." [A master who has enough to command cannot simply give a slave an order if he understands nothing of what he is commanding, rather he must, like the commander, know exactly what he is ordering and if necessary must perform it himself himself, then he has indeed done everything . . .]. Translation from the original by Mariacarla Gadebusch Bondio.

indispensable. Only if the physician is familiar with both theory and practice can he identify and correct the mistakes of pharmacists! He will be able to avoid gathering at the wrong time, buying bad or defective substances, and making mistakes in the selection of substitute substances (*quid pro quo*).[56] Typical pharmaceutical errors are connected to methods of storage: e. g. when pharmacists do not have enough stock, and as a result have to substitute one plant for another, medicines can have potentially deadly effects. De Castro complains of the tendency in his time to neglect all these disciplines and areas of medicine: physicians would leave to others what they urgently needed to master themselves (II. 5). This ignorance is censured as a deep-rooted fault, in particular when combined with arrogance (*dixit Galenus, indelebile vitium esse vehementem ignorantiam, praesertim si cum superbia coniungatur*).[57]

In comparison to his predecessors, de Castro's creation of a teaching programme for physicians is not only more comprehensive, he also calls for the appropriate stock of books of his *optimum medicus* with long lists in which the ancients (Greek and Latin authors), the Arabs and the "neoterici," the new authors, have their place on the imaginary bookshelves of a physician. Unlike Zerbi or Falcucci, he held a progressive position with regard to the authorities:

> Nobody should believe that Galen never made a mistake and that we should therefore believe him implicitly as if he were an oracle. . . . The physician should not be daunted so long as he realises that Galen wrote towards him, and who at different times and on different occasions wrote so much in which he must be contradicted. In such cases one should always determine his actual and true opinion, which is evident at many points and which is in better accord with reason.[58]

"Reason is preferable to authority" (*Rationem auctoritati esse praeferendam*), is the title de Castro gives to the eleventh chapter of his second book, where four characteristics are particularly emphasised, which, apart from presenting a credible line of reasoning, give the physician belief: prudence (*prudentia*), virtue (*virtus*), benevolence (*benevolentia*), and the highest of the four, wisdom (*summa viri sapientia*).[59] Authority is compared by de Castro to a nose made from wax which "can be easily turned in any direction," even by false argumentation, as even the most absurd mis-

---

56 de Castro 1614: II. 4 f.70.

57 de Castro 1614: II.5 ff. 72–73.

58 de Castro 1614: II.9 f. 87: ". . . Ne existimet quispiam, Galenum nunquam falli, et propterea in ipsius verbis quasi oracoli jurandum esse . . . . Ne deterreatur medicus, quotiescunque animadvertit, Galenum sibi contraria scribere, qui enim variis temporibus atque ob varias occasiones tot scripsit, in quibusdam sibi interdum contrariari necesse est: quo in casu exploranda semper erit potior ac verior ipsius sententia, qua eque pluriuso in locis legitur, et rationi magis consona videtur". Translation from the original by Mariacarla Gadebusch Bondio.

59 de Castro 1614: II.11 f. 94.

takes have their advocates (*quia auctoritas perinde est ac cereus nasus, quem facile veritas*).[60]

> For this reason the perfect physician should read books working from the assumption that even greater authors can have erred, for they sometimes waver and great men can succumb to the desire (*libido*) of their minds, can be deceived and deceive themselves. Even the most beautiful bodies bear scars.[61]

In the argument concerning those mistakes physicians make which obstruct their access to what is true and good (III, ch. 1), Galen's text on the passions of the soul and the avoidance of mistakes is, again, essential.[62] For Galen and de Castro, sadness, anger, greed, fear, and envy are the five passions of the soul which should be avoided.[63] De Castro sees these negative characteristics and dispositions as obstacles to the education of a good physician. Although Pietro d'Abano and Zerbi had also elevated medical vices (*vitia*) as a canon *ex negativo* – which they also ascribed to adverse astrological constellation of medicine – de Castro would appear to be the first to have followed through Galen's teaching here to its conclusion. Knowledge helps prevent mistakes, for this reason de Castro wanted physicians to recognise natural philosophy as the sister of ethics or moral philosophy. The latter could calm the "confusion of the spirit," which in its turn could weaken the body to a state of illness.[64] If, however, the soul is afflicted with defects, and is indeed full of bad properties, then additional (external) help should be obtained. A person should be sought to exercise criticism – a "parrhesiastes"!

---

60 de Castro 1614: II.11 f. 94: ". . . Plus fidei rationi, quam auctoritati adhibendum esse censemus: et ratio est, quia auctoritas perinde est ac cereus nasus, quem facile vertas, quocunque velis, sinistra interdum interpretatione, et nullus est tam absurdus error, qui defensorem nanciscantur" [However, this does not prevent us from judging; rational consideration requires more trust that authority. The reason for this is that authority is like a wax nose that can easily be turned whichever way one wants, even with a false interpretation, and there is no error so absurd that it can find nobody to defend it]. Translation from the original by Mariacarla Gadebusch Bondio.

61 de Castro, 1614: II.11 f. 96: "Quocirca velim medicum perfectum eo animo libros legere, ut estime, potuisse etiam magnos autore errare: nam labuntur aliquando, et indulgent viri magni ingeniorum suorum libidini, fallique et fallere possunt: et sunt in pulcerrimis corporibus sui etiam naevi." Translation from the original by Mariacarla Gadebusch Bondio.

62 See Galien, *L'âme et ses passions, les passions et les erreurs de l'âme. Les facultés de l'âme suivent les tempéraments du corps*, eds. Barras, Birchler and Morand 1995.

63 de Castro 1614, III.1 f. 110: "Porro Galenus quinque animi affectus imprimis vitandos censet, moerorem, iram, cupiditatem, metum, atque invidiam: qui sane si a reliquis hominibus, a medico multo magis sunt exterminandi, quoniam multitudini exemplo esse debent" [Galen judges that there are above all five affects of the soul which should be avoided: sadness, anger, greed, fear and jealousy. If other people must avoid these, then the physician must all the more, as he should be an example to the people]. Translation from the original by Mariacarla Gadebusch Bondio.

64 de Castro 1614: II.1 f. 56.

> For the person who is called excellent and productive by all others, the same author [Galen] wants him to choose somebody from amongst his friends who points out individual errors as an admonisher and indicator of those errors against which a physician should believe himself more protected from, than if he had physically healed a patient. But whoever regards himself as one of the best, and allows no one to pass judgement on him, of him we see that he is deluding himself in the greatest matters, as many would achieve the highest levels of competence if they had not believed they had already attained them. But whoever loves himself best is therefore most blind to himself. If we have a double satchel hanging over our shoulder, one on our chest with the nature of others, one on our back full of our own errors, we always see those of others but never our own.[65]

Returning to the question I posed at the beginning, whether fallibility itself could be considered as a motivation for physicians to concern themselves with their moral profile and to write on the topic – my answer would be positive. The awareness of medical fallibility stimulated early modern physicians to develop more or less practically oriented strategies useful in combating or at least controlling it. Authors shaped their prevention programmes starting from variously detailed educational concepts, which gradually extended to include other branches of knowledge and practices, such as anatomy, along with autopsy, optics, pharmacy, surgery, and physiognomy. The necessity of detecting and recognising the mistakes and deceptions of related practitioners (surgeons, pharmacists) was a crucial reason for extending the competencies of a physician into hitherto neglected practical fields. At the end of his *Medicus Politicus*, De Castro demonstrates the therapeutic effect of music by means of countless examples of healing that he has either collected or experienced himself he arrives at a more general warning. De Castro criticises the force of habit which leads to new, better habits being rejected. For this he blames hostility to change and the failure to admit to mistakes. Such people behave as if they had closed their ears, and persist "deafly" in their convictions. The closing thought of the *Medicus Politicus* indicates the consequences of moral rigidity. It is the perpetuation of errors, even obvious errors, which threatens society and prevents renewal.[66]

---

65 de Castro 1614: III.1 ff. 110–111: ". . . Ideo vult idem auctor ad hoc, ut unusquisque sese exhibeat dignum, qui egregious ac frugi ab omnibus appelletur, ut aliquem sibi eligat ex singularibus amicis, qui singular errata indicet, quemque censorem habeat admonitoremque ac vitiorum indicem: a quo magis medicus se servatum existimet, quam si corpora aegrotantem sanasset. E converse enim qui seipsos optimos existimant, nec de se aliis iudicium permiserunt, hos in maximis rebus hallucinari conspicimus, quia multi ad summum virtutis pervenirent, nisi iam pervenisse existimarent. At vero quidquid amat, coecum est erga rem quae amat: quivis autem nostrum se ipsum maxime diligit: idcirco erga seipsum maxime coecutiat est necessarium. Nam cum duplicem manticam e collo suspensam gestemus, unam quidem ante pectus, alieno rum: alteram a tergo, nostro rum vitiorum plenam, aliena spectamus simper, propria cernere nunquam possumus". Translation from the original by Mariacarla Gadebusch Bondio.

66 de Castro 1614, IV f. 277.

De Castro declared his need to eradicate errors and defects, to improve professionalism and establish a firm moral attitude in physicians as essential to their work. But unlike Zerbi and the other authors presented here, his intention harmonised more than ever with Galen's ideal of a *medicus philosophus*, deeply conscious of the problem of human fallibility, who would find the courage for criticism and self-criticism.

Jennifer Kosmin

When the Fetus Becomes a Child: Some Reflections from the Long Eighteenth Century

Over the course of the long eighteenth century in Europe, medico-legal definitions of fetal life increasingly shaped decision making over rights and inheritance, even as they intersected with much longer traditions of defining life and personhood through local, religious and cultural rites of initiation. During this time, a new corps of scientific experts was tasked with negotiating the limits of human life. Was the fetus sensitive – alive – while still in utero? What signs indicated vitality in a newborn baby? Were monstrous births human? How long must a neonate live/have lived in order to be considered a person with legal rights? In France and territories under the Napoleonic code, such as northern Italy, the construction of an entirely new legal category, "viability," extended the question of fetal life temporally. To fulfil its legal definition, viability required the prognosis of sufficient future existence. In other words, did the signs of life suggest the capacity for the fetus/newborn to *continue* to live independently of the mother once born, barring other complications?

This essay explores several moments during the long eighteenth century when questions of fetal life were negotiated by a variety of actors: parents, communities, the Catholic Church, medical experts, and legal authorities. As today, the negotiations surrounding fetal lives were grounded in particular local contexts, even as the participants involved referenced specific bodies of authoritative knowledge to support their positions. The construction of a class of medico-legal experts imbued with state authority represents part of a longer trajectory of biomedical ascendance in the western world, one that has obscured the intense cultural negotiations involved in determining the beginnings and ends of human life.[1] Today, as the measures of fetal life and viability become understood in increasingly technical and scientific terms and mediated by ever more advanced technologies, a historical perspective is critical for grounding a medical humanities approach to these questions.[2] In what follows, my aim is to lay bare some of the historical developments that underlie our modern conceptions of fetal life and to demonstrate that debates over fetal rights have existed in various forms over the last several centuries. One of the tenets of a critical medical humanities is that biomedicine cannot alone fully explain the nature, significance, or meaning of embodied experience.[3] Biomedicine

1 Kaufman and Morgan 2005.

2 On the medicalization of fetal life and the history of neonatalogy, see Casper 1998; Dubow 2010; van der Ploeg 2001; Lantos 2001; Baker 1996.

3 Bleakley 2014: 24.

https://doi.org/10.1515/9783110788501-004

is itself a cultural system that is embedded with particular social and political values, despite its claims to authority and objectivity. Medicalized definitions of fetal life are thus shaped by the political, social, and cultural needs of the societies that produce them, whether it be 21st-century America or late eighteenth-century Italy. I argue here that an increasingly medicalized approach to defining fetal life at the turn of the nineteenth century displaced earlier traditions – often reliant on women's knowledge and authority – that had traditionally governed when and how new lives entered into a community.

The long eighteenth century is a particularly fruitful moment in which to begin to think through the historical constructions of fetal life. In addition to witnessing the emergence of a new class of professional medico-legal experts, this period saw the interests of the Catholic Church and absolutist states projected onto the unborn in new ways.[4] Inspired by new discoveries related to embryology, many Catholic moralists began to promote the developmental model of preformationism, which would be the basis for the Catholic position that life begins at conception.[5] At the same time, states concerned with the twin goals of productivity and procreativity advanced systematic public health programs that included efforts to protect fetal and infant life while harshly criminalizing infanticide, abortion, and abandonment.[6] In Catholic countries like Italy, these interests came together in legal requirements for surgeons and midwives to perform the cesarean operation in cases where the life of the fetus was at risk, most commonly when the mother died during or before childbirth.[7] The result was a deontological shift on the part of medical practitioners to save the life of the fetus/child over or on par with that of the mother. In political discussions, the fetus became a 'future citizen', unborn but imagined as an eventual part of the body politic.

Historically, legal, medical, and religious writers have attempted to define the boundaries of fetal life for a variety of different reasons. In the early modern period, the burden of distinguishing abortions and infanticides from miscarriages and stillbirths drove the development of forensic medicine.[8] Key to such determinations was the ability to accurately establish the age of the fetus and whether it had been alive at birth. Measures of vitality were also central to the Catholic Church's official stance in this period that infants must be alive to be baptized. Theologians in the eighteenth century increasingly debated whether a fetus could be baptized in utero when its life or its mother was in distress or if it must have first 'been born'. Monstrous births, the humanity or number of individual souls within which could be difficult to discern,

4 On the history of legal medicine, see Forbes 1985; Pastore 1998; Watson 2010; Burton 2007; Clark and Crawford 1994.
5 Dal Prete 2015: 59–78.
6 Blum 2002.
7 Kosmin 2021.
8 Wessling 1994: 117–144; Jackson 1996: 84–109; Lewis 2016: 115–149; Watson 2019: 143–198.

also continued to trouble both religious and legal categories of life in this period. Additional interests related to concerns about legitimacy, inheritance, and succession. In the nineteenth and twentieth centuries, efforts to discern the boundaries of life and measures to save vulnerable fetuses/infants were entwined with public debates about resource allocation, 'quality of life,' and the place of intellectually or physically compromised persons within society.[9] Finally, since the 1970s in the United States, the conceptual categories of "fetal life" and "viability" have become virtually synonymous with highly political debates over abortion and fetal personhood.[10]

Despite the advances of neonatal medicine in the 20th and 21st centuries, however, there can be no objective scientific measure, or even estimation, of viability at any given moment. As anthropologist Astrid Christoffersen-Deb has written, there is no scientific answer to the question of, what is "viability"? There is only "the work of viability":

> Because the *viable fetus* does not solely exist within the medicalized spaces of childbirth, I have avoided discussing what "is" viability, and focused instead on the "work of viability" to emphasize that what is regarded as "viable" is condition to recalculation through social, medical and legal practices, and that through its application in different situations, the work of viability is variable.[11]

Considering the various historical evolutions in understandings of 'vitality' and the emergence of a discrete medico-legal concept of "viability" lays bare the 'work' that such categories may play in various contexts. Ultimately, it is only as a result of culturally determined negotiations about what rights (if any) an unborn fetus has, whether there is a point when a fetus has an absolute right to be delivered, and the nature of the mother-fetus relationship – questions themselves all mediated through a society's racial, economic, political, social, and religious landscape – that the *viable fetus* comes to exist as a conceptual category and legal entity.[12]

Baptism and Personhood in Historical Perspective

As the fundamental rite of initiation into the Christian community and the essential condition for eventual entrance into the Kingdom of Heaven, baptism holds a singular place in Christian ritual and theology. During the Protestant Reformation in

---

9 Obladen 2011: 563. In the late nineteenth and early twentieth centuries, these debates took place against the backdrop of the eugenics movement. See Pernick 1996.

10 Rachel Roth argues that the modern construction of fetal rights has negatively impacted women in a variety of ways beyond the typical association with the debate over abortion. Roth 2000: 1–5.

11 Christoffersen-Deb 2012: 587.

12 For a consideration of how fetal rights are determined in relation to class and race see Goodwin 2020; Roberts 1997; Davis 2019; Roth 2000; Kenney 1992; Gallagher 1987: 9–58.

the sixteenth century, when the practice of baptism was challenged, the Catholic Church responded by ardently reaffirming baptism's essentialness to salvation and taking a hardline on the fate of the souls of infants who died without baptism.[13] While theologians debated the exact justifications for this position, everyday Catholics understood baptism in immediate and personal ways. At a local level, baptism stitched together the ritual fabric of early modern communities in a way that, "transcended the crossroads between official religion and folklore tradition," and "transformed birth from a natural to a cultural fact."[14] In other words, baptism mediated the social and the physical. It was of deep, material concern to the lives of premodern Catholics, especially at a time when infant mortality was high. Traditionally, unbaptized infants were prohibited burial in churchyards and consecrated ground.[15] The wrenching emotional toil that parents might experience at the prospect of a child's soul banished to eternal damnation was no less real than the pain of the loss of the physical child.

One measure of this grief was the widespread phenomenon in parts of France, northern Italy, Switzerland, Austria, and southwestern Germany of the *répit* miracle, in which children who had died without baptism were briefly resuscitated in order that the rite could be performed. Although not unknown prior to the Reformation, the practice of parents making pilgrimages to particular sites associated with the *répit* miracle became a regular and accepted practice in the years after, and continued as late as the nineteenth century, despite repeated attempts by the Church hierarchy to suppress the unsanctioned ritual. According to Jacques Gelis, there were some twenty of these sites in the Italian Alps, and as many as 230 in France alone.[16] A notable aspect of the *répit* miracle was the consistent involvement of women. Imbued with the intrinsic capacity of generating life, women's bodies were often materially important to popular magic and religious ritual in the premodern world. In Carnia, in Italy, for instance, "the bodies of infants who had died during childbirth were brought [to the shrine] and these [women] would lay them out before the altar of the Madonna, celebrate the mass, say peculiar prayers, and suddenly shout that the Madonna had performed a miracle."[17] At the Madonna of Terlago, near Trent, women were also known to bring dead infants back to life. With the women's attestation of the miracle, a certificate of baptism could be drawn up and, most importantly, the child finally laid to rest in consecrated ground. The rituals thus served religious, communal, and also legal ends. A birth certificate

13 Poska 1998: 77–78.
14 Prosperi 2005: 197.
15 Unbaptized infants were excluded from Christian burial. Prosperi 2005: 208–210. See also Eurich 2008: 47.
16 Prosperi 2005: 205.
17 Quoted in Cavazza 1982: 551.

ensured the child could be buried in a cemetery, but also that that child could be given a name and recorded (and remembered) as having lived in the community.

The proof that a miracle had taken place was the observation of some sign of vitality in the otherwise dead child. *Répit* babies would briefly move, cry, urinate, or make some other sign of life signaling that the women could baptize them, after which they would return to death, readied for reburial in Christian soil. The miracles became so popular that, in 1755, Pope Benedict XIV issued a ruling clarifying what qualified as a "sign of life." He also reaffirmed that baptism should *not* be given to infants that did not show clear signs of life.[18] Benedict, himself deeply engaged in the scientific debates of his day, cautioned that slight movements were not in themselves sufficient to allow for baptism; rather, the child must perceptibly cry or moan as well. Benedict's ruling closely followed the logic of contemporary medical and legal writers who knew that brief movements could be both hard to detect and easy to confuse with involuntary responses to stimuli; crying, on the other hand, indicated both that respiration was taking place and that they child's lungs had developed to relative maturity.[19]

Early modern Catholics went to extreme lengths to ensure their infants were baptized because the alternative was unthinkable. To die without baptism meant to die without a soul, to be denied eternal salvation.[20] The soul was at the heart of the process of person making; it mediated between the material body and the individual, and between this world and the next.[21] The weight of the potential loss of this critical feature of identity therefore weighed deeply and heavily on the consciences of early modern parents. Even those parents who, for a variety of reasons, abandoned their infants at foundling homes invariably left them with notes detailing whether they had yet received baptism.[22] At the same time, as Benedict XIV's ruling demonstrates, the *répit* miracle was, by the eighteenth century, subject to increasingly specific medical criteria. The Church, while certainly not dismissive of concerns over fetal baptism (quite the opposite, as we shall see), *was* dismissive, if not downright hostile, to the appropriation of sacred rites by women. By formalizing the signs of life required by the Church to perform baptism, Benedict XIV not only medicalized the rite but also reinforced the fact that the sacred was the preserve of masculine authority.

18 Cavazza 1982: 110.
19 Ciancio 2017: 16–25; Woods 2009: 123.
20 Prosperi 2005: 175.
21 Prosperi 2005: 220.
22 Prosperi 2005: 192.

Theological Embryology and the Cesarean Operation

The Catholic Church's Tridentine preoccupation with baptism was reinvigorated in the eighteenth century in response to new scientific discoveries relating to embryology and the heated debates about ensoulment and animation that resulted. While some scientists and philosophers argued that the advanced organization that the animal embryo attained during development existed complete in some form from the time of conception (preformationism), others held that the embryo developed gradually from unorganized matter (epigenesis).[23] The notion behind preformationism that essentially an entire human being was present at conception, only waiting to be revealed over time during development, appealed to theologians who seized on the opportunity to harness science to support a religious worldview. To such thinkers, the preformation thesis not only demonstrated the omnipotent wisdom of God's creation, it also allowed for the argument that human ensoulment begins at conception, something that aligned well with the idea of Mary's Immaculate Conception. Nevertheless, this view represented a quite drastic revision of traditional Aristotelian and Thomistic doctrine on ensoulment, which held that animation begins at between thirty and forty days for males, and seventy to eighty for females. These figures had long provided the basis for both religious and secular legal codes. That is to say, abortion was typically only considered a crime (or only a severe crime, such as murder) if carried out after these supposed points of animation.[24] They also shaped women's own perceptions and understandings of pregnancy, as, in the early stages after conception, there was not generally seen to exist a child, but rather an unformed mass that could still be false pregnancy (*mole*), retained menses signaling some kind of ill-health, or even something more malicious or monstrous.[25]

One of the most influential religious works inspired by the debates over embryology was Francesco Emanuele Cangiamila's *Embriologia sacra: Ovvero dell'uffizio de' sacerdoti, medici e superiori circa l'eterna salute de' bambini racchiusi nell'utero* [Sacred embryology: That is, on the duty of parish priests, physicians, and officials with respect to the eternal well-being of infants still in the womb].[26] Cangiamila, a Paleman priest and jurist at the time he wrote his treatise, was deeply swayed by the logic of preformationism. He was also acutely aware of existing theological debates

23 On the debate over preformationism in Italy, see dal Prete 2015: 59–78.

24 Christopoulos 2021: 9. Christopoulos notes that despite general agreement on ensoulment, there was significant variety in how individual jurists and theologians discussed these matters. In addition, in 1588 Pope Sixtus V issued a papal bull classifying voluntary abortion at any stage of development as homicide. Sixtus' successor, Pope Gregory XIV, seems to have quickly moved to retract this bull. See Christopoulos 2021: 145–148.

25 Duden 1993a: especially 56–61, 79–82; Duden 1993b: 159–160.

26 Cangiamila, 1751. All quotes in this article refer to the 1751 Milanese edition, though Cangiamila's text was first published in Palermo in 1745.

over the fate of babies who died without baptism.[27] While proponents of probabilism, especially Jesuits, tended toward a less rigid interpretation, rigorists, held that unbaptized infants were, without exception, denied salvation. During the eighteenth century, many rigorists, Jansenists, and reformers like Cangiamila, who argued for a partnership between Catholic theology and modern science, became strident proponents of extreme measures to ensure fetal baptism.

As a work meant for a varied audience of clergy, laymen, jurists, and medical practitioners, the *Embriologia Sacra* was as much a practical handbook as it was a learned theological discussion. The text included chapters on the causes of voluntary and involuntary abortion (miscarriage), how priests might help to prevent them, why animation begins at conception, and the procedures for baptism in a variety of cases. A bestseller with wide support from the Church, the *Embriologia Sacra* was translated into Latin, Spanish, French, Portuguese, and German and remained relevant well into the nineteenth century.[28] In fact, the *Embriologia Sacra* had a direct influence on legislation in Cangiamila's native Sicily where a decree issued in 1749 made the cesarean operation mandatory for women who died while pregnant (even if there was only the suspicion of pregnancy).[29] As many as 225 postmortem cesarean sections were performed in the Kingdom of Sicily between 1760 and 1762.[30] Similar edicts followed in Venice and Milan in the 1760s.[31] The thrust of such statutes was, moreover, that relatives who impeded the cesarean might legally be viewed as assisting in murder.

Overall, Cangiamila's position on baptism and the cesarean operation was an extreme one, if not entirely new. Owing to his belief that ensoulment followed closely if not immediately after conception, the priest argued that baptism should be performed on all abortions, even those that occurred in the early days of a pregnancy.[32] The *Embriologia Sacra* explained in detail how the fetus can survive for short periods in the womb even after the mother's death, justification, according to Cangiamila, to intervene with a post-mortem cesarean at such times. Most unconventionally, he advocated that the cesarean operation be performed not only on all dead women that were suspected or known to be pregnant, but in certain cases on living women as well. While various – and often suspect – accounts of successful cesarean operations being performed on living women dated from the sixteenth century, most medical experts in the eighteenth century were extremely wary about

27 Filippini 1995: 59–63, 81–84.

28 Filippini 1995: 61.

29 Warren 2009: 652. In fact, there may have been as many as 225 post-mortem cesarean sections performed in Sicily in just the two-year period between 1760 and 1762. Warren's data here comes from Pundel 1969: 91n7.

30 Warren 2009: 91.

31 Pancino 1984: 146.

32 On ideas about ensoulment in this period see Prosperi 2005, esp. 218–299.

the mother's prospects of surviving such an operation.[33] Many jurists and public health writers similarly questioned the logic of undertaking a procedure that was almost surely guaranteed to kill the mother when the child's likelihood of survival was itself extremely low. Cangiamila's position thus only made sense when the baptism of the fetus was prioritized over the mother's life.

As Adriano Prosperi has pointed out, "in the cesarean section . . . the priest and the physician exchanged roles, and the life of the soul was the prize gained with the physical death of the mother and fetus."[34] Indeed, it was parish priests, according to Cangiamila, who would have to be ready and willing to perform the procedure themselves, as the reluctance of relatives to the dissection of the pregnant woman might prevent others from acting.[35] Families might indeed be deeply resistant to the violence that the cesarean operation enacted on the bodies of loved ones. Even though early modern Catholics held deeply to the spiritual significance of infant baptism, they also placed a high value on bodily integrity. Although some medieval religious teachings prescribed that a pregnant woman who died without delivering could not be buried in consecrated ground because of the spiritual contamination threatened by the unbaptized fetus, it is unclear to what extent such practices were still followed in the eighteenth-century Catholic world.[36] In either case, the result of the Church's insistence on baptism was the spiritual or physical condemnation of the life of the mother.

Despite Cangiamila's impassioned efforts, the official decrees on the subject of postmortem cesarean section issued in northern Italy were generally cautionary. Although officials endorsed the operation – *postmortem* – they were skeptical of priests or midwives undertaking a serious surgical operation and advocated that such procedures only be performed by trained surgeons or physicians. In Venice, for instance, a list of capable practitioners was to be posted at all apothecaries so that community members would know whom to call in such a situation.[37] Johann Peter Frank, the director of public health in Lombardy, was deeply concerned that an extreme position such as Cangiamila's might result in an overzealous priest or other untrained person operating on a woman who was in fact still alive. No doubt influenced by recent developments in artificial resuscitation and a general uncertainty about how definitively to diagnose death, Frank cautioned that the signs of death could be misleading and ambiguous. He not only outlined several tests to ensure that the mother was truly dead – such as holding a candle or glass below the woman's nose – but he also concluded that the occasional loss of a fetus was not

---

33 Blumenfeld-Kosinski 1991. esp. 7–47.
34 Prosperi 2005: 216.
35 Cangiamila, *Embriologia Sacra*, II.8, 138–142.
36 Muir 2005: 26; Korpiola and Lahtinen 2015.
37 Pancino 1984: 146.

worth a massacre of child-bearing women.[38] Moreover, Frank thought undertaking the cesarean operation in the very early stages of pregnancy was not only useless but barbaric since the fetus had no possibility of living outside the mother.[39]

Despite the differences in tone between Cangiamila and officials like Johann Peter Frank, the underlying assumptions in such discussions marked a profound shift in thinking about the relationship between mother and fetus. With the introduction of theories that potentially moved animation all the way back to conception, religious writers like Cangiamila began to think of the fetus, if not as an individual, then at least as a creature independent of the mother that was deserving of the tutelage of the state. According to Cangiamila, the fetus "becomes animated" in the first few days and therefore, "however small . . . nonetheless is living . . . and [it] is therefore never licit to murder it."[40] Frank, too, wrote with a new sensibility about the nature of the being contained in the womb. His words would have a deep impact on public health legislation not only in Lombardy, but in Europe as a whole.[41] In his *Sistema completo di polizia medica*, Frank asked, "are not the citizens still enclosed in their mother's wombs nonetheless members of the state?"[42] Influenced as much by the political arithmetic of populationism and new conceptions of the state's responsibility toward its subjects as he was by the spiritual concerns of Cangiamila, Frank nonetheless concluded that the unborn fetus was deserving of the protection of the state through laws that, among other things, would compel qualified practitioners to perform a post-mortem cesarean operation.[43] In both cases, the cesarean operation signaled something much more consequential than the development of a new medical intervention. As Nadia Maria Filippini has argued, the new sensibilities toward the fetus gestured to a "profound rupture of tradition, one that disrupted the hierarchy of moral, professional, and social ethics."[44] For the first time, the life/soul of the fetus was considered equally, if not paramount, to that of the mother.

Medical practitioners, too, seemed to consider the fetus with a new sensitivity by the late eighteenth century. Treatises from this period suggest subtle but profound shifts in the understanding of the relationship between mother and fetus and the medical professional's sense of responsibility toward each entity.[45] For instance, some proponents of the cesarean operation, such as the professor of surgery Pietro Paolo Tanaron, outlined conditions for performing the procedure both post-mortem and while the mother was still alive. If the child's passage was undeniably

38 Pancino 1984: 147–148.
39 Filippini 1995: 129.
40 Cangiamila, *Embriologia Sacra*, I.3.16.
41 Prosperi 2005: 217.
42 Johann Peter Frank 1807: 166.
43 Filippini 1995: 117–121; Prosperi 2005: 216–217.
44 Filippini 1995: 13.
45 Prosperi 2005: 252–265.

blocked yet the infant suspected still to be alive, were there men, he queried, "so barbarous, and so deprived of humanity, that they could plunge a knife into the breast of a poor, little infant (*creatura*) and cut it to pieces . . . so that it could be pulled out?"[46] The cesarean section was therefore presented as the more humane option when compared to the horrors of an embryotomy, even when the operation might put the mother's own life at risk. In fact, Tanaron went so far as to argue that the learned practitioner who failed to perform the cesarean operation in such a situation should be judged in line with any other murderer:

> Princes, and Magistrates judge to be the offenders those prostitutes, and other women, known to have caused the deaths of their children, either through a procured Abortion, or an Infanticide; so why not punish similarly those, who because of fault, or negligence, cause to perish within the womb those unfortunate infants . . . even though they could have saved them with the application of their profession? Since this question concerns [the loss of] the physical life, no less than the spiritual one, and as there should be equal consideration for the one as for the other crime, then any Practitioner (*Professore*) who out of negligence, or, even more if out of politics, or maliciousness omits [to perform] the Cesarean Operation he should receive a severe penalty, as grave as that for the perpetrator of Homicide.[47]

Although there were obvious differences in how eighteenth-century theologians, medical practitioners, and jurists thought the cesarean operation might best be applied, there was an increasing consensus on its utility, at least in post-mortem cases. Surgery and theology had combined in this instance to reimagine the nature of the relationship between mother and fetus. According to Prosperi, the prospect of the cesarean operation had dramatically "changed the social condition" of the *creature* that existed in its mother's womb; it had "become the object of great investment by powers and disciplines of all kinds, just as a special system of surveillance had been put into place over unmarried mothers."[48] By investing the fetus with a greater humanity than ever before, the changed theological and medical landscape of the eighteenth century brought women, particularly those that might be suspected of abortion or infanticide, under greater scrutiny and legal supervision.

46 Tanaron III.26. At the time, apart from the cesarean section, the only sure method for delivering an obstructed fetus was to perform a craniotomy or embryotomy and pull the baby out in pieces, a procedure typically performed by a surgeon. This was seen as the safest procedure for the mother.
47 Tanaron III.3.95.
48 Prosperi 2005: 217.

Monsters and the Limits of the Human

Monstrous births, including fetuses with congenital deformities or those with defects resulting from injury while in utero or during delivery, challenged early modern observers to consider the boundaries of the "natural" and the "human." Early modern monsters provoked a variety of emotional responses, including fear, wonder, revulsion, and pleasure, and their depiction in stories and pamphlets served as powerful metaphors for religious and political transgression. Indeed, much has been written about the significance of monstrous births as signs and portents, particularly in an atmosphere of religious turmoil after the Reformation.[49] Their origin was hotly contested, but many early modern authors continued to believe that the power of the maternal imagination or misdeeds during conception could lead to monstrous children.[50] As such, monsters represented a powerful form of control imposed over the bodies of pregnant women. At the same time, monsters were biological realities and immediate "problems" for families and their communities.

The most basic and pressing question was whether a monstrous birth should be baptized. To be baptized meant, at a basic level, to be recognized as human. As Prosperi writes, baptism had also traditionally been linked to a child's ability to inherit.[51] Thus, from the 16th century, medico-legal writers distinguished between monstrous births that lacked all resemblance to a human being and those that bore at least some recognizably human attributes. In the 16th century, the physician and author of a popular pamphlet on monsters, Giovanni Filippo Ingrassia, wrote that monsters of the "first kind" were those that carried no human form or combined human and animal form to such an extent that it was clear such creatures lacked a rational human soul.[52] Baptism was unnecessary in such cases. In cases where there was uncertainty about whether the monster was or was not human, medical and religious writers urged parents and priests to be cautious but inclusive – when there was doubt, the child could be baptized with conditions (eg "*If* you are a human, I baptize you . . .").[53] In other instances, monstrous births raised questions about where the seat of the soul was located, a debate that had existed since Plato and Aristotle.[54] If conjoined twins shared one heart, for instance, was one baptism required, or two? Ingrassia strongly favored the heart over the brain as the physical

---

49 On monstrous births as signs or portents, see Wilson 1993; Walsham 1999; Knoppers and Landes 2004; Crawford 2005; Spinks 2015.

50 On monsters and the maternal imagination see Huet 1993; Daston and Park 1998: 173–214; Finucci 2001.

51 Prosperi 2005: 215.

52 Cusumano 2012: 872.

53 For example, Baruffaldi, 1760: 40–46; Baruffaldi 1789: 19; Capecelatro 1817: 25.

54 Plato argued for the brain, while Aristotle held that the seat of the soul was the hart. Cusumano 2012: 871–872.

location of the soul, arguing that doubled limbs did not in themselves prove that there were two entities in need of individual baptisms – only two distinct hearts could demonstrate that. As Ingrassia, other early modern physicians, and many Catholic writers would argue, however, most monsters *were* in fact human, natural if unusual examples of human variety, and therefore baptism was essential, especially since many such births survived for only short periods.

Given their association with sexual and moral transgressions, however, parents of monsters were sometimes less inclined to recognize officially their offspring. In such cases, monstrous births represented "public exposures of their mothers' hidden or as-yet-unpunished desires, disorders, or crimes."[55] Parents might be left with "a permanent stigma, a halo of scandal and shame," that could also extend to the entire community.[56] As a result, many parents may have attempted to conceal monstrous births. Particularly in cases of extreme deformity – where the creature's humanity could not be detected – midwives or mothers most likely left the monster without care, to die quickly on its own. There is also evidence that monsters of all kinds were deliberately killed, though such behavior was increasingly condemned.[57] As Ottavia Niccoli notes, as late as the 18th century, the religious author Fra Diodato da Cuneo had to enjoin midwives and parents not to kill monsters of their own accord, but to consult with authorities on how to proceed.[58] The public health expert Johann Peter Frank similarly wrote that although it was most sinful, it was a "quotidian experience" to hear of monstrous newborns suffocated immediately after birth by spooked midwives or relatives.[59] Whereas the question of which births would be recognized as human had once "been subject to the final judgement of the midwife, mistress of life and death," increasingly such determinations were subject to the "stringent examinations of a religious and civil nature."[60] Yet, the admonition of Cuneo and Frank suggest changes occurred gradually over time. Even as eighteenth-century medical writers increasingly understood and wrote about monstrous births in naturalistic terms, parents and midwives continued to make decisions about who would be included and recognized as fully human members of their communities.

Although earlier associations of wonder and awe didn't entirely fall away, by the eighteenth century, monstrous births were largely accepted in learned circles as anatomical anomalies of a purely human nature. In fact, Frank wrote that such

55 Crawford 2005: 66.

56 Walsham 1999: 17.

57 On the other hand, at least some parents saw in monstrous births the potential for fame and financial return. The Turin anatomist Giambattista Bianchi recounted, for instance, the story of conjoined twins born in Pavia in 1748 whose father brought them from town to town to display them, even in poor weather, likely leading to or at the very least accelerating their death. See Bianchi 1748.

58 Niccoli 1980: 404n.13.

59 Johann Peter Frank 1807: 198.

60 Prosperi 2005: 140.

events were in a way fortuitous as the anatomical study of monstrous births would be of "great utility for the republic, and for legal medicine."[61] Much like the future citizens Frank saw enclosed in their mother's wombs, even deformed births could serve the serve the state through the investigation of their bodily exceptionality. Indeed, monstrous births became favored topics of presentations to medical and academic societies in the eighteenth century.[62] Removed from the highly charged atmosphere of their birth, in which parents and neighbors struggled with how to interpret the deformed child, such births became valued scientific objects, able to enhance the careers and reputations of those who studied or collected them.[63]

Life between Vitality and Viability

In 1812, in the small Italian town of Casalmaggiore, a woman named Angiola Chiozzi died, intestate, while seven months pregnant. A baby girl was delivered shortly after the mother's death by cesarean section. In the months following Chiozzi's death, a protracted legal battle ensued between Chiozzi's relatives and her husband over whether the baby girl had ever, in fact, lived.[64] At stake in the case was whether the mother had died childless, meaning her inheritance would be returned to her natal family, or whether that inheritance, through the fact of legitimate succession resulting from a live birth, would stay with her husband and any of his future children. Although the case hinged on a seemingly simple question – was the Chiozzi baby girl alive or dead when she was born? – what emerged was a complex and entangled set of epistemological questions about what exactly life was and how one could determine that it had started or ended.

Apart from considerations of Chiozzi's weakened state at the end of her life, most medical authorities agreed that a seventh-month fetus was potentially viable, and thus the surgical removal of Chiozzi's child by cesarean section was a relatively uncontroversial decision. Witnesses who observed the operation, which took place about half and hour after it was determined Chiozzi had died, were, however, in disagreement about whether the fetus/child had demonstrated any signs of life. The surgeon who performed the operation, Giovanni Beduschi, testified that the child had been alive when extracted, given that he had seen its right hand move and its mouth make an effort to cry, though no sound was emitted. A friend of the mother, Veronica Baccanti, recalled that the child's head was down when first

61 Frank 1807: 3.195.

62 Fontes da Costa 2004: 159; Quinlan 2009: 601–603.

63 Fontes da Costa 2004: 160–163.

64 "Chiozzi Fratelli e Sorelle c. Casazza Carlo ed Antonio," in Chiozzi 1817: 265–283. Henceforth cited as "Chiozzi."

removed and that after it was baptized under conditions, the surgeon laid it on the bed, at which time it seemed the baby's head moved, though it was hard to tell whether the motion was voluntary. Finally, a priest who had been called to deliver the last rites and baptize the child, recorded that while watching the operation, he first glimpsed a bluish arm emerge, but also that when the fetus was extracted it had made several small gasps in an attempt to breathe.[65]

All of the witnesses agreed that the infant was fully formed in all of its parts. Although the exact age of the fetus was often hard to determine in the eighteenth century, Chiozzi's doctor, who had cared for the sickly woman throughout her pregnancy, estimated that the woman was seventh months pregnant, if not some days more. This number was significant because it represented a generally agreed upon date at which a baby might be born naturally and be fully formed in its parts. A persistent commonplace originating with Hippocrates was that a seventh-month fetus was likely to survive birth but that an eight-month fetus would not. While a baby born naturally at seventh months had developed quickly, but fully, a baby born at eight months was, by contrast, a premature nine-month baby.[66]

Of course, in this case, the Chiozzi baby was not born naturally; Angiola Chiozzi did not go into labor – she died and the fetus was extracted post-mortem by cesarean section. Chiozzi's doctor, Luigi Isalberti, did, however, testify that the fetus was active – it could be felt moving both by its mother and others – up until Chiozzi's last moments of life.[67] Thus, despite the fact that Chiozzi had been sick for a protracted period of time, and in contradiction to the claims of Chiozzi's natal kin, the illness had apparently not harmed the fetus growing inside her. This was important because contemporary medical writers debated whether a disease afflicting a pregnant woman would always have a negative effect on the fetus. In fact, a number of practitioners argued that it was physiologically impossible for a fetus to survive the death of its mother at all, given the absolute dependence of the former on the latter for nutrition and respiration.[68] The matter was by no means conclusively decided, however. In the early nineteenth century, for instance, Domenico Meli, a professor of obstetrics in Ravenna, proposed that it was possible for labor to occur after the mother's death because of the uterus' unique vital characteristics. According to Meli, not only was the uterus distinctly independent compared to other organs in terms of how it interacted with the body as a whole, but that during menstruation and pregnancy the uterus actually acquired an additional capacity of 'sensible irritability'. In other words, the uterus could continue to function after the rest of the mother's own vital functions had begun to shut down, at least for a number of hours.[69]

65 "Chiozzi": 271–275.
66 McClive 2016: 168–169.
67 "Chiozzi": 272.
68 On these debates, see for instance Barzellotti 1818: 186; Corradi 1877: 1423, 1435, 1467–1470.
69 Meli 1821.

The evidence presented by the witnesses to the Chiozzi baby's delivery was determined to satisfy the court's criteria for the category of *vitality*. To be considered *vital*, the court required that the child had been born alive, and that it was born with the necessary "disposition to be able to conserve and prolong life."[70] The key indicators of this status for the Chiozzi baby were that she had been born at greater than seven months,[71] she was clearly alive in the moments before she was extracted from her mother, she gave indications of life after her delivery, and, finally, she was well composed in all of her parts. Nonetheless, Chiozzi's relatives challenged each of these points in turn. Reaching seventh months in the womb meant nothing, they argued, when the child was not born of its own inclination and instead was pulled from its mother prematurely through a violent surgical procedure. They also questioned whether signs of life in utero – when the fetus was utterly dependent on the mother –had any bearing on the fetus' capacity to live once separated from that source of sustenance. Moreover, the clear indication of the successful independence of the fetus from its mother, the ability to breath, had never truly been proven in the Chiozzi baby, who had only gasped and made indeterminate movements. As Frank had argued in the case of the post-mortem cesarean operation, the signs of death could be dangerously difficult to detect; so, too, apparently, could the signs of life.

While the indicators of *vitality* were in themselves contentious, the Chiozzi case ultimately hinged around a second, even more controversial, legal category: *viability*. *Viability* was a French legal construction enshrined in the Napoleonic code and intended in particular to apply in cases of illegitimacy, inheritance, and succession.[72] As the legal representation of Napoleonic ideals, the *Code Civil* prescribed a fundamental shift in inheritance practices aimed at undermining household heads' patriarchal authority and guaranteeing equitable transfer of property among siblings. During the early nineteenth century, the Civil Code was adopted in parts of northern Italy, including Chiozzi's Lombardy, which were then under French authority. The Civil Code was decidedly vague on defining viability, except to say that a viable child was not only born alive, but also with the capacity to develop and

70 "Chiozzi": 275.

71 This number was significant because it represented a generally agreed upon date at which a baby might be born naturally and be fully formed in its parts. As mentioned, a persistent commonplace originating with Hippocrates was that a seventh-month fetus was likely to survive birth but that an eight-month fetus would not. While a baby born naturally at seven months had developed quickly, but fully, a baby born at eight months was, by contrast, a premature nine-month baby. See McClive 2016: 168–169.

72 From 1797–1815, Lombardy, where the Chiozzi case occurred, was part of what was essentially a French puppet state under Napoleon. Criminal cases, against infanticide for instance, were concerned with vitality rather than viability. Viability as a legal concept was at this time a French particularity; no comparable legal category existed in English common law, for instance. Taylor 1873: 247–248.

reach maturity. In practice, this meant that medico-legal authorities were given a significant amount of latitude in their determinations of just exactly what viability meant. To make matters more complicated, legally, viability meant something distinctly different from vitality. The Civil Code took the stance, for instance, that life begins at conception. Thus, the abortion of a fetus at any gestations age[73] or the murder of a child born alive (vital) were crimes punished severely, even if the *viability* of the child was never explicitly proven.

Marie-Guillaume-Alphonse Devergie, a Parisian physician and one of the leading writers on forensic medicine in the early nineteenth century, defined viability as "the aptitude for extra-uterine life, characterized by the maturity of the infant, the good conformation of the principal organs of the economy, and the healthy state of those organs at the epoch of birth."[74] Although Devergie's definition of viability differs little from the Italian court's concern with vitality in the Chiozzi case, further considerations of the former category reveal it to be epistemologically quite distinct from the latter. For one, a number of medico-legal writers understood viability as a status that could only be designated retrospectively. That is, while a living child's rights of succession and inheritance could not be denied, a child who died early in life could retroactively be declared non-viable. In such cases, the determination of non-viability was expected to be predicated upon anatomical investigation. Only medical experts, it was argued, could make the necessary assessments to determine if the child had some underlying condition or poor conformation of parts – originating prior to birth – that would have impeded its ability to mature to adulthood.

François-Emmanuel Fodéré, a French physician trained in Turin who became one of the leading writers on the legal concept of viability, argued that, "if the newborn (*neonato*) is not viable, it is the same thing as if he had never lived, since it is in the hope of the life that his birth anticipates" that makes him eligible for succession.[75] For Fodéré, a child born with "grave defects of conformation" should be considered, from its birth, as "interdicted," that is, legally forbidden from succeeding or inheriting.[76] In other words, a physical impairment that seemed to forecast a baby's imminent death could be taken as legitimate grounds to deny that child – even while 'alive' – his or her right of succession and inheritance. The Scottish obstetrician

73 This was a departure from Roman law, Canon law, and the Carolina, all of which tended to be concerned with abortion only after quickening, when the mother sensed the fetus moving within her body and the point at which a fetus was traditionally seen as 'animated.' See Burton 2007: 147–153; Riddle 1997: 87–94. On the Carolina, see Lewis 2016: 136–137.

74 Marie-Guillaume-Alphonse Devergie, quoted in Guy 1844: 175.

75 "Riflessioni del Professore Barovero sull'articolo *viabilité*, scritte dal Professore Fodéré, ed inserite nel vol. 57 del *Dictionnaire des Sciences Médicales*," in Fodéré 1822: 192. Henceforth cited as "Riflessioni del Professore Barovero."

76 "Riflessioni del Professore Barovero": 193.

G.L. Bonnar suggested that, in cases of premature births, at least eight days must pass before the child could be declared viable. This was the figure he derived from his own tabulated observations on the outcomes of premature births and which he believed represented the necessary amount of time for the child "to have outlived the immediate obstacles to independent animal existence."[77] In other words, the premature child was, for certain legal purposes, neither alive nor dead for the first week of its life.

The legal liminality of the sickly newborn is brought into even sharper relief when one considers that Fodéré and his followers also believed that viability could often not be definitively determined except by anatomical investigation. As much as external conformation, viability depended on the proper development of the internal organs, especially the brain, liver, heart, and lungs.[78] Only a post-mortem investigation could accurately chart the color, texture, and maturity of these crucial organs and to know if the child had been born healthy or had been likely to die since birth. Particularly in cases like Angiola Chiozzi's, where a child was delivered by a post-mortem cesarean operation, it was critical to approach the question of its viability with the exactitude that only an internal examination could provide. It was a failure of the courts, charged Fodéré, if judges were moved by the suffering of a surviving husband to simply declare that the dead child had been born viable.[79]

Not all medico-legal writers agreed with Fodéré's rigid legal definition of viability. The Italians Giacomo Barovero and Giacomo Tommasini both strongly argued that in the Chiozzi case, the baby girl had in fact demonstrated the quality of viability. Tommasini, taking into consideration the perfect conformation of the baby's parts at birth, the motions of its hands and mouth, and subtle inhalations, wrote passionately that the child had clearly lived, and that no specific defect could be pointed to that would have impeded it from developing to maturity. The child's death was not, therefore, a fate determined before its birth, but rather a result of *having lived*. Furthermore, all of the child's movements after birth "indicated the action and influence of the nervous system on the organs of respiration, and on the exterior muscles; and this nervous principle that is the same thing as the vital principle," would have, in more opportune conditions, prolonged and perfected the child's life.[80] Tommasini also picked apart the inherent contradiction in Fodéré's and others' condition of viability that a child must not only live but also have the capacity to live *for some length time* (however quantified). The legal acknowledgement of viability could thus only be determined retroactively *or* be reliant on an incredibly difficult estimation of a child's probability of survival, neither of which served the ends of equitable and efficient legal practice.[81] For Tommasini and

77 Bonnar 1865: 593.
78 "Riflessioni del Professore Barovero": 193.
79 "Riflessioni del Professore Barovero": 193.
80 Tommasini 1836: 159–60.
81 Tommasini 1836: 165.

Barovero, the Chiozzi case had a clear verdict: the child, born fully formed, had demonstrated life independent from its mother and was thus deserving of its civil rights.[82]

In matter of fact, the legal life of the Chiozzi case was just as dramatic and surprising as the circumstances that gave rise to it. Initially, the tribunal in Cremona where the case was heard ruled that the Chiozzi baby girl had indeed been born alive, with the qualities of both vitality and viability. Using much of the same reasoning as Tommasini and Barovero – the baby had been born after seven months and was lively in utero up to its delivery, it had been fully formed in its parts, had moved, and made efforts to breathe – the court ruled against Chiozzi's living relatives in their plea to deny the child a right to succession and inheritance. In 1819, however, after a lengthy review in the appellate courts, the case was retried in Milan. Reviewing the evidence, the court ultimately declared that the signs of life were indeterminate – the witnesses' statements disagreed on which parts of the child had moved, for instance. The court's experts also questioned the impact of the mother's sickness on the fetus, and the amount of time that lapsed between Chiozzi's death and the cesarean operation.[83] Rather than ruling with the presumption of life, the Milanese court instead approached the case with skepticism. Life was increasingly a property that had to be proven by forensic experts with substantial evidence.

Conclusions

While Fodéré's argument that an infant born with little prospect of reaching adulthood could be declared, in a legal sense, to have never lived at all, may strike us today as cruel and unjust, it underscores the fact that conceptual categories like "vital" and "viable" are historical and cultural constructions rather than biological realities. In early modern *répit* shrines, dead infants were temporarily granted the quality of life in order that they might assume a place in the larger Christian community, both on earth and in heaven. Life in this instance was a negotiated status that served to restore order and cement shared communal values at a time when the premature death of a child disrupted the social equilibrium. Pope Benedict XIV's ruling on the miracles, however, underscored a shift toward medico-legal rationalism, even among the Church, by the eighteenth century. While granting that some infants taken to miracle shrines might demonstrate the signs of life – in particular, crying – he reasoned that in such cases it was not actually the result of a

82 Tommasini 1836: 162–164. Tommasini also pointed out that many seemingly robust babies born naturally at nine months also quickly succumbed to death for reasons that were hard to discern.
83 "Chiozzi": 280–283.

miracle, but of the child's nearly imperceptible breathing that had simply gone undetected in the moments after its birth.

Secular judges, too, were turning increasingly to medico-legal experts as arbiters of the legal status of life. It was no long parents, midwives, or even priests that could make legal judgments about the perfection of the newly born child or the way any imperfections might cast doubt on that child's ability to participate as a legally recognized member of the society. The legal category of viability in early nineteenth-century France and Italy emerged to regulate matters of inheritance and succession, By being deemed viable, an individual was granted the "right to claim rights" and therefore a recognition of personhood.[84] In the context of fetal life, as Sara Dubow writes, "the idea that the maternal-fetal relationship is primarily a moral one in which the mother transmitted values and character to her unborn child was rearticulated as primarily a physiological and biological one in which the mother was primarily a source of nutrition and shelter for the fetus."[85] The capacity to live was divorced from concerns about salvation and morality, and instead defined in explicitly biological terms. The process of transforming from fetus to child, of becoming a person integrated into society, was medicalized.

This history also reveals a long tradition of patriarchal institutions – religious, medical, and legal – acting as arbiters of women's bodies. The moment when the fetus came to be seen as ensouled and alive, for instance, had once been dependent on a woman's subjective sensation of her own body; increasingly during the early modern period, however, medico-legal experts aimed to usurp this embodied authority with mathematical calculations of gestational age and experimentally determined analyses of embryonic development. Likewise, whereas authorities had traditionally agreed that the mother's right to live always superseded that of the fetus in utero, the Catholic Church's preoccupation with infant baptism increasingly positioned the mother and fetus as competitors for a limited number of resources in a high stakes game of eternal salvation. Finally, populationist concerns during the eighteenth century engendered particularly harsh attitudes toward infanticide and abandonment, often involving the strict surveillance of and presumption of guilt surrounding women's bodies. The legacy of such measures is visible today in the increasing criminalization of pregnancy, particularly among women whose racial, class, or citizenship status makes them especially vulnerable.

Since the twentieth century, these trends have only accelerated. The authority of the medical establishment in making determinations about what it means to be "alive" has expanded to influence the legal landscape surrounding disability, gender, sexuality, mental health, and, of course, reproductive rights. These medico-legal determinations both shape and are shaped by the Western cultural values

84 Christoffersen-Deb 2012: 589.
85 Dubow 2010: 37.

placed on binary gender identity, whiteness, heteronormativity, able-bodiness, "neurotypicalness," and so forth. Thus, while eighteenth- and nineteenth-century jurists believed putting legal determinations in the hands of medical experts would bring them closer to the 'truth,' the fact is that there was never any biological reality of 'viability' to ascertain, only, to use Deb-Christoffersen's words, "the work of viability."

Manuela Gallerani

Ethical Responsibilities in the Curing/Caring Relationship

Introduction

If we adopt an epistemological multi-perspective approach[1] and attempt to go beyond the illusion of human's inexhaustible efficiency or a "cult" (culture) of efficiency we can see that both *health* and *illness* inevitably trace their mark (and accompany) throughout every person's life.[2] If we *comprehend* these existential aspects at their very core, will we think of life as a life-long personal *project* that is both authentic and makes true sense.[3] This *existential project* should, in the first place, give a meaning and *value* to every instant of a person's life, under any conditions and at any moment, whether defined by health and well-being, or by pain and suffering.[4] Secondly,

1 Or rather a systemic (cf. Balint 1957, *The Doctor, His Patient and the Illness*; Charon 2006a, *Honoring the Stories of Illnes*; Greenhalgh 1999, *Narrative based medicine in an evidence-based world*; Bert 2007, *Medicina narrativa. Storie e parole nella relazione di cura*), multidisciplinary (Benedetti 2011, *The Patient's Brain. The Neuroscience Behind the Doctor – Patient Relationship*; Good 1994, *Medicine, Rationality, and Experience. An Anthropological Perspective*; Foucault 1963, *Naissance de la clinique: une archéologie du regard médical*; Lévinas 1991, *Entre Nous. Essais sur le penser-à-l'autre*; Bruner 1990, *Acts of Meaning*; Borgna 2017b, *Le parole che ci salvano*) and complex approach (Morin 1999, *L'Intelligence de la complexité*; Bocchi and Cerruti, eds., 2007, *La sfida della complessità*), i.e. a global approach that brings together the results obtained with idiographic methodologies (from the Greek ιδιος-γραφικος = ídios and graphikós, which means "describe the particular") aimed at the study of single specific cases (therefore not generalizable) and nomothetic methodologies based on large collections of data and on 'evidence' that allow generalizations.

2 Within this work, the term *illness* is used to describe anything and everything that alters a person's state of general well-being. So, this concept covers everything that a person perceives in terms of the associated phenomena, going back to the deep personal experience of whoever (user, patient) feels the need for care, whether because of bodily discomfort, malaise or pain (the symptoms of illness) or because of feeling afraid, worried and in need of help. The term *disease* is instead used to describe organic illnesses and symptoms caused by a pathology requiring surgery.

3 Wittgenstein 1953, *Philosophische Untersuchungen*; Costa 1987, *Binswanger. Il mondo come progetto*; Foucault 2008, *Le gouvernement de soi et des autres: Cours au Collège de France (1982–1983)*; Bodei 2002, *Destini personali. L'età della colonizzazione delle coscienze*; Damasio 1999, *The Feeling of What Happens: Body and Emotion in the Making of Consciousness*; Gallerani 2015, *Prossimità inattuale. Un contributo alla filosofia dell'educazione problematicista*.

4 Jonas 1985, *Technik, Medizin und Ethik: Zur Praxis des Prinzips Verantwortung*.

Note: For purposes of clarity and to better understand the interdisciplinary approach of the essay, in this text the titles of the works referred in the notes are not omitted.

https://doi.org/10.1515/9783110788501-005

it should lie at the foundation of a *culture* of *cure* and *prevention* – the prevention of illness, to smoothen the way for a person's global well-being and good health – by which we mean the *copying strategies* we use, in a constructive-critical way, to counter both a desire to distance ourselves or our *removal* (in the words of, among others, Hans Jonas, Norbert Elias, Michel Foucault and Edgar Morin),[5] from the illness or pain and our *indifference* towards human fragility. A fragility understood as inherent in the intimate essence of human existence.[6] Following the above outlined approaches and in accordance with the tenets of the Medical Humanities movement (henceforth referred to as MHs), setting an *etiquette* of suffering for professional health carers means thinking of a new way to interact both with the *patient* (who is first and foremost a unique and non-replicable *person*), with that person's caregivers as well as with that patient's *illness*.[7]

The Curing/Caring Relationship

Starting from these premises, one may observe that the curing/caring *relationship* as well as the manner in which this relationship is co-constructed, communicated and experienced, take on a crucial role.[8] This relationship must, primarily, be based on *actively listening* to the Other (the not-self) and on *taking care* to establish a significant relationship with the Other and the Other's world (in the meaning of Heidegger's *In-der-Welt-Sein*, Being-in-the-world; *Dasein*, the peculiar experience of human being), to transform the curative/caring relationship into one defined by *ethical commitment*, *deontological responsibility* and the ability to co-construct common meanings or common, *collective narratives* (in the words of the psychologist Jerome Bruner).[9] It is well known from statistical evidence that, to a large measure, the efficacy of any therapeutic intervention depends on the *type* of curing/caring

5 Jonas 1979, *Das Prinzip Verantwortung: Versuch einer Ethik für die technologische Zivilisation*; Elias 1982, *Ueber die Einsamkeit der Sterbenden in unseren Tagen*; Foucault 2001, *Biopolitica e liberalismo: detti e scritti su potere ed etica, 1975–1984*; Idem 1994, *Poteri e strategie. L'assoggettamento dei corpi e l'elemento sfuggente*; Morin 1970, *L'homme et la mort*.

6 Agamben 1998, *Homo Sacer: Sovereign Power and Bare Life*; Goffman 1963, *Stigma: Notes on the Management of Spoiled Identity;* Sontag 2003, *Regarding the Pain of Others*.

7 Halpern 2004, *The Etiquette of Illness. What to say when you can't find the word*.

8 The interesting etymology of the word *relationship* comes from the Latin *relatio* and, in turn, from *relatus*, the past participle of the verb *referre*, in the meaning of *establishing a bond*.

9 Within this essay, the word 'Other' is written with a capital O, following the Lévinas perspective, cf. Lévinas (1980), *Totality and Infinity* (originally 1961) and Idem (1991), *Entre Nous. Essais sur le penser-à-l'autre*. On the meanings of the *collective narratives* see, among others: Bruner 2002, *Making Stories. Law, Literature, Life*. On this issue see also: Nussbaum 2016, *Not for Profit: Why Democracy Needs the Humanities*; Gottschall 2012, *The Storytelling Animal: How Stories Make Us Human*; Peters and Biesta (eds.) 2008, *Derrida, Deconstruction, and the Politics of Pedagogy*.

relationship that is established between doctor and patient.[10] Furthermore, when the relationship is truly meaningful and thus based upon reciprocal co-constructed trust (which the interlocutors build up together), then whoever is sick and being cared for is seen as a *person*, fully and totally, that is with all his/her physical, psychic, cognitive, emotional and imaginative aspects, and not merely as a *patient* or a customer/user; and by extension, as a person with *needs, emotions* and *desires* that transcend the condition of being ill.[11]

This essay is an attempt to analyze the meaning of *responsibility for what is said* in a curing/caring relationship, relating to both *care* (in its two-fold meaning of caring and of *curing* in a medical sense). By a caring relationship, we mean the set of actions taken or not taken by the caregiver that have the purpose of *extrapolating* or *receiving* the patient's life experience.[12] All of this occurring within a perspective of *co-responsibility* (the *co-* means shared, reciprocal, agreed with the Other) and respect for the *times* and *ways* of elaborating the traumatic life episodes of the person being cared for. *Cure-care*, in this sense, relates to both in its Latin meaning of *cura sui* (take care of your Self) and as *taking care* of a relationship, which implies mastery over a wide linguistic repertoire, ranging from *verbal* and *non-verbal language* to *paraverbal communication*, which encompasses all the messages we transmit through verbal sounds, glances, gestures, pauses and silences.[13] From here on, we can infer how the curing/caring relationship is directed, firstly, to *receive* with respect and rigor – what is said and done, and also what is not said and *not* done – by someone who is too fragile or ill to express his/her need. This need has deep roots and a *desire* to be acknowledged.[14]

Reaching a *diagnosis* is preliminary to any later intervention, both when it is part of the actual process of curing/caring for a patient and when it addresses the wider community or society, for example when promoting the right lifestyle in terms of diet or preventive care.[15] But only when there is the capacity to co-construct a relationship based upon mutual trust (between doctor and patient) can the curing/caring process actually start. In this way, the sick person and the doctor can establish the *right distance* (between each other) and the *right mental outlook* about the path that both, in

---

10 Bert 2003, *Parole di medici, parole di pazienti: counselling e narrativa in medicina*; Lingiardi 2018, *Diagnosi e destino*; Orletti and Iovino 2018, *Il parlar chiaro nella comunicazione medica. Tra etica e linguistica*; Orletti 2000, *La conversazione diseguale. Potere e interazione.*

11 On these aspects see Ekman and Friesen 1976, *Pictures of Facial Affect*; Ekman 1999, *Handbook of Cognition and Emotion*; Panksepp 1998, *Affective Neuroscience: The Foundations of Human and Animal Emotions;* Nadel and Camaioni, eds., 1993, *New Perspectives in Early Communicative Development*; Billig 2008, *The Hidden Roots of Critical Psychology*.

12 Launer 2002, *Narrative-Based Primary care: A Practical Guide.*

13 Cf. Watzlawick, Beavin Bavelas, Jackson 1967, *Pragmatics of Human Communication: A Study of Interactional Patterns, Pathologies and Paradoxes.*

14 Lacan 1991, *Le séminaire, Livre VIII. Le transfert.*

15 Lingiardi and McWilliams 2017, The *Psychodynamic Diagnostic Manual. PDM-2.*

their own roles, will have to undertake. As a consequence, the professional scope of physicians includes their capacity of activating *relational* and *decisional* pathways in their patients who will thus consider their doctor's words as an indispensable *co-adjuvant* of a drug. On the other hand, *dialogue* and *relationship* between doctor and patient are regulated through (verbal and non-verbal) behavior, by their respective roles and also through their own *social representations* of the illness.[16] Communicating a diagnosis in an effective way is important, first, to explain the *symptoms* of an illness, its causes and possible cures while discouraging patients from turning to unproven sources of treatment, such as may be found on the internet or social networking sites. A correct diagnosis must be delivered to each person being cured/cared for using words and tones that are *personalized* for that person. Doctors and caregivers must identify the right words for each patient according to his/her age, culture, sensitivity and the expected emotional response to the information received.

Let us proceed with order and take a step back. A few years before the MHs emerged within scientific debate – in the United States around the mid-1960s – the physician Michael Balint published a seminal book entitled *The Doctor, his Patient and the Illness* which can be considered, under certain aspects, the programmatic manifesto of the MHs movement.[17] Balint considered the complex dynamics that play out in the clinical relationship between doctor and patient and claimed that, in medicine, the *doctor* himself/herself is the most frequently prescribes *medication*, although we have no list of all the possible contraindications that may emerge

16 Cf. Goffman 1963, *Stigma: Notes on the Management of Spoiled Identify*; Goffman 1986, *Frame Analysis: An Essay on the Organization of Experience.*

17 A well-known *method* is to work in groups (cooperating) and in training groups that go under the name 'Balint Groups' (see Balint 1957, *The Doctor, His Patient and the Illness*). The conceptual premises of Balint's approach can be summarized thus: a) the most frequently prescribed *medication* is the *doctor* himself/herself, but the relative pharmacology (therapeutic action, posology, toxicity and collateral effects) is virtually unknown; b) the doctor, in general, pays considerable attention to the illness and symptoms, less to the *patient* and far less to the doctor-patient *relationship*; nevertheless, a dysfunctional relationship can generate more than just anxiety and lead to therapeutic or diagnostic errors; c) the modern system of cure/care requires doctors to have a new emotional and relational skillset, without which their work can become less effective but more draining (with the risk of burnout). The conceptual nucleus of this approach is *transferable* to other caring professions, because it in centred on: 1) the analysis of the *relationship* between professional and user (with careful attention to the emotional-relational skills in play); 2) actions taken by the work team as an enabler of thought (looking carefully at the best understanding of group processes, the dynamics of communication and organization of the system); 3) learning based upon *caring/curing practices* and on *experience*, as well as on scientific knowledge. This, therefore, sets in motion a process of systematic caring/curing, at the heart of which are four essential *objectives*. The first is to improve the caring *relationship* (care in listening to the patient's needs and emotions). The second is that this team work involves reciprocal debate, and the third is that there is supervision over the *role* of the person providing the care/cure (and that of the person receiving it). The fourth is that this process induces a serious *reflection* about caring/curing practices, to create good practice.

during the relationship between doctor and patient. Pursuing his analysis, Balint highlighted the fact that the persons who cure/care often tend to exercise a sort of *conviction* or persuasion over their patients, akin to converting the patient to the doctor's own approach to the illness, healing and medical expertise. Nevertheless, paraphrasing the words of the American philosopher John Dewey, we must not forget that we do not learn from experience: we learn from reflecting on experience.[18] The rigour of this Deweyan thought is even more fitting when we are talking about a curing/caring process. People who are being cared for must be helped by cure/care professionals (not only doctors) to *think about* the *sense/meaning* of their illness experience every day, as it represents a phase, a step in their life path. This critical phase is part of the complexity and plurality of life experience and, while painful, is just one phase among many Others – and often the source of deep changes brought about by the very processes of reflection and self-analysis.

In this sense – despite the increasingly pressing rhythms dictated by healthcare facilities – the time necessary to establish a relationship of trust between the caregiver and the person cared for can be used to encourage well-being (individual and for the team) and to curtail anxieties and reciprocal misunderstandings, as well as avoiding possible errors in therapy. Caregivers must have developed *emotional, relational and empathetic* skills (and other life skills), without which their work could easily be less effective and more exhausting (with the risk of burnout). On the other hand, the capacity of *reflecting* on one's experience also plays a large part in promoting *best practice*, in the sense that it leads to improvements both in individual and team decision-making processes and the dynamics of organization and communication, through continuous interaction between thinking *in*-action (when acting) and *on the* action (after acting).[19]

In the United States, ahead of Europe, the MHs approach introduced a shift in paradigm within the world of medicine and professional cure/care, from a predominantly *biomedical* model to a *biopsychosocial* approach, where the *person-patient* and that person's pathology/illness are studied with equal attention. In the *biomedical model*, attention is centered on illnesses understood as biological entities. By studying the effects of the chosen interventions, an illness can be observed through rigorous clinical and laboratory procedures used to investigate both its pathogenesis and its progress. The basic assumptions in the biomedical model (nomological approach) are confirmed in *Evidence-Based Medicine* (hence to fore, EBM), where scientific evidence establishes the guidelines in the fields of diagnosis and therapy.

---

18 On this issue, see Dewey 1996, *The Collected Works of John Dewey, 1882–1953: The Electronic Edition.*

19 On this aspect see, for example: Argyris and Schön 1974, *Theory in Practice: Increasing Professional Effectiveness*; Schön 1984, *The Reflective Practitioner: How Professionals Think in Action*; Schön 1987, *Educating the Reflective Practitioner: Toward a New Design for Teaching and Learning in the Professions*; Schön, ed., 1991, *The Reflective Turn.*

Clinic and research work is conducted using statistical and qualitative criteria, and interest converges above all on pathology, while the sick person seems to remain in the background, having become completely identified with the illness.

A clear difference emerges between the *nomological* approach (pursued through the biomedical model that, because of its epistemological status, tends to formulate 'general laws' that refer to the illness rather than the sick person) and the *idiographic* approach envisaged by the MHs movement. This approach is interested in learning about the sick person and that person's individual and family history, as these elements are essential to create a comprehensive clinical picture and, as a consequence, lead to a more effective therapeutic intervention. It follows that patients are recognized as subjects with rights and so become an active part of the therapeutic pathway, with full respect of their entitlement to an informed consent explaining their treatment. In this approach, there is the recognition that, within clinical practice and in the relationship with the sick person, factors external to the illness come into play. Among these are the psychological, cultural and social factors that interact heavily and at various levels of awareness, affecting both the patients and their internal world (their fears, anxieties, sense of fragility and loss of independence) and their doctor. From an ethical point of view, the introduction of an 'informed consent' recalls the four keystone principles *(basic ethical principles)* of bioethics: establishing that all patients have the right to be fully informed about their state of health and the potential risks and benefits of the treatment they are receiving.[20] Patients can also decide to delegate another person to make decisions on their behalf. This right falls within the principle of personal *independence* and is strictly linked to the other bioethics principles: a) respect for the patient's *autonomy* and right to self-determination (*voluntas aegroti suprema lex*); b) *beneficence*, knowing how to make decisions in the best interest of the patient (*salus aegroti suprema lex*); c) non-*maleficence* (*primum non nuocere*) which in turn is balanced against beneficence, as the effect of the two principles together can produce a *double effect*, explained as the *consequences* deriving from a given set of actions) and d) *justice*, which states that everyone has the right to receive suitable, fair and correct treatment according to the principle of equality without discrimination. In a broader sense, these ethical principles are reconnected to Amartya Sen's and Martha Nussbaum's *capability approach* and the ethics of a substantial freedom (based upon principles of equity and justice, agency and substantial freedom or capability, quality of human life and well-being). Taking up the thread of Amartya Sen's thought, we can say that caring means to give meaning to the human experience and the

20 Beauchamp and Childress 2001, *Principles of Biomedical Ethics*; Walter and Klein, eds., 2003, *The Story of Bioethics. From seminal Works to contemporary Explorations*; Leone and Privitera (eds.), 2004, *Nuovo Dizionario di Bioetica.*

quality of life. And even better caring means ensuring good quality of life, substantive freedom and capabilities.[21]

The Medicine as *Scientia et Humanitas* (and *recte loqui*/speak rightly and properly)

Another approach slotting into the tracks of the *biopsychosocial* model, and in particular within the context of MHs[22], is that of *Narrative Evidence-Based Medicine* (NEBM). Elaborated by Rita Charon, this approach states that the relationship between doctor and patient should be defined by empathy, reflection, professionalism and trust (we shall return to this point later).[23] In these approaches (and perspectives), people are recognized as complex dynamic systems where their personal experiences and cultural conditioning have a bearing on the mental (and social) representations of their illness and also, at a deeper/inner level, on 'their feeling of Self' as a sick person. Starting from these premises, to help us gain a better understanding of the person in his or her totality, rather than gathering anamnestic data through the usual questions in a clinical report or protocol, we will proceed through a language that *intentionally* places the person being cared for at ease, *escorted* during the process of formulating an organic and truthful story of the Self. In other words, the person cared for is helped to compose a *responsible* story that allows the caregiver to activate and co-construct a relationship that makes sense. The caregiver, in this case, is a *reflective professional* with *emotional intelligence* who believes in the value of the relationship itself, which is perceived as the first element on which the cure/care is based.[24] This relationship must be built on and is fueled by words said and not said, emotions, gestures, silences and glances that can create a climate of participation, commitment and reciprocal project-making.[25] In other words, there is a shift from a doctor and disease-centered medicine to a patient-centered medicine, where, in the first instance, the fundamental aim is the specialist expertise of the doctor treating the

---

21 Sen 1999, *Development as Freedom*; Idem 2005, "Human Rights and Capabilities"; Sen 2010, *The Idea of Justice*; Nussbaum, *Creating Capabilities: The Human Development Approach*; Nussbaum and Sen, eds., 1993, *The Quality of Life*.

22 *Scientia et Humanitas* are two words that appear on the *logo* of the eminent *Bologna's Medical Surgical Society* and underline the double *role* of the doctor: as scientist and humanist (in the *classical* meaning of the word *humanitas*). The Medical Surgical Society was founded in Bologna in 1802. This historical Italian medical association has been publishing the Medical Journal *Bullettino delle Scienze Mediche* since 1829.

23 Charon 2001, "Narrative Medicine. A Model for Empathy, Reflection, Profession, and Trust."

24 Goleman 2007, *Social Intelligence: The New Science of Human Relationships*.

25 Argyris 1982, *Reasoning, Learning, and Action. Individual and Organizational*.

illness/disease, which is often decontextualized from the person suffering from this ailment. In the second instance the patients' *life experiences* are re-evaluated and, alongside their own personal stories, their illness reveals the interiorized cultural models that unconsciously guide the painful experience itself.[26]

Inevitably, different styles of doctor/caregiver-patient relationships, spring from these two models. In the first, the predominant element is the controlling behavior of the doctor, who, as the disease expert, can become paternalistic and authoritarian. The doctor manages the relationship, setting the times and ways for consultations, asks questions designed to investigate the symptoms, builds diagnostic hypotheses, informs and prescribes the treatment, but all in a framework where clinical interest is concerned with the illness. As if to say, the need for a precise clinical anamnesis prevails over attention or concern for the subjective component in the form of the sick person. It may be the case that when they have to communicate dismal diagnoses or other "bad news," doctors use reassuring expressions to protect the patient from anxiety and anguish but, without immersing themselves into the patient's individual and unique life experience. In these proceedings, it is as if the doctor is placed above the true needs of the person being care for, who must be stimulated into fighting a *battle* against the illness without giving up. But, unless they are shared and co-constructed with the person being cared for, certain apparently reassuring words (which have entered medical semantics) are fated to fall on deaf ears without producing any positive reaction in the person to whom they are spoken. Patients, on the contrary, may feel crushed by the doctor's words, if they are difficult to understand and interpreted as alien to them and their life experience, because they are a source of disappointment and disorientation. It is well known, for example, that the diagnosis of a serious pathology can affect the patient's behavior, mood and self-perception, engendering *dysphoric,* emotional and physiological responses described in the literature as an anxiety-depressive disorder. Patients must be helped to understand and give a new meaning to the doctor's words, through shared words, re-elaborating them and relocating life experiences linked to the illness within the entirety of their life experience, in the knowledge that their life does not coincide *only* with their illness. It is much more than that.

In the patient-centered medicine, doctors receive the sick person's storyline (as the expression of the patient's needs, desires and fears) and they *observe and learn* about the sick person, in a multi-dimensional perspective that combines biological knowledge of the illness and of the ill person. This knowledge is initially reached through "active listening" (in the words of Thomas Gordon, the American clinical psychologist colleague of Carl Rogers), *syntony* and *empathic care*, as these are skills that amplify the caregiver's own ability in interpretation and prediction. In this sense, doctors manage to combine their *social awareness* (a wide range of

26 Augé 2014, *Une ethnologie de soi: Le temps sans âge.*

perceptions that include perceiving another person's mental state and emotions and the ability to grasp and understand complex situations), with their *social* skills (synchrony, concern, capacity to guide and orient another person) and *clinical expertise.* Going into greater detail, *primary empathy* means being able to feel and *endure the pain*[27] of other people, and know how to sense their world and accept their uniqueness (*as if* trying to feel like those people, with them, while keeping the "correct distance," i.e, the requisite clarity of mind and strength to evaluate a situation without being overwhelmed by a patient's emotions or words). *Syntony* means listening carefully and being receptive to the words of the other person, described as being on the *same wavelength.* Lastly, *empathic listening* is the ability to recognize and comprehend[28] the thoughts, feelings and intentions of the other person.[29]

The Gentle Listening

In a sick person, the benefit of knowing that one is listened to carefully and without judgment produces further emotional and cognitive benefits. In telling their story, patients can quench – at least in part – their inner tension that can set in motion defense mechanisms such as negation or removal, and other types of refusal behavior that can be self-destructive and self-harmful. Translating suffering into (spoken or written) words releases the patients' intrapsychic anxiety because of the cathartic effect of presenting their inner life experience (their internal world), and, more than that, it also provides the means for the person to tell a story that may be recounted

27 *Endure pain* is used here in the sense of sharing the emotion of suffering. The verb comes from the Greek πάθος (*pàthos)*, deriving from the theme παθ- of the verb πάσχω, which in turn can be traced back to the verb πάσχειν (in vox media or the middle voice, a Greek construction that is neither in the active nor in the passive voice) meaning *to suffer* or *to become emotional*, and to συμπάθεια, in the literal sense of 'affinity of feeling'.

28 *Comprehend* (from the Latin *cum* and *prehendere*) is a verb of great semantic density meaning contain, include, understand. It can be traced to knowing how *to understand* and *grasp*, therefore reorganizing one's balance of knowledge. In other words, to understand what one can make one's own, so that it becomes part of one's way of being, perceiving, thinking. The etymological connotation underlines that this process takes place through the intellect and emotions and also through the senses and body.

29 Newman, Danziger, Cohen, eds., 1987, *Communication in Business Today*; Goleman 2006, *Emotional Intelligence*; Adolphs 2001, "The neurobiology of social cognition"; Baron-Cohen 2003, *The Essential Difference: The Truth about the Male and Female Brain*; Baron-Cohen, Golan, Wheelwright, and Hill 2004, *Mind Reading: The Interactive Guide to Emotions*; Gallese 2003, "La molteplice natura delle relazioni interpersonali: la ricerca di un comune meccanismo neurofisiologico"; Gallese, Goldman 1998, "Mirror Neurons and the Simulation Theory of Mind-Reading"; Gallese, Keysers, Rizzolatti 2004, "A unifying View of the Basis of Social Cognition"; Gallese, Fadiga, Fogassi, Rizzolatti 1996, "Action recognition in the premotor cortex"; Rizzolatti and Craighero 2004, "The Mirror-Neuron System"; Rizzolatti and Sinigaglia 2006, *So quel che fai. Il cervello che agisce e i neuroni specchio.*

in more or less ordered syntactic sequences in time and space, but can be expressed and shared. However, for this to occur, it is necessary for patients to be *motivated* to search for the right words to express their *pain* or *suffering* and for them to have *trust* in the person listening, who, therefore, is engaging with this story telling and *listening gently*. In the words of the Italian psychiatrist Eugenio Borgna, *gentle listening* implies that the listener opens up to the Others, to acknowledge and respect them, with humanity and participation, as well as with trust, consideration and gentleness.[30]

In a relationship based upon sharing, caregivers can gain information from the story told by the persons being cared for that can help them formulate a diagnosis. And more, the caregiver can gain entrance into that person's inner world, now helping to strengthen defenses, now suggesting other interpretations of suffering and illness apart from those which, due to dejection, that person is unable to perceive.

On the other hand, it is well known that in any social context story telling (narrations) is an ideal *medium* defining, as it does, a space and time that encourages the meeting between thoughts, cultures, persons and social representations. As mentioned earlier, respect for the patient's words and stories lies at the basis of *narrative medicine* or, better, of *Narrative Evidence-Based Medicine* which weaves *evidence-based* studies of molecular medicine with humanities.[31] According to this approach, *curing/caring* does not merely coincide with external treatment (drugs, therapies) to be administered to the patient, but it is also based on a therapeutic alliance and involves totally reconsidering the role played by the intersubjective *relationship*, across two facets. Focus is directed, on one side, to the relationship between the caregiver (doctor, nurse) and the person being cared for (patient), concentrating on the story-telling of illnesses (spoken and written), and life experiences of discomfort and suffering. On the other side, this narrative fine-tunes the careful evaluation of the relationship between curing/caring *theory* and *practice*, in order to understand the origins of the disease and the painful experience, and to derive from it the hypothesis of a personalized intervention shared between doctor and the receiver of cures and care. In the curing/caring relationship and in the search for shared meanings, the person who is curing/caring will keep an asymmetric position relative to the person receiving the cure/care, and so avoid generating intimating divergencies or inappropriate transversal dynamics (known as transference or in French and Italian as trasfert). The former joins in the story of the latter, trying to re-elaborate and understand this life experience and, as far as possible, to establish empathy with it. For this very reason, because of each patient's unique experience, the usual repertory of clinical phrases offered to the sick person can be

30 Borgna 2017b, *Le parole che ci salvano*; Idem 2017a, *L'ascolto gentile* (gentle listening); Idem 2020, *Speranza e disperazione*.

31 Bucci, ed., 2006, *Manuale di medical humanities*; Charon, Montello, eds., 2002, *Stories Matter: The Role of Narrative in Medical Ethics and Psychoanalysis and Narrative Medicine*.

unsatisfactory or barely pertinent, especially when all the patient's possible resources must be called into play, and all his or her energies employed to elaborate the experience of suffering from this disease. In this sense, if the assumptions at the foundations of EBM and NEBM can be interwoven, this can trigger a genuine and virtuous hermeneutic process.[32] This last statement is confirmed in experimental research and clinical observations – think of the studies on the placebo effect – which demonstrate how the *mind-body problem* can be identified with the *mind-brain problem*. The brain, notoriously, is the seat of thought, conscience, emotions, intelligence: it is where our mental events occur. So, recent studies have demonstrated that mental activity is always correlated with the activation of a cerebral region. This means that thought is closely linked to the nervous activity emerging from the brain, and this can be recorded through positron emission tomography (PET scans) or functional magnetic resonance imaging (fMRI scans). Both these methods can be used to observe which cerebral regions are activated in any given moment of a person's life. Despite this, science today is still unable to explain whether, within the complex and dynamic mind-brain system, it is thought (mind) that determines the activation in this cerebral region, or if it is the other way round. Experimental studies tend to go *beyond* the molecular and physiological description of the illness, overcoming an idea that re-proposes rigid mechanistic schemes of how the brain works.[33] This is also propounded in MHs, with the elaboration of a more global and eco-systemic vision of the illness, placing themselves sometimes in dialogue with and sometimes in opposition with other approaches or functionalist theories, without for this refuting their originality.

In any case, the true Gordian knot lies in settling the question of *conscience*, because scientists are not yet able to confirm whether conscience and mental events are global characteristics that emerge from the combination of cerebral regions, neurons and molecules (and their connections) or whether they are, instead, responsible for guiding and influencing human behaviour. With regards to this dilemma, MHs propose, in first instance, that a person's global conscience is the prerequisite for a relationship that is both curing and *caring*, of *being at the side* of patients – even when they are suffering from serious, incapacitating, chronic or oncological pathologies – in their path towards independence. In other words, when they are searching for well-being that can give back meaning to their life and to the experience of the illness itself. The curing process, therefore, develops along a complex pathway of *comprehension* directed towards re-connecting the "clinical" perspective of the cure/care giver (with its set of exams, diagnostic checks, medicinal therapies) and the perspective of the person being cured or, better still, that

---

32 See among others: Charon 2006a, *Honoring the Stories of Illness*; Bert 2007, *Medicina narrativa*; Frank 1985, *The Wounded Storyteller*; Brody 1987, *Stories of Sickness*.
33 Benedetti 2013, *Il caso di G.L. La medicina narrativa e le dinamiche nascoste della mente*; Idem 2008, *Placebo Effects: Understanding the Mechanisms in Health and Disease*.

person's self-cure mechanisms. By applying *technologies of the self* (remembering Michel Foucault's work),[34] such as *reflective writing* and the *telling of oneself*, the patient reassembles the threads and the texture/weave of his or her life story. The caregiver has the task of providing the structure to the story when there is none, recognizing and illuminating the *plot woven* within the *hidden* storyline. There will never be one single interpretation of the sick person's story, as the viewpoint will be personal and relative, so it must be received and *interpreted* whatever its *form* (linear or confused) or its content.

The Role of Active Listening

The process of curing/caring is based on a relationship involving both *active listening* (in the words of the psychologist Carl R. Rogers) and on a willingness on both sides to learn from each other. This brings to light, in all its urgency, the role of the body, because *knowledge* is always *embodied*, by which it is meant the knowledge of one's own body, which is often tacit.[35] The ranks of caregiving professionals are filled with many teachers, doctors, nurses, social care workers who know this well. Within their daily work and practice, they are called upon to put into play their *competences* and *life skills*. In other words, they must engage with their *knowledge* (their expert, specialized knowledge), their ability to *know what to do* (their skills and practice), to *know how to be* (knowing how to make decisions, knowing how to interact with Others and with complex problems) to *know how to communicate* (knowing how to cooperate and work effectively in teams) and to *know how to learn* (life-long learning).[36] Indeed, the body is not only a means to gain knowledge, but it is itself at the basis of knowledge. People know factual reality through their bodies and so develop their abilities in the sensory, cognitive, emotional and sentimental sphere. At the same time, they elaborate an "embodied knowledge," which is experiential, produced *by* and *within* our "being as a body."[37] A body that is never unhistorical, but is programmed for social interaction within the socio-historical *hic*

---

34 Cf. Martin, Gutman and Hutton, eds., 1988, *Technologies of the Self: A Seminar with Michel Foucault*. See also: Foucault 2008, *Le gouvernement de soi et des autres: Cours au Collège de France (1982–1983)*; Foucault 2001, *Biopolitica e liberalismo: detti e scritti su potere ed etica, 1975–1984*; Foucault 1994a, *Poteri e strategie. L'assoggettamento dei corpi e l'elemento sfuggente*.

35 Maturana and Varela 1987, *The Tree of Knowledge: The Biological Roots of Human Understanding*; Damasio 1994, *Descartes' Error: Emotion, Reason, and the Human Brain*.

36 Kannen 2012, "'*My body speaks to them': Instructor reflections on the complexities of power and social embodiments*"; Kolb 1984, *Experiential Learning: Experience as the Source of Learning and Development*; Lawrence 2012, "Intuitive Knowing and Embodied Consciousness."

37 Leder 1992, *The Body in Medical Thought and Practice*.

et nunc (here and now) and is, therefore, regulated by the political and cultural system of the *milieu* to which every person belongs.

To understand the role of *active listening* in the sense of a welcoming, non-judgmental *attitude*, one may uncover how the *person-centered approach* developed by humanistic psychology scholars (starting from the lesson of Carl R. Rogers) helps to explain how effective listening is in terms of improving practice and relationships in both the personal and professional spheres.[38] This is especially so when listening is combined with empathy (see above), *unconditional positive regard* and *authenticity* (transparency, clarity and *congruence* in proposing oneself). *Unconditional regard* is the capacity of appreciating and accepting the other person's existential reality (what the person being cured/cared for has lived through his or her experiences) refraining from any form of interpretation and/or judgment. Acceptance, nevertheless, does not mean sharing or approving other people's ideas, opinions or feelings *tout court* but, rather, it is acknowledging that the other person is free to have these ideas, opinions and feelings. It is more a form of deep *respect* for the other person and is also the means for developing reciprocal *trust*, which is an indispensable part of a curative relationship (see Carl R. Rogers and the person-centered therapy, also known as person-centered psychotherapy, person-centered counselling and client-centered therapy).[39] *Authenticity*, on the contrary, denotes being spontaneous, consistent and transparent in one's relationships. People are authentic when they express and behave as they really are with their own feelings and perceptions, without using stereotypes or platitudes, clichés in their speech, and when they remain in empathy with Others.[40]

Following the footsteps of Jacques Lacan, the Italian psychotherapist Massimo Recalcati states that, rather than in the function that defines and bonds the *significant* and the *signifier*, the unconscious dimension of an artwork lies in the *cleft*, the fracture that separates them. An artwork, in other words, defines a *cut* (a separation, a cleft, a resistance). It is, therefore, the impossibility of fully reconducting the "significant storyline" to its primary meanings. More correctly, the *unconsciousness of the artwork* consists in its not being translatable, its eccentricity compared to any hermeneutic interpretation.[41] Therefore, an art image, like a visual metaphor or a poem, enjoys its own intrinsic semantic density, which is impossible to translate unless symbolically. Starting from these premises, in the next part of this essay, I will try to extract the deepest meaning of the curing/caring relationship by proposing a few peculiar examples from literature (chosen among thousands of possibilities,

---

38 Rogers 1951, *Client-Centred Therapy: Its Current Practice, Implications, and Theory*.

39 Rogers, "Client-Centered Therapy" in Arieti, ed., *American Handbook of Psychiatry*, vol. 3.

40 See among others: Rizzolatti and Sinigaglia 2006, *So quel che fai. Il cervello che agisce e i neuroni specchio*; Rizzolatti and Sinigaglia 2019, *Specchi nel cervello. Come comprendiamo gli altri dall'interno*.

41 Recalcati 2011, *Il miracolo della forma*, and Recalcati 2016, *Il mistero delle cose*.

leaving aside the most famous authors, those akin to Francesco Petrarca, Giovanni Boccaccio, Michel de Montaigne, Luigi Pirandello, Italo Svevo, Umberto Saba and Carlo Michelstaedter, because these would require another and different work, more space and are beyond the scope of this essay). As in a trick with mirrors, where the past dialogues with the present, we take up the paradigmatic thoughts of several writers who question themselves about curing and caring, disease and death. They take it upon themselves to describe them, sometimes through metaphors[42] (often seeing curing as a *fight* or battle) and other times trying to trace, within their poems the most intimate meanings of life. Among these, for example, is the ideal 'doctor and philosopher' described by the ancient Greek physician Galen (Pergamon 129 CE – Rome c. 201 CE). He reflects on whether to tell his patient the true facts, stating that a doctor is the 'friend of truth,' because he tells his patients the truth, but truth that is measured (or in a Latin word, *temperantia*).

Moreover, how is it possible not to remember the life experience and the effect produced by the words spoken by the doctor consulted by the quiet, calm and patient *Ivan Ilyich*? – in Tolstoy's famous 1886 novel – and his understanding: ". . . Maybe I didn't live as I should have done?" came the sudden thought. "But how can that be when I did everything properly?"[43] "This conclusion left Ivan Ilyich with a sickly feeling, filling him with self-pity and great animosity towards the doctor who showed so much indifference to such an important question. And in the light of the doctor's confusing pronouncements the pain, that dull, nagging pain that never went away, was taking on a new and more serious significance."[44] The illness worsens, despite the treatment and he continues to consult doctors, but "The time for fooling himself was over: something new and dreadful was going on inside Ivan Ilyich, something significant, more significant than anything in his whole life. And he was the only one who knew it; the people around him didn't know, or didn't want to know,"[45] until the pain became such that "He took in some air . . . stretched out, and died."[46]

Virginia Woolf in *On Being Ill* also considers the important role of *language* and observes that ". . . but let a sufferer try to describe a pain in his head to a doctor and language at once runs dry. There is nothing ready made for him. He is forced to coin words himself, and, taking his pain in one hand, and a lump of pure sound in the other . . . so to crush them together that a brand new word in the end drops out" (Woolf, 1926, p. 34).[47] This permanent search for meaning and the search for

---

42 Lakoff and Johnson 1980, *Metaphors We Live By*; Lakoff and Johnson 1999, *Philosophy in the Flesh: The Embodied Mind and Its Challenge to Western Thought.*

43 Tolstoy (1886) 2016: 80.

44 Tolstoy (1886) 2016: 41.

45 Tolstoy (1886) 2016:42.

46 Tolstoy (1886) 2016: 90.

47 Woolf 1926, *On Being Ill.*

the precise words to express it is a distinctive trait of Woolf's narrative style. And, nevertheless of an existence that alternates moments of profound discomfort with "moments of being":[48] that is, those particular epiphanies, those instants (or involuntary memories) able of illuminating, just for a moment, the daily routine.

The words used – in the Seventeenth Century – by Molière in *The Imaginary Invalid* (1673), also translated as *The Hypochondriac*, are totally different in tone, register and basic purpose. In this case, the satirical topic concerns the invalid's hypochondria and also the incompetence of the doctors caring for him. All of them are characters sporting pompous and ridiculous names, Monsieur Purgon the physician, Monsieur Fleurant the apothecary, Monsieur Diafoirus the physician. Towards the end of the play, the character Béralde, expresses his opinion saying: "I don't see anything more ridiculous, than for one man to undertake to cure another." This reveals a certain truth, because it is indeed difficult to cure the sick person who does not wish to take on the personal responsibility or commitment to cure him or herself.

At this point of the analysis, based on what has been said so far, it is interesting to recall the value of the "circumstantial paradigm" (in the words of the Italian historian Carlo Ginzburg). This is an interpretive method that re-evaluates marginal data or phenomena (usually considered unimportant and undervalued) by considering them, instead, as revealing clues: they provide the key to accessing the most complex levels of existence and most hidden movements of human experience.[49] Moreover, Edgar Wind argues that the true human personality is revealed, above all, in non-verbal language and in those small, unconscious gestures that more fully reveal the true character of each person than any formal attitude, carefully thought out and externally expressed.[50] Hence, the meaning of the words of both the scientists and the artists mentioned in the present analysis is even better understood.

Peculiar Words of Care

Moving our attention from the meaning of the words of contemporary writers and poets, one may discover how each one was able to present an exemplary picture of his or her identity and his or her experience of *disease* and *cure*, providing care for

48 Jensen 2007, "Tradition and Revelation. Moments of Being in Virginia Woolf's Major Novels."
49 Cf. Ginzburg 1986, *Miti emblemi spie. Morfologia e storia*. On these aspects (and the 'circumstantial paradigm') see also: Romano 2019, "Freud, Morelli e la nascita del paradigma indiziario in psicoanalisi"; Gallerani 2021b, "Images, Thoughts and Words on Disease in the Works of Michelangelo"; Eadem 2021a, "Disease in Art and Art(ist) in Disease: Reflections on Paradigmatic Works by Leonardo da Vinci and Michelangelo Buonarroti."
50 Cf. Wind 1968, *Arte e anarchia*, p. 62 (author's translation).

loved ones or received this care (see the poem by Francesco Tomada, as it will be addressed in the following pages).[51] It is as if they transformed a personal experience into high literature, or better still, into a universal message that can send very diverse messages to different people. These words resound in each one of us differently, and we can reflect in (or distance ourselves from) them. But they are unlikely to leave us *indifferent*, because even when they connect to deep, shared meanings, they can always be read in a different way, or be interpreted newly. They evoke vision, intuition and sudden powerful realizations that surprise us and throw wide windows open into reality, *illuminating*, observing and analyzing it as never before. And so it is when we address an aesthetic experience, in *real presence, living* a very personal experience that, for this very reason, we are often unable even to *explain*. We cannot explain it because, that emerging moment of emotion – stimulated by words, images, colours, sounds or the natural environment – has the power of rapidly transforming both our interior reality and the reality around us, and place a question mark over everything that first appeared obvious, reasonable and serious.

Any aesthetic experience (in a Dewey's meaning) will consult, question, surprise, make us think and, unexpectedly, let us discover new meanings where none appeared to be. And conduct us further, beyond conventions, norms or precise expectations.[52] For this reason, the words of the poets are peculiar and precise. Precise in the sense that they are chosen with care and attention: they are selected from common and less common words, to compose original verses able of express an idea or a thought, to identify an emotion and to make them understandable in a concise narrative text, that is *essential*. Every poem can be considered a real *distillation* of words, because every single word refers to deep meanings: it refers to uncommon emotions, feelings, moods, images, colors (and much more), because uncommon are the people who read those words and make them live in themselves. Indeed, in Nancy's words: ". . . if in some way we have access to a threshold of meaning, this happens thanks to poetry."[53]

Therefore, one may note the sensitivity, incisiveness and pained irony that runs through Raymond Carver's poem entitled *What the Doctor Said*.[54] And how, in his

---

51 On the question of identity and how does a mind build itself, see among others: Freeman 2001, *How Brains Make Up Their Minds*; Ammaniti and Gallese 2014, *The Birth of Intersubjectivity. Psychodynamics, Neurobiology, and the Self*; Rizzolatti and Gnoli 2016, *In te mi specchio. Per una scienza dell'empatia*; Northoff 2019, *La neurofilosofia e la mente sana. Imparare dal cervello malato.*

52 Clover, Sanford and Butterwick (eds.) 2013, *Aesthetic Practices and Adult Education*; Reid and West (eds.) 2014, *Constructing Narratives of Continuity and Change. A transdisciplinary Approach to Researching Lives*; Merrill and West 2009, *Using Biographical Methods in Social Research*; Leavy 2021, *Method Meets Art. Arts-Based Research Proctice.*

53 Nancy 2017, *La custodia del senso. Necessità e resistenza della poesia*, p. 21 (author's translation); Gallerani 2012, *L'impegno lieve. Il razionalismo critico e l'ideale estetico.*

54 Carver 1996, 'What the Doctor Said,' *in All of Us: Collected Poems*. See also: Pivano, *Pagine americane. Narrativa e poesia 1943–2005.*

final poem (*Things I'll Not Do*), Allen Ginsberg instead sheds light on the life of a man who responds to a diagnosis of terminal illness (cancer) with the will and courage to continue writing, despite everything. So, while the poet is saying farewell to the world, he dwells once more on his attachment to life, as if to observe once again that, while he did not look after his body much (or his health), he certainly took care of "looking after his mind." He writes a farewell to life (and farewell to all the things he will not do, again or ever) which is without doubt melancholic, but in a certain way also soothing. This is the last act in an existence that was both tormented and overflowing with experiences, with his writing acting as a unifying theme throughout. He also expresses his mood and life experience through the emblematic verses of two poems entitled *Things I'll Not Do* (now in *Death & Fame*, 1997) and *When I die*.[55]

The choice to consider these authors is intentional and deliberate, searching for revealing clues about the meaning and significance of an artist's work, thanks to the circumstantial paradigm. It is a motivated choice, first of all, to highlight the words of care and the personal narrative style (the sensibility) of each of them. Secondly, to observe how writers and poets – so different from each other – have been able to use writing as an *instrument* of self-knowledge and, perhaps, of self-care. In any case, every storytelling clearly expresses the urgency to narrate and share experiences of illness (or caring) that are both intimate and painful. A narrative able of recomposing a privileged relationship with the one's own inner world and with external world.

What The Doctor Said by Raymond Carver

On the meanings of the narrative and to better highlight what has been said so far, Carver's verses are very eloquent (i.e., vv. no. 3–5, vv. no. 12–13, vv. no. 20–22):

> He said it doesn't look good
> he said it looks bad in fact real bad
> he said I counted thirty-two of them on one lung before
> I quit counting them
> I said I'm glad I wouldn't want to know
> about any more being there than that
> he said are you a religious man do you kneel down
> in forest groves and let yourself ask for help
> when you come to a waterfall
> mist blowing against your face and arms
> do you stop and ask for understanding at those moments
> I said not yet but I intend to start today
> he said I'm real sorry he said
> I wish I had some other kind of news to give you

---

55 Ginsberg 1999, *Death & Fame: Last Poems etc 1993–1997*, pp. 98–99 and Ginsberg 2006, *Collected Poems, 1947–1997*, p. 1130.

I said Amen and he said something else
I didn't catch and not knowing what else to do
and not wanting him to have to repeat it
and me to have to fully digest it
I just looked at him
for a minute and he looked back it was then
I jumped up and shook hands with this man who'd just given me
something no one else on earth had ever given me
I may have even thanked him habit being so strong.

Things I'll Not Do (*Nostalgias*) by Allen Ginsberg

To better highlight both the mood of poet and impact of illness and death on himself (after a long terminally ill), the Ginsberg's verses appear very touching (i.e., v. no. 1, v. no. 11, vv. no. 35–36 and also the icastic final verse). This particular lyric is very interesting, because it represents both a sort of testament and the last lines of one's autobiography:

Never go to Bulgaria, had a booklet & invitation
Same Albania, invited last year, privately by Lottery scammers or
recovering alcoholics,
Or enlightened poets of the antique land of Hades Gates
Nor visit Lhasa live in Hilton or Ngawang Gelek's household & weary
ascend Potala
Nor ever return to Kashi 'oldest continuously habited city in the world'
bathe in Ganges & sit again at Manikarnika ghat with Peter,
visit Lord Jagganath again in Puri, never back to Birbhum take
notes tales of Khaki Baba
Or hear music festivals in Madras with Philip
Or return to enter to have Chai with older Sunil & the young coffeeshop poets,
Tie my head on a block in the Chinatown opium den, pass by Moslem
Hotel, its rooftop Tinsmith Street Choudui Chowh Nimtallah
Burning ground nor smoke ganja on the Hooghly
Nor the alleyways of Achmed's Fez, nevermore drink mint tea at Soco
Chico, visit Paul B. in Tangiers
Or see the Sphinx in Desert at Sunrise or sunset, morn & dusk in the
desert
Ancient collapsed Beirut, sad bombed Babylon & Ur of old, Syria's
grim mysteries all Araby & Saudi Deserts, Yemen's sprightly
folk,
Old opium tribal Afghanistan, Tibet – Templed Beluchistan
See Shanghai again, nor cares of Dunhuang
Nor climb E. 12th Street's stairway 3 flights again,
Nor go to literary Argentina, accompany Glass to Sao Paolo & live a
month in a flat Rio's beaches & favella boys, Bahia's great
Carnival

Nor more daydream of Bali, too far Adelaide's festival to get new song sticks
Not see the new slums of Jakarta, mysterious Borneo forests & painted men & women
No more Sunset Boulevard, Melrose Avenue, Oz on Ocean Way
Old cousin Danny Leegant, memories of Aunt Edith in Santa Monica
No more sweet summers with lovers, teaching Blake at Naropa,
Mind Writing Slogans, new modern American Poetics, Williams
Kerouac Reznikoff Rakosi Corso Creeley Orlovsky
Any visits to B'nai Israel graves of Buba, Aunt Rose, Harry Meltzer and Aunt Clara, Father Louis
Not myself except in an urn of ashes.

March 30, 1997, A.M.

Death & Fame

Words of care also are hidden in the ironic and sarcastic verses of this Ginsberg's poem about death (see in particular the following verses):

When I die
I don't care what happens to my body
throw ashes in the air, scatter'em in East River
bury an urn in Elizabeth New Jersey, B'nai Israel Cemetery
But I want a big funeral
St. Patrick's Cathedral, St. Mark's Church, the largest synagogue in Manhattan
First, there's family, brother, nephews, spry aged Edith stepmother 96, Aunt Honey from old Newark
[. . .]
'Father Death comforted me when my sister died Boston 1982'
'I read what he said in a newsmagazine, blew my mind, realized others like me out there'
Deaf & Dumb bards with hand signing quick brilliant gestures
Then Journalists, editors's secretaries, agents, portraitists & photography aficionados, rock critics, cultured laborors, cultural historians come to witness the historic funeral
Super-fans, poetasters, aging Beatnicks & Deadheads, autograph-hunters, distinguished paparazzi, intelligent gawkers
Everyone knew they were part of 'History' except the deceased
who never knew exactly what was happening even when I was alive.

February 22, 1997

These words (and verses) seem to remind us, once again, that disease does not coincide with *diagnosis*, but can bring to light the entire world of relationships that envelop the sick person. This world is lain bare, telling, alongside one person's story, also the story of that person's family, revealing the familiar illnesses, the types of cure received, the relationships co-constructed between people (parents, friends and partners). All this is tightly linked to a capacity for self-care and looking after Others. Philip Roth returns to these topics and unfurls them in his autobiographical book entitled *Patrimony: A True Story* (edited in 1991).[56] Here, he explores the relationship between a terminally ill father and his son, who becomes his carer (and caregiver), pulling away the veil from a family's story and its arrangements, its dynamics and, within these dynamics, inhibiting the things that are said and not said. And so it is possible to observe life memories emerging, of how the writer was cared for by his parents and how the roles will be exchanged when it is him, the son, to care for his elderly father. In a circular physiological process that is part of the circle of life, but which prompts a sense of powerlessness and defeat when faced with a relentless disease. The incipit of *Patrimony: A True Story* is very expressive: "My father had lost most of the sight in his right eye by the time he'd reached eighty-six, but otherwise he seemed in phenomenal health for a man his age when he came down with what the Florida doctor diagnosed, incorrectly, as Bell's palsy, a viral infection that causes paralysis, usually temporary, to one side of the face." Then, in a later chapter, he writes an impressive description of his father's tone of voice (during a medical visit) not meant to be one of anger but, rather, it was more anxiety and bewilderment. This was all new, and he felt powerless to find a way to help him. They went through this experience together.

Finally, it is interesting to consider – in order to complete and conclude this analysis – the poetics of a contemporary Italian poet Francesco Tomada. Many illustrious Italian poets have written poems dedicated to their mothers or to a symbolic idea of mother, such as, for example, Dante Alighieri (referring to a Heavenly Mother, the Virgin Mother), Giovanni Pascoli, Salvatore Quasimodo, Pierpaolo Pasolini, Giuseppe Ungaretti and Eugenio Montale, just to mention some of the most important. Nonetheless, the choice to evoke Tomada's words of care is explained by the unexpected adherence of his poetic (sense and sensibility) to the analysis and the theses advocated in this essay.

In his profound poetic *voice* (words) Tomada shows an intimate feel/sensitivity and, at the same time, a deep attention into take care of the Other and, in particular, to one's mother. A deep emphaty expressed through gestures of *respect* and *care* towards the mother. As is well known, the mother represents the first *significant adult* or the primary best caregiver for each child: all this, without wishing to carry out any psychoanalytic investigation of the writer's Oedipus complex, but only with the aim to observe the point of view of a man who takes care of a woman, his mother. And more, *how* and *why* in doing so he feels the urge to write and share his personal experience with Others. That is to say, he feels the need to express himself with writing (poetry) and to use words, at first to describe emotions, thoughts and experiences

56 Roth 1991, *Patrimony. A True Story.*

both of his inner world and his life. Secondly, to fix their experiences on the blank page as well as in their own memory and consequently to share this sensibility with Others (i.e. the readers). This last point is also noteworthy.

In his poem entitled *Saved souls* Francesco Tomada expresses simple and delicately, the ways in which he cares for his fragile elderly mother.[57] In his case, the act of caring sanctions, on the one side, an intergenerational passage and a clear inversion of roles, but under the form of reciprocal mirroring of the one (the now adult son who cares) in the other (the elderly mother who needs care). Care, therefore, is transformed into an act of restitution and gratitude, the son's *gift* of caring for his mother as she had cared for him in times past.

The style and ways for caring of a son/daughter for a mother (or father) broadly reflect the set of caring practices learnt in the bosom of a family, from parents or other carers (caregivers). Having care (for oneself or Others) is linked to the human being and is truly a way of being human (*à la* Heidegger).[58] And more. Having care, thus interpreted, presents some interesting analogies with the Latin's meaning of *pìetas* that is intended as a feeling that does not perfectly coincide with piety as we understand it, but is much more complex and nuanced: it is a mixture of devotion and respect towards both high moral values and ethical ideals but also towards Others (i.e. the family, humankind).

As is well known – starting from the studies of Sigmund Freud, Anna Freud and Melanie Klein – how indispensable is, in early infancy, to give the child a harmonious, well-balanced upbringing and instil resilience, but it is still a primary need in other ages, throughout life. As a pre-social skill, caring encourages well-being and is a precious lifelong act that people can share and co-construct together. Hence, learning the 'ethics of care' means developing not only a personal disposition (or attitude)

---

57 Francesco Tomada's poem is here reported in original Italian version to highlight the carefully research of each word and meaning. So that it is possible to *hear*, feel and make experience both the sound and *rhythm* of the verses in two different languages (Italian and English). *Anime Salve* (Saved Souls): "Dieci anni fa cambiavo i vestiti ai miei bambini, / lavavo la loro nudità e lo sporco / Prima di averli, pensavo che mi avrebbe impressionato, e invece no / Oggi faccio lo stesso con te / E quel pudore assoluto che ci ha sempre accompagnati / non esiste più, / Non c'è vergogna in nessuno dei due / Ho imparato prima ad essere padre / e solo dopo figlio / Appena in tempo, mamma, / Ma ce l'ho fatta adesso puoi andare."

58 Being-in-the-world. A fundamental basis of our being-in-the-world is, for the German philosopher Martin Heidegger, not matter or spirit but care. The core of human beings lies in their basic structure of care. This is expressed by Martin Heidegger in his *History of the Concept of Time: Prolegomena* (the text is that of a course of lectures that Heidegger gave at the University of Marburg during the summer semester 1925) where he states that human beings should be "shepherds of Being," meaning that they are such through what they do and through what they take care of. Cf. M. Heidegger, *History of the Concept of Time: Prolegomena*, Bloomington, Indiana University Press 2009.

but also appropriating practices (actions) that can be shared socially: not only with family or friends:[59]

> ***Saved souls*** by Francesco Tomada
>
> Ten years ago I changed my children's clothes,
> I bathed their nakedness and washed away their dirt
> Before having them, I thought this would bother me,
> But it was not so
>
> I do the same for you today
> And the complete modesty once always present between us
> Is there no longer,
> There is no shame for either of us
>
> I learnt first to be a father
> And only later to be a son
> Just in time, Mother,
> But I got there and now you can
> go . . .[60]

Conclusions

As it has been suggested and argued in the analysis throughout this essay, active listening is the prerequisite for an effective relationship of care. A care that is understood as *measured, respectful and equitable*: the three keywords of the *Slow Medicine* approach promoted by Carlo Petrini and other founding members.[61] If we analyze the meaning of each of these three words, we can observe that the Slow Medicine recognizes that "doing more" does not always mean doing better. In fact, the dissemination and use of treatments or new diagnostic procedures is not always necessary (or accompanied by benefits for patients). Moreover, it should not be

---

59 Gilligan 1982, *In a Different Voice. Psychological Theory and Women's Development*; Noddings 2013, *Caring. A Relational Approach to Ethics and Moral Education*; Manning 1992, *Speaking from the Heart. A Feminist Perspective on Ethics*; Tronto 1994, *Moral Boundaries: A Political Argument for an Ethic of Care.*
60 *Saved Souls* (*Anime Salve*), in Francesco Tomada 2014, *Portarsi avanti con gli addii* (no page numbers), English translation by Manuela Gallerani.
61 The first Italian National Conference on Slow Medicine (held in Turin in November 2011) attended by Alberto Dolara and Carlo Petrini set the stage for a broadly represented organization for a medical *reform*. Indeed, hyperactivity in clinical practice is often *unnecessary*. In the *slow medicine* network, healthcare professionals, especially the doctors and nurses, are given the time they need to run through their patient's personal, family and social problems exhaustively, to help them overcome excessive anxiety when faced with non-urgent diagnostic and therapeutic procedures, to assess new methods and technologies *carefully* and to offer adequate emotional support to terminally ill patients and their loved ones.

overlooked that economic interests encourage both an excessive use of drugs or health services and an expansion of people's expectations on health (and life span). In this sense, a *measured* medicine implies the ability to act with moderation, in a gradual and essential way (and without waste). A measured medicine is engaged in the respect of both the environment and the integrity of the natural and social ecosystem. A *respectful* medicine is able to take into consideration the orientations and the values of each person in different moments of life and, at the same time, the professionals of caring act with empathy and respect. *Consequently*, an *equitable* medicine asks appropriate care and a good *quality* of curing/caring, for both patients and health professionals. Therefore, an equitable medicine facilitates people to access both to social and health service network and is against inequality treatments.

In this framework, care becomes a *rule of life*, embracing the underlying assumption of (ethical) *responsibility* that can be traced to the notion of *epimeleia heautou* (care of Self and of the Other) and to *parresia* (to speak candidly or to ask forgiveness for so speaking), which is also a way of speaking to a person's *soul* and is an act relating to *how* this soul will be formed (in the words of Michel Foucault). In this sense, the care becomes an ethical and also *political* action, because the purpose is to find well-being and the *good* for a good life or at least *good enough*. That means with dignity, at any age, time or condition (including frailty) one finds oneself in.

In other words, in a respectful, measured, equitable *curative* relationship, caregivers must be most careful to say the right things, at the right time – from diagnosis to cure – and in the right way to that unique, unreplaceable person who is in front of them, face-to-face.[62] To take up the words of Paul Ricœur ethics are defined through three levels: "I would define the ethical perspective with these three terms: hope for a good life, with and for Others, within the right institutions."[63] From here, the contribution of Medical Humanities appears essential in an ethics of curing/caring that encourages good health as the grounds for a good life.[64] By this it is possible to mean a life directed towards self-fulfillment and empowerment, individually and socially, as well as towards *good* and *happiness* (in the meaning of *eudaimonia*), pursued because it is the intrinsic aim of a human and engaged life project.

---

62 Cassidy, Werner, Rourke, Zubernis and Balaraman 2003, "The Relationship Between Psychological Understanding and Positive Social Behavior"; Gallerani 2016, "Narrative Medicine and Stories of Illness: Caring for the Sick requires active Listening"; Gallerani 2020, "The Responsibility in speaking of Care."

63 Ricœur 2007, *Etica e morale*, p. 34 (author's translation). On this theme see also: Bauman 2008, *Does Ethic have a Chance in a World of Consumers?*

64 Veatch 1981, *A Theory of Medical Ethics*; Nussbaum and Sen, eds., 1993, *The Quality of Life*.

Haling or Hale: The Body in the Arts and Literature

Stephen C. Meredith

Disease and the Problem of Evil in the Novels of Thomas Mann

Disease and the Problem of Evil: An Overview

An obvious point to start: disease, most of the time, is associated with suffering of the afflicted and anyone who loves the afflicted. Even when asymptomatic, a disease carries the threat of suffering and/or death. It might then seem clear that disease should be considered as an evil. There are, however, many subtleties that tend to undermine this statement. Consider, for now, only those subtleties – and not even all of these – that are relevant to the fiction of Thomas Mann. The first is the question of whether a disease can ever be elevating or spiritualizing. Certainly (as discussed below) some of Mann's characters considered disease as a pathway to the spirit. In this respect, these characters are in a lineage that can be traced back to Novalis (among others) or further. The following quotes, for example, are from Novalis:

> Likely, diseases are the stimulus and the most interesting subject for our meditation and activity . . .
>
> Every disease is a musical problem, healing is a musical solution . . .
>
> If a man began to love disease or sorrow, the most exciting pleasure would penetrate him . . .
>
> Could disease not be a means of higher synthesis?[1]

Although the meaning of these quotes is rather opaque, one can imagine some of Mann's characters saying similar things.

The second is whether an individual can or should be blamed for getting or having a disease. For (as also discussed below) in much of Mann's fiction, disease is in some way desired or sought after. Did Mann, then, side with Job's interlocutors who, seeing Job "afflicted . . . with loathsome sores from the sole of his foot to the crown of his head,"[2] said that surely, he deserved this fate?

And the third is this: even when disease is taken as an evil, *simpliciter*, something grim and horrible, *how* does one think about disease as an evil? For it is the common view among Christians[3] – after Augustine and Thomas Aquinas and many others –

1 Quotes are from Novalis 1997. See also Biasin 1975: 8; Krell 2007: 289–309; Stone 2008: 141–164.

2 Job 2:7. Though often called leprosy, whatever Job had almost certainly was not what we now call leprosy (i.e., Hansen's disease, caused by *M. leprae*).

3 Evil is a problem for all religions but may be a particularly poignant problem for Christians like Augustine and Thomas Aquinas, who believed in one loving and omnipotent God. The question became for them how such a God could allow evil to occur in the world. The *problem* of evil was stated

https://doi.org/10.1515/9783110788501-006

that evil is the privation (not simple negation[4]) of the good in a subject: *privatio boni in subjecto*. To cite Thomas Aquinas's oft-used example, blindness is an evil in a human being, because it is the loss of a particular goodness – sight – that properly would belong to a human being. If disease is the privation of the good of health, then one might conclude that it is the privation of our *nature*. But how can disease be a *privation* of nature when, truly, nothing could be more *natural* to us than to get a disease and die? This grim reality is also central to Mann's fiction. At the end of this essay, I will argue that Mann did, in the end, come to an Augustinian / Thomistic view of evil as *privatio boni in subjecto*, but this is a position towards which he evolved.

Little Herr Friedemann and an Overview of Mann's Obsession with Disease

With these questions in mind, we turn now to the fiction of Thomas Mann. Mann's fiction gives the sense of an author with a morbid disposition and an obsession with disease. This obsession was his avenue for examining the broader Problem of Evil. After a brief overview of the diseases portrayed in his stories and novels, I will turn to some early short stories and three novels to discuss in more detail: *Buddenbrooks* (which has clear autobiographical aspects), *The Magic Mountain*, and his late masterpiece, *Dr. Faustus*.

The survey of Mann's works about disease starts with the bitter, almost brutal anti-romantic irony of the short fiction. It was as if Mann, sickened by saccharine household fiction, sought to turn their schmaltzy tropes on their heads. Mann's first published work, *Little Herr Friedemann*, in many ways typifies his work of this period.[5] The "protagonist" of this story (or novella) becomes a hunchback at the age of one month when his alcoholic nurse drops him. After an early disappointment in love – the girl he loves prefers another – he renounces love and devotes himself to a life of resignation, taking pleasure only in literature and music. This resigned and deprived life, distanced from the emotions of love, lessens his inner turmoil, but his calm is disrupted when he falls in love in spite of himself with the friendly and sociable,

---

succinctly in Boethius's *Consolation of Philosophy*, in a verse that is consonant with Augustine's writings: "If God exists, whence comes evil? If He does not exist, whence comes good?" [Si quidem deus est, unde mala? Bona uero unde, si non est?] The first part of this verse is often quoted without the second, but the second is at least as important.

4 A privation is the lack or absence of something that, by nature, ought to be there. A negation, on the other hand, is the simple absence of something, whether it ought to be there or not. Negations are not always evil: as Thomas Aquinas put it poetically, it is not evil for a human being not to have the swiftness of a roe or the strength of a lion.

5 *Little Herr Friedemann* (*Der kleine Herr Friedemann*) was first published in 1896 in *Die neue Rundschau*, and then was included in a book of Mann's short stories collectively entitled *Der kleine Herr Friedemann* (1898).

though notably not beautiful wife of a military commandant in the town, Frau Commandant von Rinnlingen. He confesses his love to her, saying how she has made him realize that his life of renunciation has been a lie: though he had tried to convince himself that he was happy in that life, in reality he was not. She praises his bravery in making such a confession, and starts to make a similar one to him – until she abruptly stops and laughs in his face:

> He had touched her hand with his as it lay beside him on the bench, and clung to it now, seizing the other as he knelt before her, this little cripple, trembling and shuddering; he buried his face in her lap and stammered between his gasps in a voice which was scarcely human: "you know, you understand . . . let me . . . I can no longer . . . my God, oh, my God!"
>
> She did not repulse him, neither did she bend her face towards him. She sat erect, leaning a little away, and her close-set eyes, wherein the liquid shimmer of the water seemed to be mirrored, stared beyond him into space. Then she gave him an abrupt push and uttered a short, scornful laugh. She tore her hands from his burning fingers, clutched his arm, and flung him sidewise upon the ground. Then she sprang up and vanished down the wooded avenue.

This is not mere rejection: it is a special cruelty by which she seems to sympathize but then brutalizes Herr Friedemann. It is also a revelation. One assumes that Herr Friedemann knew what he was, but the narrator clinched it by adding "this little cripple" emphatically. She, at the same time, is revealed as ugly and shallow. Distraught and despondent, Herr Friedemann drowns himself in a river, which the unsubtle narrator describes starting with a rhetorical question, followed by less than omniscient speculations:

> What were his sensations at this moment? Perhaps he was feeling that same luxury of hate which he had felt before when she had humiliated him with her glance, degenerated now, when he lay before her on the ground and she had treated him like a dog, into an insane rage which must at all costs find expression even against himself – a disgust, perhaps of himself, which filled him with a thirst to destroy himself, to tear himself to pieces, to blot himself utterly out. On his belly he dragged his body a little farther, lifted its upper part, and let it fall into the water. He did not raise his head nor move his legs, which still lay on the bank. The crickets stopped chirping a moment at the noise of the little splash. Then they went on as before, the boughs lightly rustled, and down the long alley came the faint sound of laughter.

The disgust that Herr Friedemann feels towards himself is mirrored in the derisive laughter with which the story ends, preceded by "a luxury of hate" and "an insane rage." In order to drown himself, he drags his body on his belly, thereby joining himself with other literary insects or worms like Gregory Samsa and Dmitri Karamazov.[6]

---

6 In Dostoevsky's novel, Dmitri repeatedly refers to himself as an insect or worm. For example, in the poem on "Joy" that he wrote and recited to his brother Alexei, he stated (and later repeated), "To insects – sensuality." He then makes this point even more explicitly by saying, "I am that very insect, brother, and those words are precisely about me." In Dostoevsky 1990: 108. As for Gregor Samsa, although he is widely taken to be an insect, in the text of "Metamorphosis" Kafka used the word "Ungeziefer", meaning "vermin", not necessarily an insect.

The story is thematically similar to a later story, "The Dillettante,"[7] and also has clear resonances with *Buddenbrooks* (discussed below) where Thomas Buddenbrooks discovers Schopenhauer's pessimistic writings. The sublimation of Herr Friedemann's sexuality into literature and art, prompted by his physical repulsiveness, was in some ways as Freud proposed: unacceptable impulses get transformed along socially acceptable pathways. In the case of Herr Friedemann, however, this type of defense mechanism ultimately proved unsuccessful. In his early life, whenever he sought female companionship, a competitor repeatedly denied him this pleasure. Herr Friedemann, in any case, seemed to be seeking a mother rather than a lover. As an infant, his mother had not cared for him, leaving his care to the alcoholic nurse, to devastating effect. In an early stage of the encounter with Frau von Rinnlingen, she sexually provoked him (she stuck out her lower lip and put her breasts forward) – to which Herr Friedmann replied by sucking his finger and fondling his own breast. When later she suggested that he play the violin in a duet with her – literally as well as metaphorically "making music" – he merely blushed and declined, at which point she turned hostile. In these ways, he was denied even maternal caresses, let alone mature intimacy with a woman.

For our purposes, the main point is that the deprivation of the protagonist – the inability to engage in the pleasure of "life" – started with a physical ailment. His resignation to a life enriched only by the intellect resulted directly from his physical deformity. Sporadic attempts to reverse this pattern by seeking female companionship, however psychologically immature, only disrupted the uneasy compromise that had allowed him a somewhat calm existence. In the end, an attempt to end his social and sensual isolation leads to his suicide: no longer able to continue a life of renunciation, he succumbs to disappointment.

Before turning to the three novels, here, then, is a brief survey of diseases in the works of Thomas Mann. The list is not limited to diseases, strictly defined; in some cases (as in the case of Herr Friedemann) the issue is physical (or mental) deformity.

- "Tobias Mindernickel" is mentally ill.
- "Little Lizzy": the protagonist, Lawyer Jacoby, is morbidly obese.
- "The Way to the Churchyard": the protagonist, Praisegod (*Lobgott*) Piepsam, has alcohol-induced neurological disease.
- *Buddenbrooks*: Thomas Buddenbrook dies of septicemia after having a "bad tooth" extracted. Christian Buddenbrook has phobias and somatic delusions. Hanno Buddenbrook, the last in the family line, dies of either typhus or typhoid fever (see below).

7 Original published as "Der Bajazzo," 1897. This story is a rare Mann story written in the first person.

- "Death in Venice" takes place in the midst of a cholera epidemic – from which the protagonist dies.
- *The Magic Mountain* takes place in a tuberculosis sanitarium; all of the characters, even the doctors and nurses, are sick.
- "Tristan" also takes place in a tuberculosis sanitarium.
- *The Black Swan*: the central character dies of uterine carcinoma.
- *Dr. Faustus*: the protagonist (Adrian Leverkühn) has migraines, and later contracts and dies from neurosyphilis, after a period of 24 years. The narrator (Serenus Zeitblom) has an essential tremor. Nepomuk ("Echo") Schweigstill, Adrian's cherubic nephew, dies of bacterial meningitis.

This is a partial list. The unifying theme of these conditions is not a medically rigorous definition of "disease," but rather the grotesquerie of these conditions: the aberration of nature, the radical unwellness of the misfit's body. The slippery term "decadence" might also apply: some of the novellas and stories (e.g., "Disorder and Early Sorrow") demonstrate the decline of a family from earlier vigor and status. Although "grotesque"[8] has come to mean a type of sardonic humor that is improper, distorted, and unnatural, this meaning derives from an earlier one. As Sir Thomas Browne wrote in *Religioso Medici* (1643), "There are no Grotesques in nature," implying that where grotesques are found, they are somehow against nature. And yet to the many characters in Mann's fiction, the grotesque is also appealing.

In the remainder of this article, I will examine three of Mann's novels in more detail. I will argue that Mann's use of disease in these novels is in part metaphorical, in that a physical disease that also serves as an outward manifestation of a disease of the soul. There are four features that I think define Mann's concept of disease, typified by tuberculosis in *The Magic Mountain* and tertiary syphilis in *Doctor Faustus*:

1. *Diseases are hidden, secret – not only in showing the state of the soul, but also in their physical manifestations they are a hidden process.* Thus, a disease gets uncovered and discovered through the course of the novel. In *The Magic Mountain*: Hans Castorp is genuinely amazed – as well he should have been – by the ability to look inside himself through the use of x-rays – quite a novelty in 1907. Indeed, he carries an "inner portrait" – a chest x-ray – of his some-time girlfriend Clavdia Chauchat. These x-ray pictures – "internal portraits" as Hans Castorp calls them – are also metaphors, which reveal an inner and secret disease. X-rays are thus likened to the psychoanalytic process of unearthing unconscious thoughts and feelings. Indeed, this comparison is made explicitly

8 The word arose in early modern French, with parallels in other languages, to refer to certain crude paintings made on unpolished wall, i.e., "painting appropriate to grottos" (Oxford English Dictionary).

through Dr. Krakowski, a psychoanalyst at the sanitarium. Mann's affinity with psychoanalysis was described in his essay on Freud.

2. *Diseases are sought after: however gruesome the disease, it is also something the character wants, actively or passively seeks, and something that fulfills a wish.* Disease for Mann is quite similar to neurotic symptoms and dreams for Freud. As Freud stressed, the neurotic patient is not malingering, but the neurotic symptom nevertheless fulfills some unconscious wish, however harmful that symptom might ultimately prove to be. Similarly, dreams are not always conventionally pleasant, but are also somehow the fulfillment of a wish. Thus, one goal of psychoanalysis is to bring the wish from unconsciousness into consciousness. For Mann's diseased characters, disease has this characteristic of neurotic symptoms: that whatever harm the disease does, it is also sought after, and fulfills some unconscious wish.
3. *Disease is a type of decay.* That is, *the characters* view their diseases as a manifestations of decay, usually from a healthier and more virtuous past state. Disease, therefore, becomes a type of fall from grace, and indicates an inner, *moral* rottenness. In *Buddenbrooks*, Thomas Buddenbrook, apparently at the apex of his life and career, intuits that the family's decay and fall is already well advanced (see citation, below). In *The Magic Mountain*, Mann plays with the heights of the Alps by declaring them also to be a descent – into Hades, where Radamanthus (the nickname for the head physician, Dr. Behrens) has dominion. It is a fall also of the stolid North German burgher – a budding engineer in a shipping firm – into the slovenly ways of art and philosophy. Indeed, much of Mann's earlier fiction – for example, *Tonio Kröger* – tells of the artist who comes from an upper middle-class background – as Mann did – and has a bad conscience about the dissolute ways of artists. There is a sense of shame in these artists, who are what a female artist-friend of Tonio Kröger called him: *bourgeois manqué*. Early in *The Magic Mountain*, Hans's doughty uncle Tienappel visits Hans at the sanitorium – and then flees in panic. He flees, it is implied strongly, so that he does not fall into the seductive decadence of the mountain.
4. Finally, *disease is also something for which the character feels guilty*: it occurs through a failure of vigilance and discipline, it is something "succumbed to," "slovenly," as any child raised on "Struwwelpeter" would well understand. For Mann, this would include or even center on what he referred to as his "homoerotic urges" (his term).

Buddenbrooks

Thomas Mann was born in 1875 in the city of Lübeck on the Baltic Sea. He was the second son of a socially prominent grain merchant and senator in the "free city," which was also called a "Hanseatic city."[9] In the glory days of the Hanseatic League (12th century), Lübeck was once one of the largest and wealthiest German cities, the hub of a trade system that stretched as far as England, Scandinavia, The Netherlands, and Russia. Despite the decline of the league starting in the 15th century, and its official dissolution in 1869 when Kaiser Wilhelm I established the German Empire, Lübeck retained the words "Hanseatic City" in their official titles – one of only three cities to do so.[10] To Mann, this term might have been redolent of decadence and past glory.

By all accounts, Mann's father was a rather stolid man of the middle class, but his mother was "exotic" – the author's word for her. She was born in Brazil of mixed German, Portuguese-Creole ancestry, and had artistic inclinations – in particular, she was musically talented. When Mann's father died at the age of 51 (Thomas was not quite 17) they were forced to liquidate the family firm and sell the opulent family home.

Although Thomas completed his schooling in Lübeck – and he was, at best, an indifferent student – the family moved to Munich. Thomas joined his family in Munich, where he worked for a time as an apprentice in an insurance firm; but he was utterly uninterested in the work, and was sullen and withdrawn, preferring to spend his time reading, painting, writing poetry – and he started writing *Buddenbrooks*, at age 22, finishing it when he was 26. He was in university for one year, but was mainly self-taught – a pattern which continued through the rest of his life. Nevertheless, he had an uncanny knack for penetrating to the heart of a certain philosophers.

Buddenbrooks, originally entitled *Downfall*, incorporates these and other biographical details. The novel ends with the remains of the Buddenbrook family preparing to embark for another city. In the end, Mann settled on the subtitle, "The Decline of a Family." Although the novel follows the decline and deaths of one family, it is a quantum leap in complexity and depth from the short stories that preceded it, and it also analogizes this private decline to the more public decline of Lübeck. This coincided not only with the unification of Germany and its rise as a modern industrial state, but also with its emergence as a colonial empire (albeit a modest one compared to England and France). The historical arc of *Buddenbrooks* begins in the prenational Germany of the first Johann Buddenbrook, a patriot who

---

9 The Hanseatic League (from Old High German *Hansa* for "convoy") was a defensive confederation of merchant guilds and, later, merchant towns in Northern Germany. After being a dominant force in Northern German trade in the late 1100s, they declined and lost power as these cities were subsumed by German states, and as English, Dutch and Swedish traders rose, starting about 1450.

10 Hamburg and Bremen were the others.

is firmly rooted in Lübeck, but who is worldly and well-travelled, old fashioned (he wears knee-breeches and a powdered pigtail), a tradesman who builds a large and thriving business, but still loves classical, humanistic education. The contrast to the last male member of the Buddenbrook line (the emphasis of the novel is definitely on the males) is stark. Hanno (Johann IV) is the antithesis of this virile patriarch: he is sickly, weak, gloomy, introspective, effeminate (probably homosexual) and, of course, a budding artist. He fits in poorly with the vulgar militarism of his school. This militarism reflects the rise of the German empire, and contrasts with the enlightened secular humanism of Johann Buddenbrook.

For this section of the essay, I focus on the demise or denouement of the penultimate generation – Thomas, Christian, and Antonie (Tony) – and of little Hanno. Tony, with her daughter and granddaughter, survives the male members of her family.[11] She is a sad figure: she married twice, and divorced twice, and has seen her son-in-law sent to prison. She renounced the only love of her life (Morton Schwarzkopf) at her family's behest, in order to marry a man of greater wealth and social prominence. With this first husband, Bendix Grünlich, she had a daughter (Erika). Unfortunately, however, Grünlich was an unscrupulous businessman who falsified his accounts to hide debts, and had married Tony only for her family's money. Tony's father refused to bail him out of his debts, and instead brought his daughter and granddaughter home with him, letting Grünlich go bankrupt, after which Tony divorced him. Her second husband, Alois Permaneder, was a provincial bumpkin, but at least was also a seemingly honest (at least, compared with the first one) hops merchant from Munich. The couple moved to Munich, but Tony hated the southland (people had funny accents, there was no seafood, and so many of them were Catholics – "gibberish. And then there's Catholicism; I hate it, simply have no use for it, as you know."[12] – and so forth . . .). She became pregnant, but the baby was stillborn. The marriage collapsed when Tony discovered her husband drunkenly trying to rape the maid. She sued for divorce, and somewhat surprisingly, her husband accepted the terms and apologized for his behavior – and even returned her dowry. Tony moved back to Lübeck with her daughter, but this daughter later suffered a fate similar to her mother's: she married a manager at a fire insurance company (Hugo Weinschenk), who was sent to prison for insurance fraud. She did, however, have a daughter, Elizabeth, who survives at the end of the novel. But the surviving females are all mateless, a damned state in the context of provincial Lübeck, and Tony, once an arrogant and imperious beauty, is described as dumbfounded that she should end her life in so mean a state: "Her grandfather had driven all over the country in a coach-and-four."[13] And physically, she is described

11 Christian, though alive, is as good as dead: by the end of the novel, he has disappeared, forever confined to an asylum.

12 Mann 1993: 271.

13 Mann, 1993: 646.

thus: “there were even several hairs on her upper lip – Tony Buddenbrook’s pretty upper lip.”[14]

At first glance, the two brothers, Thomas and Christian seem polar opposites. As a child, Christian was irresponsible and lazy in his studies, and a hypochondriac (of which, more below). Rather than fulfilling his father’s demand that he “[remain] with Mr. Richardson for another year [to] acquire mercantile experience” (presumably, in order to enter the family business in due course), he preferred to travel, “out of a love of adventure.” He was unable to hold a job in Lübeck, and rather than associating with the merchant class while in his hometown, he preferred the company of the déclassé: artists, actors, and prostitutes. Long before this, however, we are given an epiphanic incident from his childhood:

> They are sitting at the dinner table; the fruit has been served and is being enjoyed amid genial conversation. Suddenly Christian places a peach he has bitten into back on his plate, his face turns white, and his round, deep-set eyes grow larger and larger above his oversize nose.
>
> “I shall never eat another peach,” he says.
>
> “Why not, Christian? What sort of nonsense is that? What’s wrong?”
>
> “Just imagine what would happen if . . . just by accident . . . I swallowed this big peach pit and it got stuck in my throat . . . and I couldn’t get my breath . . . and I’d jump up and choke and die a horrible death, and then you’d all jump up . . .” And suddenly he adds a short groan, an “Oh!” of total terror, sits up erect in his chair, and turns to one side as if he is about to bolt.
>
> His mother and Mamselle Jungmann do in fact leap to their feet.
>
> “Good heavens – Christian, you haven’t swallowed it, have you?!” For it really does look as if that is what has happened.
>
> “No, no,” Christian says, gradually calming down, “but what would happen if I did?”
>
> The consul, who has also turned pale with fright, begins to scold him, and even his grandfather raps indignantly on the table, forbidding him any such foolish pranks. But for a long time afterward, Christian does not eat a single peach.[15]

His reaction to the peach was phobic, i.e., fear beyond actual risk, a point he apparently understood at a later age, though it did not relieve all of his anxiety. The inclusion of this passage is striking: it indicates Christian’s oddness, and sets the stage for where he will end up: in an asylum, of which more shortly. The fear can be interpreted symbolically, standing in for the more general one of being unable to breathe, that he would “jump up and choke and die a horrible death” – not only being unable to breathe, but even worse, making a big scene out of it, and breaking the stuffy decorum of the dinner table, because if he “jumped up,” everyone else would jump up, too – as, indeed, they did on this occasion. The fear of being unable to breathe played itself out in an ostensibly comical scene in which Christian burned some medicinal green powder in his office to aid his breathing. The burning

14 Mann 1993: 646.
15 Mann 1993: 59.

made everyone around him cough and rush from the room, unable to breathe. This scene more overtly shows the wishful aspect of the symptom: he, himself, had set fire to the green powder, thereby making himself and everyone around him "jump up," as if to declare, "you see! No one can breathe in this atmosphere."

Christian's solution to his phobia was simple enough: avoid peaches. But in Christian, this phobia was part of a larger pattern of failing to adapt to the demands and social requisites of his class. In contrast to the (apparently, temporarily) successful Thomas, Christian failed at almost everything he did: he was a poor student, a poor employee, and a spectacular failure in his marriage to a lower-class mistress – a courtesan who was on the rebound from one of his cronies, and who ultimately had Christian committed to an asylum, after which she was able to resume her former habits.

In his early days, fearing that he would choke to death, so to speak, in Lübeck, he traveled to London and faraway San Francisco in the U.S. and Valparaiso, Chile. He returned eight years later, outlandishly "dressed in a yellow plaid suit that certainly hinted at the tropics and carried a swordfish sword and a long stalk of sugarcane. He looked half embarrassed and half preoccupied as he stiffly returned his mother's embrace."[16] He still had trouble swallowing food, however, and feared that he would choke on it:

> Suddenly he said, "You know, it's strange – sometimes I feel like I can't swallow. No, now don't laugh. I'm being quite serious. The thought occurs to me that I can't swallow, and then I really can't. What I've eaten is clear at the back of my mouth, but these muscles here, along the neck – they just won't work. They won't obey my will, you see. Or, better, the fact is: I can't bring myself to actually will it." Quite beside herself now, Tony cried, "Good heavens, Christian, what silly nonsense. You can't make up your mind whether to swallow. No, you're just being ridiculous. What are you talking about?" Thomas said nothing. But his mother said, "It's your nerves, Christian. It was high time you came home. The climate out there would surely have made you ill."[17]

He then shocked his family with tales of violence, his companionship with unsavory individuals, and, it is implied, his own sexual adventures, which sets the stage for the later *brüderkrieg*. Christian got drunk while dining at the club, and for a while he is giddily upbeat, regaling the group with stories and songs:

> He waxed enthusiastic; he was speaking in tongues now. He told his stories in English, Spanish, Plattdeutsch, and Hamburg dialect; he described knifings in Chile and robberies in Whitechapel, hit upon the notion of rummaging through his collection of comic songs, and now recited some and sang others, mugging like a professional comedian and displaying an especially whimsical talent for hand gestures.

16 Mann 1993: 228.
17 Mann 1993: 23–232.

A walkin' down the street,
Just lazin' through my day,
I chanced to spot a lass
Up ahead a way.
In Paris togs she was . . .
I offered her my arm.
She turned and eyed me hard
And spurned my manly charm:
"Go home, my boy, and tend your farm . . . ."

Christian's manic mood continued for a while, but:

> Then, very suddenly, he fell silent – the expression on his face changed, his arms and legs went slack. His little, round, deep-set eyes began to wander restlessly in all directions, and he passed his hand down his left side – it was as if he were listening for something inside him, where strange things were happening. He drank another glass of liqueur, which restored his spirits a little; he tried telling another story, but then, looking rather depressed, he got up to leave.[18]

Mann gives an extended excerpt from the song, "That's Maria," an English music hall song, which was considered risqué at the time, and which Mann had used earlier in his story, "Little Lizzy." It could almost be considered Christian's anthem: the trivial content tells Christian's story of courting, and ultimately being rejected by a stylish but risqué (read: déclassé) woman. In fact, this was the second time he mentioned the mangy dog in San Francisco, and as for "Maria," she got four mentions, mostly ribald but in one case (the above quoted discussion of his difficulty swallowing), the mention was coupled and laced with anxiety. In this gathering, he sang the song but became unable to sustain the manic mood, his forced gaiety evaporating as rapidly as it had arisen. Indeed, Christian has his leitmotifs, another one being an ostensibly chance, glib remark, which provoked and incensed his brother, Thomas. Christian declares – twice, both times when he is in his cups: "Seen in the light of day, actually, every businessman is a swindler."[19] The problem, of course, is that the statement has more than a grain of truth in it[20] – demonstrated, for example, by Tony's and Erika's husbands – but Thomas reacts to this statement with righteous indignation, and pours out a lifetime of anger at his irresponsible brother, demeaning him as "a laughingstock with your love affairs, your harlequinades, your diseases, and your remedies . . . as a growth, a fester, on the body of our family," thereby likening Christian to another disease, cancer. This incident leads to Christian's dismissal from the business and exile to Hamburg, where he works in a lowly import-export business.

18 Mann 1993: 397–398.
19 Mann 1993: 280.
20 It is also hinted that some of Thomas's business transactions are less than completely honest.

Whatever Thomas felt on Christian's departure, the main loss was to the club and theater in Lübeck, which had lost a boon companion.

The "diseases" for which Thomas castigated Christian – and which Thomas bravely resisted – are never specified and were presumably hypochondriacal. His legs were described as somehow spindly and crooked; barring the unlikely occurrence of rickets in his social class, this awkwardness is of a piece with his large beaky nose. Christian complained that his nerves are too short on his left side. These symptoms might have been real to Christian, but the cause was likely psychological, not physiological.

Christian literally disappears from the novel. Exiled to Hamburg and from the family business, he spent money he didn't have on Aline Puvogel, who lived with her two children, though it is not clear that she is or was ever married to their father(s), and she is later referred to as "Fraulein Puvogel,"[21] The name, when pronounced quickly, sounds like "Puffvogel," which would mean "brothel bird."[22] As the narrator notes – and this is well known to the gossip channels – "Christian Buddenbrook was not the only merchant in Hamburg who maintained intimate and costly relations with this personage."[23] Of course, his family would revile anyone like Aline Puvogel, and Toni did not disguise her feelings in a letter to Aline. But the marriage to Aline was just a final touch in the family's decline – or rather, the stage in the process, which is finally completed with the death of Hanno. As to Christian, his end is dismissed in a matter-of-fact paragraph:

> There was bad news about Christian. His marriage, it seemed, had not improved his health. His morbid fantasies and fixations had grown worse, and more frequent, and at the urging of his wife and a physician he had entered a sanatorium. He did not like it there and wrote plaintive letters to his family . . . At any rate, it allowed his wife, quite apart from the material and social advantages that she derived from her marriage, to continue her former independent life without embarrassment or regard to others.[24]

Although the text does not quite say explicitly that Christian was committed, clearly this is the case since he is not permitted to leave. The chapter ends euphemistically: to remain locked up "really was the best thing for him" – "best" for whom? one wonders – and Aline was thereby able "to continue her former independent life without embarrassment" – meaning, presumably, as the mistress of some other businessman.

21 Mann 1993: 600.
22 I am indebted to Dr. Jonathan Zerweck, a post-doctoral fellow in my laboratory, for this suggestion.
23 Mann 1993: 342.
24 Mann 1993: 604.

If there is one chief character in the novel, it would probably be Thomas Buddenbrook, who shares the author's first name.[25] Like Thomas Mann's father, Thomas Buddenbrook succeeded in business, built a grand house and was elected to the town's Senate. As a child, Thomas Buddenbrook was properly behaved, a dutiful student, driven to succeed, and likely to inherit the family business. He dressed conservatively and tastefully, even somewhat fussily. His whole demeanor bespoke discipline and diligence. He married a beautiful woman – though, as it turned out, the marriage was not truly a happy one. Christian never adapted to the demands for propriety of the Lübeck business class; Thomas, however, was only apparently better suited for it. He wore a brave front, but maintaining it became increasingly a trial as the novel progresses. If Thomas is the man Christian "should have been," then Christian is the man Thomas feels he must avoid becoming at all costs. Thomas's propriety and success are shams – self-delusions.

Thomas's two rows with Christian centered on the latter's impropriety, but Christian, whatever his faults might have been, did not shrink from a rebuttal: he recognized his brother's hypocrisy. The first row, concerning Christian's joke (so he called it, afterwards) that every businessman is a swindler, paled in intensity to the second, which came immediately after the hard death of their mother. While dividing up their mother's possessions, Christian surprised Thomas by asking for dishes and linens – because he planned to marry Aline. Thomas, who had become Christian's guardian, recoiled in fury, and declared that he would not allow Christian to marry Aline and degrade the family name. In this way, Thomas justified his vindictiveness, for his anger went even further than warranted even by Christian's irresponsibility. Christian's quirks notwithstanding, he recognizes the source of Thomas's reaction:

> "Your heart is so icy and full of malice and disdain," Christian went on, and his voice was somehow muffled and squawking at the same time. "As far back as I can remember, I have felt such icy contempt coming from you that I've always been frozen to the bone in your presence. Yes, it may be a strange way to put it – but if that's what I really felt? You rebuff me just by looking at me, and you almost never even do that. And what gives you the right to act that way? You're human, too. You have your weaknesses."[26]

Christian's first name, like Thomas's, seems significant. Giving Christian his name often seems ironical, but here, he is faulting his brother with an icy heart and a lack of charity. The alert reader might recognize the last statement in the above citation as a veiled allusion to something about which Christian had already reminded Thomas on several occasions. Thomas, like Tony, had renounced his one chance at

25 Although Mann's name was Paul Thomas Mann, the name "Paul" was seldom used.
26 Mann 1993: 500.

true love when he declined to marry Anna, "the little flower-shop girl with the Malaysian face."[27] (Mann mentions her "Malaysian" face three times in the novel.) Like Alina, Anna is physically healthy – denoted above all by remarkably healthy teeth (a trait Gerda also shares), in contrast to Thomas's decaying teeth. Thomas maintained remote contact with Anna through the years, and Anna attended Thomas's funeral.

Thomas, however, reacted to Christian's rebuke not with greater rage, but with a confession:

> "I have become what I am," Thomas said at last, with emotion in his voice, "because I did not want to become like you. If I have inwardly shrunk away from you, it was because I had to protect myself from you, because your nature and character are a danger to me. I am speaking the truth."[28]

And yet, he ended by threatening Christian:

> But if you challenge me, push things to extremes, we shall see who comes out the worst for it. I'm telling you – be careful. I shall be ruthless. I'll have you declared incompetent, I'll have you locked up, I'll destroy you![29]

Which, in any event, Aline did for him. As Tony had given up the very tamely revolutionary Morten Schwarzkopf (who later became a respectable physician) and had entered into a disastrous marriage with Bendix Grünlich, so Thomas had entered into a nearly loveless marriage with Gerda Arnoldson. After bidding Anna a sad farewell as he left for Amsterdam earlier in his life, he reminded her that he never promised to marry her – and she recognized that this could never happen because of their class differences. It was in Amsterdam that he met Gerda.

Mann depicts Thomas's marriage to Gerda as superficially ideal. She is beautiful, albeit in a slightly alien way: she has telltale blue shadows under her eyes (which Hanno also has), and her beauty is repeatedly called "exotic" or "strange," for which one can read "foreign" or "dangerous" – or, like Mann's own mother, "Southern." Worse still: she has a Stradivarius violin, on which she plays duets with her father, who "played the violin like a gypsy, with savage passion."[30] And still worse: she loved Wagner. But music might be one of the few things about which Gerda was passionate. Thomas and Gerda left for a two-month honeymoon but protracted it to seven months. Thomas built her an unnecessarily opulent and expensive house – significantly, across from Anna's flower shop. In contrast to Anna, who is pregnant with her third child at Thomas's funeral, Thomas and Gerda have only one child, the sickly Hanno. After returning from their honeymoon, they

27 Mann 1993: 206.
28 Mann 1993: 502.
29 Masnn 1993: 503.
30 Mann 1993: 260.

soon grew apart, and ultimately slept in separate bedrooms. Gerda is nervous, snappish, and cold; she rarely smiles; she suffers from migraines; it is hinted that she might be sexually frigid.[31] She makes no friends in her new town, with the exception of one young and musical Lieutenant von Throta, as discussed below.

Thomas shows only occasional insight into his own situation. One example, as noted, was his recognition that he became what he is in order to avoid becoming like Christian. Thomas is most forthright and at ease with his main ally in life, his sister Tony. Despite marrying, having a child, and building a grandiose new house, he recognizes that this outward show of glory really belies an *inner rot* that signals the family's decline:

> The last few days I've been thinking about a Turkish proverb I read somewhere: 'When the house is finished, death follows.' Now, it doesn't have to be death exactly. But retreat, decline, the beginning of the end. Do you remember, Tony," he went on, slipping his arm under his sister's and lowering his voice even more, "when Hanno was christened, how you said to me, 'It's as if a whole new era is beginning'? I can still hear it quite clearly, and it seemed to me then you were right, because then came the election for senator, and I was lucky, and this house rose up here out of the earth. But 'senator' and 'house' are superficialities, and I know something else that you weren't even thinking about that day, something I've learned from life and history. I know that the external, visible, tangible tokens and symbols of happiness and success first appear only after things have in reality gone into decline already. Such external signs need time to reach us, like the light of one of those stars up there, which when it shines most brightly may well have already gone out, for all we know."[32]

The irony, of course, is that he has failed to heed his own insight: he has built the house and death will come. Except: he perhaps built the house *so that* death would come. Thomas had aged prematurely; Gerda, vampire-like, did not seem to age at all:

> The only change that time had brought could be found in the difference in their ages, which, although it was quite small when measured in years, had began to show itself in obvious ways.
>
> People looked at them and saw a rather stout man who was aging quickly and a young woman at his side. People thought that Thomas Buddenbrook looked, as they said, rather tumbledown. Indeed, although his vanity – which had become almost comical by now – kept him propped up, "tumbledown" was the only word for him. Whereas Gerda had scarcely changed at all in the last eighteen years.[33]

31 At least, with her husband. Although the time she spends with Lieutenant von Throta is suspect, it is not clear whether they have sexual relations.

32 Mann 1993: 378–379.

33 Mann 1993: 557.

Worse: Gerda found a lover, but not simply a lover, perhaps a kindred spirit: "René Maria von Throta, a Rhinelander by birth, was a second lieutenant in one of the infantry battalions stationed in the town."[34] Thomas frets not that she has a handsome young sexual partner, but on the contrary, that this lieutenant was "a most unmilitary man"[35] who played several musical instruments "exceedingly well":

> What was he afraid of? . . . Oh, he would almost have been happy if he could have called him that [a beau], thought of him as that, if he had been able to despise him as a shallow, empty-headed, and vulgar young man who was working off a normal dose of youthful energy with a little music in order to win ladies' hearts.[36]

Thomas worried about his wife and the lieutenant, but he declines to display any jealousy, as that would be unseemly. As he worried, however, a rare tender moment occurred between Thomas and his son, breaking their usual awkwardness:

> And then, quite suddenly, Hanno heard something being said above him that had no connection at all with their conversation – in a soft, anxious, and almost imploring voice that he had never heard before. But it was his father's voice, and it said, "The lieutenant has been in there with Mama for two hours now, Hanno."
>
> And at the sound of that voice, little Johann raised his golden-brown eyes and fixed them – larger, clearer, and more loving than ever before – on his father's face, with its reddened eyelids beneath pale brows and its white, slightly puffy cheeks behind the long tips of the stiff mustache. God knows how much he understood. But one thing was certain, and they both felt it, that at that moment, as their eyes met and held, the estrangement and coldness, the constraint and misunderstanding between them fell away . . .[37]

Here, the father breaks his emotional isolation, becoming a child to the child, anxiously confessing his worries to his son – and the son responds as if he had been waiting for such a moment all along. But rather than building on this moment, Thomas "drill[ed] Hanno in the practical things that would be important in his future life and work; he examined his intellectual abilities, pressed him for decisive statements about his love of the occupation awaiting him, and burst into rage at every sign of resistance and languor."[38] The reason is clear: "The fact was that Thomas Buddenbrook, at forty-eight, had begun to count his days more and more, to reckon with the approach of death."[39]

34 Mann 1993: 557.
35 Mann 1993: 557
36 Mann 1993: 559.
37 Mann 1993: 561–562.
38 Mann 1993: 562.
39 Mann 1993: 562.

It is in this setting that Thomas discovers – literally – a book by Schopenhauer:

> Pausing in the smoking room for a cigarette after second breakfast,[40] he had found it tucked in a corner of the bookshelf, behind a row of sturdy tomes, and recalled that he had bought it casually on sale at a bookshop years ago – a rather thick volume, poorly bound and badly printed on thin, yellowed paper, just the second half of a famous metaphysical system.[41]

Although this book pops up in a dream-like fashion and is not identified by name, the one chapter he read is entitled "Concerning Death and Its Relation to the Indestructibility of Our Essential Nature,"[42] which identifies it as coming from *The World as Will and Idea*, and in any case, its significance is clear enough. All along, Thomas had been fighting the brave fight, defending against the allure of death, resisting its pull, contrary to his true inclinations. Disappointed in marriage, disappointed also in his only son, losing ground in his business about which he no longer cared – if he ever really did – he declares to himself that "before it was too late, he must either achieve some clear readiness for death, or die in despair."[43] Somehow, this chapter of Schopenhauerian pessimism about life's pointlessness and death's allure puts him into a state of "intoxication," and (after dinner, which no doubt was sumptuous), he asked himself:

> What was death? The answer to the question came to him now, but not in poor, pretentious words – instead, he felt it, possessed it somewhere within him. Death was a blessing, so great, so deep that we can fathom it only at those moments, like this one now, when we are reprieved from it. It was the return home from long, unspeakably painful wanderings, the correction of a great error, the loosening of tormenting chains, the removal of barriers – it set a horrible accident to rights again. . . .
>
> Was not every human being a mistake, a blunder? Did we not, at the very moment of birth, stumble into agonizing captivity? A prison, a prison with bars and chains everywhere! And, staring out hopelessly from between the bars of his individuality, a man sees only the surrounding walls of external circumstance, until death comes and calls him home to freedom.[44]

---

40 Like Hans Castorp in *The Magic Mountain*, Thomas Buddenbrook eats very well: a second breakfast!.

41 Mann 1993: 564.

42 Arthur Schopenhauer, *The World as Will and Idea, Supplements to the Fourth Book*. Chapter XLI, transl. E. F. J. Payne. New York: Dover Publications, 1966.

43 Mann 1993: 564.

44 Mann 1993: 566–567.

He had hoped for a type of immortality in his son, but now thinks to himself: "And I hoped to live on in my son? In another personality, even weaker, more fearful, more wavering than my own? What childish, misguided nonsense!"[45]

He reviews the spiritual history of his family. His grandfather seemed in many ways a straightforward soul, building the business and having no real regrets. As for his father,

> Dogmatic faith in a fanatical biblical Christianity, which his father had been able to couple with a very practical eye for business and which his mother had then adopted later as well, had always been alien to him.[46]

Thomas "had toyed with Catholicism all his life" – a damning trait in this setting, in line with his penchant for an exotic wife. But like Hans Castorp's vision in the snow, Thomas's is short-lived and faded as he digested his dinner and sank into the cushions of his bed. He settles back into what he had been reared in, and into what, despite his occasional tepid, halting protestations, he had never left:

> And so Thomas Buddenbrook, who had stretched his hands out imploringly for high and final truths, sank back now into the ideas, images, and customary beliefs in which he had been drilled as a child.[47]

And it is in this setting that Thomas Buddenbrook will die of a bad tooth. Thomas Buddenbrook is only one of Mann's many characters with decaying teeth, the symbolism of which is apparent. Hanno's teeth are crowded and brittle, and both father and son suffer in the hands of the dentist, Herr Brecht, who is probably incompetent. The extraction is described in graphic detail. Herr Brecht first tried to extract the entire tooth, but broke the crown, so that he was forced to extract each of the four roots, one by one. Thomas, at that point, had had enough, and begged to continue the next day.[48] After leaving the dentist's office, Thomas walked a short distance, but then reeled "as

45 Mann 1993: 567.

46 Mann 1993: 563.

47 Mann 1993: 569.

48 If Herr Brecht used anesthetic inhalants, this was not mentioned explicitly in the text. Before the extraction, Herr Brecht stated, "'I'll paint it [the tooth] a little,' he said. He began at once to put this decision into action, daubing the gum freely with a pungent liquid." (Mann 1993: 654.). After the aborted extraction, Thomas Buddenbrooks is taken home (although it was not stated explicitly by whom of how). In the next chapter while he is lying in his bed recuperating, the text states, "The odors of carbolic acid, ether, and other medicines wafted toward them." (Mann 1993: 659), so it is possible that Herr Brecht had used inhalants. It is also possible, however, that he incompetently underestimated how difficult the extraction would be; another possibility is that he considered it "weak" for Thomas to receive anesthetics. Inhalant anesthesia had been used before the time in which the novel takes place. Dr. James Simpson, Professor of Obstetrics in Edinburgh, had introduced chloroform in 1847. In 1877, Dr. Joseph Clover produced the first portable regulating ether inhaler. Chloroform or ether rendered a patient unconscious or semi-conscious for a short period of time, so that brief procedures could be performed. Both of these inhalations had if someone had taken hold of his brain and with incredible force started swinging it in wide concentric circles that grew smaller and smaller . . . ."[49] He had fallen in the street. "Since the street sloped steeply downhill, his head lay a good deal lower than his feet. He had fallen face-down, and a puddle of blood immediately began to form around his head. His hat rolled off down the street a little way. His fur coat was splattered with muck and slush. His outstretched hands in their white kid gloves had come to rest in a puddle."[50] The extraction of the "bad tooth" led to septicemia: the seemingly localized rot had spread through the entire system. This cause of death connects Thomas Buddenbrook to Mann's father, who also died of septicemia.[51]

Death, especially at an early age, ought to be a sad thing, but Mann depicts Thomas's death as a grotesque, even comical thing. After collapsing in the street, Gerda, shuddering, says,

> You can't believe how he looked when they brought him in. No one has even seen a speck of dust on him, he never allowed that, his whole life long. What vile, insulting mockery for it to end like *this*.[52]

Indeed, it *was* an insulting mockery for this vain and fastidious man to collapse face down in the mud, but it was even more of a mockery that his wife focuses on this point. Two doctors come and go, and as usual in Mann's fiction, they achieve nothing, though Mann does note (twice) that Dr. Langhals has "beautiful eyes." The illness lasted long enough for Christian to return home from the club, and for other visitors to arrive. After Thomas dies, the family sits together for the propriety of addressing envelopes for death notices. Christian, notably, "had not shed a single tear and was a little ashamed of the fact. . . . Now and then he would sit up straight and rub his hand across his bald head and say in a choked voice, 'Yes, it's terribly sad.' He said this to himself, as a kind of stern reprimand, and forced his eyes to moisten a little."[53] The crowning indignity comes when Hanno starts to laugh because "[h]e had come across a name with a curious sound to it – it was just too irresistible."[54] The grotesquerie of Thomas's death becomes a scandal:

---

serious side effects, however, even sometimes causing death. "Brief procedures" could include tooth extractions, but the extraction of Thomas Buddenbrook's tooth was not quick. After some widespread use, dentists returned to an older and, as it turned out, safer inhalant, nitrous oxide, which had been discovered in 1772 by Joseph Priestly, and started being used in dentistry some 20 years later.

49 Mann 1993: 585.

50 Mann 993: 570.

51 Mann's father developed septicemia after an operation for what turned out to be inoperable bladder cancer.

52 Mann 1993: 587.

53 Mann 1993: 592.

54 Mann 1993: 592.

> A tooth – Senator Buddenbrook had died of a toothache, that was the word around town. But, confound it all, people didn't die of that! He had been in pain, Herr Brecht had broken off the crown, and afterward he had simply collapsed on the street. Had anyone ever heard the like?[55]

The inappropriate and wild laughter of Hanno recalls that of Adrian Leverkühn as a child, modeled (by legend) on that of Beethoven. Hanno is dismissed and sent to bed, but the laugh is a fitting if sardonic comment on the end of Thomas Buddenbrook.

Hanno's death is the end of the line for the Buddenbrook family, defined, as it was, by the male members. After Hanno's death, Gerda decides to return to Amsterdam, and says, briefly, "I'm sorry that I won't be able to say goodbye to Christian."[56] The reader learns, briefly, in the final chapter of the novel, that although Christian probably could make a life outside of the institution, his wife had locked him there forever, and probably "was in league with the doctor," which no one in the Lübeck family minded. But Hanno's death was of a completely different order. When, in this chapter, the conversation turns to Hanno's death, Tony weeps openly, and makes the surprising claim:

> Suddenly Frau Permaneder broke into tears. "I loved him so," she sobbed. "You don't know how much I loved him, more than any of you – yes, forgive me, Gerda, and you're his mother."[57]

This statement ought to be insulting to Gerda, but apparently was not. Tony, ever plain-speaking, was merely stating a fact: Gerda is a bit of a cold fish, or seems so, despite having dutifully stayed in Lübeck after Thomas's death, as long as Hanno was alive. But the reader must wonder what, exactly was the nature of Tony's love for Hanno, for Hanno was not necessarily quite the angel he was said to be after his death. For Tony, he was, perhaps above all else, the one who was supposed to have carried on the (male) Buddenbrook lineage, and now that he was dead, that lineage was gone forever.

The penultimate chapter describes Hanno's death in what appears, at first, to be graphic, clinical terms, but this appearance is deceptive. The opening of the chapter is: "Typhoid runs the following course" – a line that is repeated two pages later.[58] Mann divides the disease into three stages,[59] but while there are many medical details, Mann was much more concerned with the spiritual state of this particular patient. For example, the description of the "course" begins as follows:

---

55 Mann 1993: 593.

56 Mann 1993: 647.

57 Mann 1993: 648.

58 Mann, 1993: 723, 725.

59 Whatever sources Mann used for this chapter, more modern medical textbooks give four stages, each about one weeklong. As discussed, the medical facts were used only as needed to fit the literary needs of the novel.

> In the incubation period, a person feels depressed and moody, and this quickly grows worse, to the point of acute despondency. At the same time, he is overcome by physical lassitude, which affects not only his muscles and tendons, but also the function of all internal organs, and most especially the stomach, which rebels and refuses to accept any food. There is a great desire for sleep, but, despite extreme weariness, such sleep is restless, shallow, and nervous and leaves the patient unrefreshed. He experiences headaches, and his mind feels numb, edgy, and dazed, with occasional spells of dizziness.[60]

This is not incorrect, exactly, but it certainly emphasizes the mood of the affected individual rather than physical signs and symptoms, as a typical medical description would. It speaks to despondency, lassitude, weariness, restlessness – an exacerbation of the Weltschmerz that had been undermining the Buddenbrooks all along, but had accelerated in the two last generations. Only after this paragraph does Mann mention fever and other physical phenomena that actually characterize this disease.[61] The discourse then turns medical, but the doctor equivocates about the diagnosis:

> For there is one thing that Dr. Langhals does not know, one question that he cannot answer, and so he gropes in the dark. Until the third week, until the decisive crisis, the question of either-or hovers in the air, and he cannot possibly tell whether in this case the disease he calls "typhus" is an inconsequential mishap, the result of an infection that might perhaps have been avoided and that can be combated with the resources of science – or if it is quite simply a mode of dissolution, the garment in which death has clad itself, though it could just as easily have chosen some other disguise, and for which there is no known remedy.[62]

---

60 Mann 1993: 642.

61 It is not clear whether Mann meant "typhus" or "typhoid fever," though it might not matter. The German text states, "Mit dem Typhus ist es folgendermaßen bestellt." The German text uses "Typhus" three times, once each in this sentence and its repetition a few pages later, and once when Dr. Langhals equivocates about the diagnosis ("Er weiß nicht, ob die Krankheit, die er »Typhus« nennt, in diesem Falle ein im Grunde belangloses Unglück bedeutet . . . "). In the English translation, the translator used "typhoid" for the repeated sentences, and "typhus" during Dr. Langhals' equivocation. The problem is that "Typhus" is the German word for both diseases. In English, "typhus" and "typhoid" have similar sounding names, but these names denote separate disease entities. "Typhus" comes from the Greek, τῦφος (tûphos), meaning both "fever" and "delusion," both of which are common features of the disease. Typhus is actually a group of infectious diseases caused by Rickettsiae, organisms discovered in 1906 by H.T. Ricketts after whom this class of organisms was named. These organisms are carried in insect reservoirs and vectors, such as fleas and ticks. In contrast, typhoid (or typhoid fever), meaning "resembling typhus," is caused by by *Salmonella enterica* and related bacteria. These organisms grow in the intestines, Peyers patches, mesenteric lymph nodes, spleen, liver, gallbladder, bone marrow and blood. There is no insect vector or reservoir; rather, typhoid is spread by eating food or drinking water contaminated by the feces of an infected person. Although the symptoms are similar, there are distinctions: for one, the onset and course of fever in typhus is typically more acute than in typhoid fever. Because they do share symptoms, they were recognized as separate diseases only in the early 19th century.

62 Mann 1993: 644.

Dr. Langhals wonders whether "the disease" is an "inconsequential mishap, the result of an infection that might perhaps have been avoided and that can be combated with the resources of science." This is how the doctor would describe a mere bacterial infection, which is serious enough; but the worry is that what Hanno suffers from is a far more serious alternative: "quite simply a mode of dissolution, the garment in which death has clad itself," and "for which there is no known remedy," The issue, in other words, is whether or not Hanno is tied to life, or wishes to succumb to death and its rest.

Mann's second iteration of the "course" or typhoid is a remarkable passage that gets down to the real business at hand:

> As he lies in remote, feverish dreams, lost in their heat, the patient is called back to life by an unmistakable, cheering voice. That clear, fresh voice reaches his spirit wandering along strange, hot paths and leads it back to cooling shade and peace. The patient listens to that bright, cheering voice, hears its slightly derisive admonishment to turn back, to return to the regions from which it calls, to places that the patient has left so far behind and has already forgotten. And then, if there wells up within him something like a sense of duties neglected, a sense of shame, of renewed energy, of courage, joy, and love, a feeling that he still belongs to that curious, colorful, and brutal hubbub that he has left behind – then, however far he may have strayed down that strange, hot path, he will turn back and live. But if he hears the voice of life and shies from it, fearful and reticent, if the memories awakened by its lusty challenge only make him shake his head and stretch out his hand to ward them off, if he flees farther down the path that opens before him now as a route of escape – no, it is clear, he will die.[63]

This is what Hanno Buddenbrook dies of: a failure to heed the "bright, cheering voice" and "slightly derisive admonishment" of life, failure of "a sense of duties neglected, a sense of shame, of renewed energy, of courage, joy, and love." It is clear, by the end of the paragraph that Hanno is not succumbing to an infection: he is fleeing from life, into death.

The Magic Mountain

A great deal has been written about Thomas Mann's portrayal of disease and life in the tuberculosis Sanatorium Berghof,[64] and it is neither necessary nor possible to try to recapitulate all of it. Here, the focus will be on how Mann's treatment of disease relates to a view of the Problem of Evil.

---

63 Mann 1993: 644.

64 This includes an essay by Meredith 1999. In this article, I discussed more fully Hans's amazement that sick people could also be stupid (e.g., Frau Stohr): he swallows entirely the trope of the ill as more spiritual and sensitive than the healthy – a point about which Settembrini continually chides him.

As is well-known, *The Magic Mountain* takes place in a tuberculosis sanitarium in the Swiss Alps, somewhat modeled after the sanatorium at Davos, at which Mann's wife stayed for 6 months. Mann visited her there for some weeks, and, as he wrote in his postscript, "The Making of The Magic Mountain," his early impressions of the place were much like those of Hans Castorp, and like Castorp, he was examined by the physician who advised him that he too had tuberculosis and should take the six months cure. Hans Castorp is from Hamburg, not Lübeck, and goes to the International Sanatorium Berghof in Davos-Platz to visit his seriously ill cousin, Joachim Ziemssen, not his wife. Hans intends the visit, which starts in 1907, to last the three weeks until he is scheduled begin his job as an engineer in the ship building-firm of Tunder and Wilms. He stays for seven years, and when the novel ends, he is last seen on a battlefield in the Great War, his fate unknown.

What kind of disease is tuberculosis? This is not to speak of microbiology or pathophysiology. To Mann, tuberculosis was also, again, a bodily manifestation of a disease of the soul, a characterological trait, with the four characteristics named above. Hans Castorp is first described as a mediocre ("einfacher," "mittelmassig") young man. He first denies that he belongs at the sanatorium, then gradually becomes accustomed to, even dependent upon the ways of the mountain. This mediocre young man will prove to be not entirely mediocre: he will even have some flashes of brilliance, though not often or for long. One such "flash" is in the chapter "Walpurgisnight," in which, in the carnival atmosphere, he is permitted to speak not in ordinary language, but entirely in French to the girl he has been pining for: a "slovenly," artsy, Slavic woman, Clavdia Chauchat. Mann plays with a Spenglerian construct:[65] history "moves" to the west; hence, Germans are people of the middle, between the ancient and morally lax east in Asia,[66] where Clavdia is from, and the "stern west," typified by Spain of the Jesuits. Hans shudders at her loose ways –

---

65 Oswald Spengler (1880–1936) was a German historian and philosopher of history, best known for *The Decline of the West* (*Der Untergang des Abendlandes*), a work in two volumes published in 1918 and 1922 to wide popular success but mixed scholarly reviews. He argued that world culture is a "superorganism" with limited, predictable life stages and spans. In its late stages, a "Culture" became a "Civilization," a word imbued by him with negative connotations: controlled by technology, a mass society, imperialistic, petrified and fossilized. World cultures had passed through several stages; for example, he considered the first millenium to have been the time of the Near East. The West, as the title of his book suggests, was currently in its decline. Mann read Spengler, and compared reading him for the first time to reading Schopenhauer for the first time. Yet Mann also wrote a far from positive review of Spengler's work: he was reading Spengler shortly after the state of his own "conversion" to democratic ideals. One would like to believe, therefore, that Mann used this construct mainly to set up the dynamic between Hans and Clavdia, and between Settembrini and Naphta, among others in the novel – and to stand for an ideological struggle then going on in Germany.

66 She is from Kirghiz, so not very far East, but Mann racialized her, repeatedly referring to her slightly slanted, "Kirghiz eyes."

she even bites her fingernails!! – but he is attracted to her. Well, fair enough, why not? But when Hans falls for her, it is portrayed as a giving in to decadence and waywardness. Yet Mann, the constant ironist, gives us frequent hints that she is really not as much of a slut and not as immoral or amoral as Hans takes her to be. In fact, we need to be honest about our hero, Hans Castorp. He's a bit of a prig – rather pompous, and, as we'd say nowadays, a real tightass. Is it really necessary to change one's shirt twice a day, and have them sent to Hamburg to be ironed, because no one here can do it right? Must one's socks be monogrammed?

Hans eventually manages to fall and "lighten up," yielding to that artist-lover part of himself that *wants* to "fall." But in another sense, he truly is the mediocre man of the middle – in the sense of being quite average. Early in his stay, he discovers that he, too, has a spot in his lung – this token of the artist-lover. In addition, he is at the *International* Sanatorium Berghof, a microcosm of the world; this, too, tends to universalize this one middling young man. The spot of tuberculosis, therefore, also becomes a token of a trait tying Hans to the broader humanity, much like original sin. As with the Biblical fall, Hans's is a volitional act, a willful sin, stemming from a sickness that is embedded within his ordinary human nature.[67]

Hans's desire to "fall" had at least two particular precedents, which, though distinct, converge in Mann's focus on disease. The first is that Clavdia awakens in Hans a long repressed memory of an earlier, homoerotic love, also of someone of Slavic nationality, Pribislav Hippe.[68] The second is Hans's early experiences of death: that of his parents, and perhaps even more importantly, of his grandfather. When, at the age of seven, his parent died and he is sent to live with his grandfather, one of his favorite events was the examination of the baptismal bowl, with his name, and that of seven generations of Castorps engraved on it.[69] When his grandfather dies, Mann tells us about Hans's view of death, which is closely related to his view of disease on the mountain:

> Analyzed and put into words, his feelings might have been expressed as follows: there was something religious, gripping, and sadly beautiful, which was to say, spiritual about death and at the same time something that was the direct opposite, something very material, physical, which one could not really describe as beautiful, or gripping, or religious, or even as sad. The religious, spiritual side was expressed by the pretentious lying-in-state, by the pomp of

67 Hans's volitional acts, however, occur in the post-lapsarian world. To say that sinning was "embedded" in human nature or that it was necessary before the fall would be a clear departure from the Augustinian or Thomistic notion of original innocence.

68 As is well known by now, the publication of Mann's unexpurgated diaries in 1979 showed what was clear in any case: that Mann had what he called intense "homoerotic" urges. During his youth, Mann had two "crushes": one with Armin Martens (who was the source for Hans Hansen, Tonio Kröger's first love) and one with Willri Timpe, who actually served as the model for Hans Castorp. At age 25, while in Munich, Mann developed an intense emotional attachment to Paul Ehrenberg, who was later the inspiration for Adrian Leverkühn.

69 Mann 1995: 21.

> flowers and palm fronds – which he knew signified heavenly peace – and also, and more to the point, by the cross between the dead fingers of what had been his grandfather, by the blessings a copy of Thorvaldsen's Christ extended from the head of the coffin, and by two towering candelabra on either side, which on an occasion like this also took on an ecclesiastical character. The explicit and well-intended purpose of all these arrangements was apparently to show that Grandfather had now passed on forever to his authentic and true form. But they also served another purpose – one that little Hans Castorp likewise noted, if not admitting it to himself in so many words; in particular, the masses of flowers and more especially the very well represented tuberoses were there for a more sobering reason – and that was to gloss over the other side of death, the one that is neither beautiful nor sad. but almost indecent in its base physicality, to make people forget it or at least not be reminded of it.

The tuberoses (what else would they be in this novel?) were there to cover the stench of death, which is not merely unpleasant in a sensory way, but is considered "indecent." Mann was fond of doubles. *Felix Krull*, the hero of Mann's unfinished late comic masterpiece, especially dealt in doubleness, as when he espied a brother-sister pair "an open balcony of the *bel étage* of the great Hotel Zum Frankfurter Hof." They were "[s]lightly foreign in appearance, dark-haired, they might have been Spanish, Portuguese, South American, Argentinian, Brazilian – I am simply guessing – but perhaps, on the other hand, they were Jews . . . " but in any case, "Both were pretty as pictures." Yet what Krull admired most about them was their doubleness:

> Dreams of love, dreams of delight, and a longing for union – I cannot name them otherwise, though they concerned not a single image but a double creature, a pair fleetingly but profoundly glimpsed, a brother and sister – a representative of my own sex and of the other, the fair one. But the beauty lay here in the duality, in the charming doubleness. . . . Dreams of love, dreams that I loved precisely because – I firmly believe – they were of primal indivisibility and indeterminateness, double; and that really means only then a significant whole blessedly embracing what is beguilingly human in both sexes.[70]

The words translated as "double creature" (*Doppelwesen*), "duality" (*Doppelten*) and "charming doubleness" (*lieblichen Zweiheit*), all speak to Felix's stock and trade: the duplicity of the con man, but for Mann, this is even more the duplicity of the artist, who, in creating an artistic world, fools his audience. As a young boy, Felix had entertained the adults by greasing the bow of a violin so it would not squawk, and then pretending to play a Hungarian dance. Later in the novel, he referred to his shady life as a "double life."[71] He later described his own life as a double life and double existence (*Doppelleben*); doubleness runs through the entire novel, and indeed most of Mann's novels.

Felix Krull is a conman – he is not in it for money, really, though that would be part of it too; but rather, he is in it for the sake of the art, for the beauty of it: crime is, for Krull, aesthetic. *Felix Krull* was Mann's last novel, by which time he

70 Mann 1992: 70–71.
71 Mann 1992: 207.

was, perhaps, finally comfortable enough in admitting that yes, he was an artist and not a ship builder like Hans Castorp's forebears, or a grain merchant like his own.

In an early scene in the unfinished novel, which had also appeared in the short story, "Felix Krull," published in 1911, these themes are joined with disease as well. As a boy, Felix's father took the boy to the theater. Felix loved it, and as it happened, his father knew the star, with the absurd name Müller-Rosé. After the performance, Felix's father took him back stage to meet Müller-Rosé:

> Müller-Rosé was seated at a grubby dressing-table in front of a dusty, speckled mirror. He had nothing on but a pair of grey cotton drawers, and a man in shirt-sleeves was massaging his back, the sweat running down his own face. Meanwhile the actor was busy wiping face and neck with a towel already stiff with rouge and grease-paint. Half of his countenance still had the rosy coating that had made him radiant on the stage but now looked merely pink and silly in contrast to the cheese-like pallor of his natural complexion. He had taken off the chestnut-wig and I saw that his own hair was red. One of his eyes still had deep black shadows beneath it and metallic dust clung to the lashes; the other was inflamed and watery and squinted at us impudently. All this I might have borne. But not the pustules with which Müller-Rosé's back, chest, shoulders, and upper arms were thickly covered. They were horrible pustules, red-rimmed, suppurating, some of them even bleeding; even today I cannot repress a shudder at the thought of them.[72]

These pustules suggest a finding that occurs in secondary syphilis. In any case, the point is that Müller-Rosé is a fraud: a half-and-half, a duplicity – like his face and name – and that's a good thing! The fraud was good. This is art: a kind of fraud, a duplicity, in which the real world is duplicated in a duplicitous artistic rendering. Fiction, indeed, is a fake – and that's good!!

To return to *The Magic Mountain*, then, death and disease are not the physical *or* the spiritual; they are the physical *and* the spiritual. Disease, then, consists of both its spiritual and physical aspects.

The main "double" (or duplicity) in *The Magic Mountain*, however, is the pair of potential mentors vying for young Hans Castorp's mind and soul: Ludovico Settembrini and Leo Naphta. Settembrini is the self-appointed spokesman for enlightenment and reason, but he is also a bit of a cartoon, who reminds Hans Castorp of an "organ-grinder,"[73] and whose simple song, a paean to "progress," is forever the same. He is also a bit of a prig: he assumes the mentor's role to Hans, though he is only 35 years old, not that much older than Hans. He has other unsavory traits: he is a misogynist (he's all against Clavdia) and yet he behaves lewdly, and is a would-be womanizer. Naphta is his intellectual opponent. Naphta's métier is darkness – a *kind of* depth, perhaps, but in the end, an impossible and indigestible mess. He is contradiction itself, and there could be no

72 Mann 1992: 28.

73 This might be more racializing on Hans's part – or perhaps also on Mann's part.

other end for him but suicide. He is a Jewish Jesuit, a luxuriant ascetic, a reactionary communist, and above all, a proto-fascist. He rejects umbrellas as the paragon of bourgeois mediocrity and softness – and yet he is ill, too weak to live even the physically undemanding life of the academic. And as for this illness, he seems to delight in having it, and in living in a "silken cell." He is another grotesque – one of many in *The Magic Mountain*. Naphta arrived on the mountain almost immediately after Clavdia Chauchat departed, as if one takes the place of the other. Mainly, he underscores the appeal of this grotesque and the irrational for Hans and other Germans.[74]

The *appeal* of darkness and evil is a central theme to Augustine, illustrated, for example, by his well-known discussion of the theft of the pears in *Confessions*. Mann, who obviously was not a systematic theologian, nevertheless makes this point as well. But he takes Augustine's view of evil (as privation of the good in a subject) even further. Settembrini's and Naphta's battle for Hans Castorp, as Mann makes clear, stands also for the battle for Germany: would it choose the (perhaps in some ways superficial) path of reason and enlightenment, or the forces of darkness and irrationality. (Mann was hardly an optimist on this point.) Settembrini had introduced Castorp to Naphta reluctantly, and after one particularly repulsive tirade from Naphta, he warns his young charge against Naphta. At first, Settembrini's warning is somewhat restrained, but here, he had one of his moments of clarity and perspicacity, when he rose above himself, as it were:

> Gentlemen – I would like to warn you . . . about the person whose guests we just were . . . and to whom I introduced you, very much against my will and intentions. As you know, chance wished it otherwise. I could not help it, but I bear the responsibility and it weighs heavily on me. It is at the very least my duty to point out to you as young people certain intellectual risks you run in associating with that man and to beg you, moreover, to keep your relations with him within certain prudent limits. His form is logic, but his nature is confusion.[75]

Castorp had feigned ignorance or innocence about the danger in Naphta's views, ("Warn us about what?" he had said), but Hans's coyness derived from his attraction to that very darkness. The chapter title, "The City of God and Evil Deliverance," is a nod to Augustine. Settembrini continues:

> "Gentlemen," Herr Settembrini continued now, stepping very near the young men, spreading the thumb and middle finger of his left hand into a wide fork, as if to concentrate their attention, and raising the forefinger of his right hand in warning, "imprint this on your minds: the intellect is sovereign, its will is free, it defines the moral world. If it isolates death in a dualistic fashion, then by that act of intellectual will, death becomes real in actual fact – *actu*, do you understand?"[76]

---

74 This point becomes central in "Mario and the Magician," and above all in *Dr. Faustus*.
75 Mann 1992: 399.
76 Mann 1992: 404.

Death, Settembrini argues, would no longer be only the absence of life – as evil is privation of the good; it would become *actu* – *actual* in the Aristotelian sense, which is to say, a Manichean[77] representation of evil. Settembrini continues:

> It becomes a force of its own opposed to life, an antagonistic principle, the great seduction . . .

The issue in this citation is the nature of evil – or rather, *whether* evil has a nature. Settembrini, in discussing death, finds himself, intentionally or not, in agreement with Augustine and Thomas Aquinas, in arguing that evil is not a self-subsistent thing, but is, rather, the privation of the good in a subject. To see death as "a force of its own opposed to life, an antagonistic principle" is not merely a serious metaphysical error; it is also, as Settembrini says, "a great seduction."

As time goes on, and weeks on the mountain turn into months, which turn into years, and novelty turns into ennui, the debates between Naphta and Settembrini grow more strident and raucous, and eventually – if one observes carefully – on occasion, they even swap positions. It all ends, grotesquely, in a duel. Settembrini fires harmlessly into the air – symbolically, the harmlessness, even impotence of rational enlightenment. But when it is Naphta's turn to fire, he shoots himself in the head. Even if one finds Settembrini weak, shallow, and an ineffectual "organ-grinder," he is shown to be preferable to the self-destructive darkness of Naphta.

The Augustinian view of evil that Settembrini espouses is terribly counterintuitive, because both the force of evil and even its appeal are real, certainly to us. So it is not certain, by any means, that Hans would be convinced by Settembrini, and despite the basic decency of the young man, when we see him last, at the end of the novel, he, like his fellow countrymen and other Europeans, are rushing headlong into the darkness of World War I, his fate unknown. To speak of "the darkness of WWI," which was, indeed, incredibly dark, is to say nothing of World War II.

Doctor Faustus: The Life of the German Composer Adrian Leverkühn, Told by a Friend

From this novel of enormous thematic richness, this essay will focus on one part, Chapter 25, in which the protagonist, Adrian Leverkühn, confronts the devil. Like other Faust stories, this one is also a critique of knowledge. But in this novel, the Faust figure is no longer a magician, as in the chapbooks, nor a weary scholar

---

77 Augustine had been a lector (but not a novice) with the Manicheans. Whether or to what extent Manichean philosophy is what Augustine portrayed it to be, for Augustine and for the late Medievals (e.g., Thomas Aquinas), the Manicheans became an all-inclusive category for any kind of dualistic philosophy, i.e., one in which there was not a single creator-God, but two opposing gods or forces, one of light and the good, one of darkness and evil.

seeking adventure and timelessness in experience, as in Goethe's *Faust*: he is a musician – an amalgam of several 20th century, mainly German composers. But before he committed himself to a life in music, he had studied theology for a while at the University of Halle, where he learned the homely, straight-talking religion of Martin Luther. His main teacher there, Professor Ehrenfried Kumpf, spoke in the boisterous, hearty style one imagines for Luther, and like Luther, he sees devils everywhere. He stays vigilant and aware of them, he says, in order to combat them.

Why a musician? Music, to Mann, was emblematic of Germany's impasse in the early 20th century, and it also had a dual (or *duplicitous*) nature. On one hand, music is coldly analytical, almost a mathematical puzzle, and indeed, Mann's protagonist, starting in childhood, seemed more inclined to mathematics than music. On the other hand, it is inchoate, inarticulate and wordless, a thing of dense emotionality. These two sides correspond, roughly, to the Apollonian and Dionysian sides of music in Nietzsche's analysis, of which Mann makes much use.

But music, perhaps most of all the arts, was facing a problem: tonal music seemed to be spent, at its end: to prolong it in older styles was to crash into the Scylla and Charybdis facing the 20th century artist: parody and kitsch. Leverkühn, using the 12-tone scale (which Mann appropriated from Schönberg by way of Theodor Adorno's interpretations) had managed to innovate, but achieved only a sterile intellectuality; hence, Leverkühn believed his music to be in need of the Dionysian element. But in the context of his native Bavarian and Lutheran background, such an element would veer towards the demonic. Leverkühn sought a *breakthrough* – a term redolent of Nietzscheanism. It hardly mattered that this breakthrough would pass beyond humane bourgeois values and even beyond the paganism of Wagner. It would go further: to *barbarity*. This was the wild dithyramb – and yet, not exactly the dithyramb of the Greeks, a hymn to Bacchus. Rather, it was a type of paganism that was suffused with the Lutheran-demonic.

The narrator of the novel is the artist's boyhood companion and life-long friend, Serenus Zeitblom – significantly, part of the Catholic minority. Zeitblom had been "retired" by the Nazis, even as his sons fought in the war. Although living in Germany, he is somewhat peripheral to the horrors of the Third Reich.[78] Ending in 1945, Zeitblom interweaves Leverkühn's story with a horrified vision of the events leading to the moral collapse of Germany. He speaks in a Biblical cadence, foretelling, like Ezekiel, the coming ruin of his nation.

As Zeitblom relates, Leverkühn's foremost trait, from his childhood on, was his coldness, even with his family and childhood friend. While Zeitblom spoke to his friend *in Du*, Leverkühn invariably replied *in Sie*. With the exception of his culminating last piece, "The Lamentation of Dr. Faustus," Zeitblom describes Leverkühn's

78 This, of course, becomes an important question in the novel. Is someone who distances himself from horrors really free of responsibility for them?

music as brilliant but cold, dry, even lifeless – like the rest of his life. To break through and remedy this sterility, he took the symbol-laden step of *intentionally* contracting syphilis from a prostitute – an episode modeled after a similar story about Nietzsche's life, which Mann, perhaps, seemed to take as true, or perhaps only as useful for his novel. Leverkühn, upon his arrival in Leipzig for his musical studies, was led unaware to a brothel, rather than to a restaurant, as he had requested, by a suspiciously red-haired guide. In his sexual purity, he was shocked and dismayed about where he had been brought, and he resisted the prostitute's touch – then, and for an additional year. Finally, unable to resist, he returned, this time voluntarily, to the brothel. Upon finding that she had left the city, he sought her out where she had traveled (Hungary). Although she nobly warned Leverkühn of her disease, he nevertheless (as Zeitblom prudishly put it) received "all the sweetness of her womanhood." The progression of Leverkühn's disease is apparent to the alert reader. Mann inserts several meticulous medical descriptions into the novel. Consider this description of Leverkühn near the end of the novel:

> in his eyes, which previously had usually been half-veiled by the lids, but which now were open wider, almost exaggeratedly so, allowing one to see a band of white above the iris. This could appear somehow menacing, all the more so, since such a wide-eyed gaze had a kind of rigidity about it, or should I say, one noticed a kind of standstill, over whose nature I puzzled for a long while, until I realized that it was caused by the pupils, which were not perfectly round, but somewhat irregularly lengthened, and always stayed the same size, as if they were not subject to the influence of any change in light.[79]

This is a textbook description of the Argyll Robertson pupil, originally described as a sign of tertiary syphilis.[80]

In Chapter 25 Leverkühn converses with the devil – or seems to do so. Zeitblom speaks to the reader after Leverkühn's death – and, *a fortiori*, near the catastrophic end of the Third Reich. He reproduces a document in Adrian's handwriting, apparently a dialogue between Leverkühn and the devil. Zeitblom "defaults" to a naturalistic interpretation of this dialogue: this scene is, to Zeitblom, a hallucination, likely the result of Leverkühn's incipient madness from tertiary syphilis. Thus, both voices in this apparent dialogue are assumed to be Adrian's, written down as a work of fiction, but betraying a horrifying degree of cynicism, despair and mockery.

79 Mann 1997: 508.

80 The Argyll Robertson pupil was described in 1868 by the Scottish ophthalmologist and surgeon, Douglas Moray Cooper Lamb Argyll Robertson. Argyll Robertson pupils are small and irregular, and unresponsive to light, but they do constrict briskly to near targets (light-near dissociation or "accommodation"). The finding can be unilateral or bilateral. This finding is also sometimes associated with atrophy of the iris. Although first described in association with tertiary syphilis, other forms of damage to the rostral midbrain can result in similar findings. In particular, the Argyll Robertson pupil results from lesions of the dorsal aspect of the Edinger-Westphal nucleus (necessary for iris sphincter relaxation and pupil dilation), which interrupt the pretectal oculomotor light reflex fibers.

The other obvious alternative Zeitblom barely considers – and then only to dismiss it:

> But if he, the visitor, did not exist – and I am horrified to admit that such words would allow, even if only conditionally and as a possibility, for his reality! – it is gruesome to think that the cynicism, the mockery, and the humbug likewise come from his [Leverkühn's] stricken soul . . . [81]

That the devil could be ontologically real is barely admitted into the realm of possibility.

For Mann, a sensitive reader of Dostoevsky, the scene is an homage to the chapter of *The Brothers Karamazov* in which Ivan Karamazov confronts the devil. In addition to the similarities of Mann's and Dostoevksy's chapters, however, there are also some notable contrasts.

Ivan Karamazov was trained in natural sciences at the university. His thrust was to deny the ontological reality of the devil at all costs, to maintain that the supernatural world, including and above all, God, does not exist. For this the devil mocks him:

> Spiritualists, for example . . . I like them so much . . . imagine, they think they're serving faith because devils show their little horns to them from the other world. 'This, they say, 'is a material proof, so to speak, that the other world exists.' The other world and material proofs, la-di-da! After all, who knows whether proof of the devil is also a proof of God?[82]

Ivan is parried mercilessly into the impossible position of insisting that he is mad and hallucinating; for only if he is mad can the devil before him be ontologically unreal.[83] In contrast, Leverkühn lives in an inspirited world of God, albeit a *hidden* one, and ubiquitous devils; his denial of the devil's ontological reality is feinted only briefly and then abandoned. The devil exists and is sitting there before him; indeed, he is far more concerned with what *name* to call the devil than in the question of whether devil is really there.

Leverkühn's devil, like Ivan Karamazov's, is a trivial, paltry devil. This is a lovely Augustinian touch: sin and evil do not have a nature or essence of their own, but stand as impoverishments of nature. Dostoevsky's devil is a faded gentleman with a nose-cold, wearing out of style, threadbare clothes, and speaking a Russian peppered with French, with a lisp said to resemble Turgenev's aristocratic one, and (most damningly to Dostoevsky) a former serf owner. Also like Dostoevsky's devil, Mann's devil brings forth an atmosphere of icy cold – which to both Mann and Dostoevsky is the reigning characteristic of hell, rather than heat. The chill of hell

---

81 Mann 1997: 237.

82 Dostoevsky 1992: 636–637.

83 The devil, if he did exist, would be a spiritual creature; hence the issue is whether the devil is ontologically, not physically, real. In making this distinction, I am following the example of Frank 2002: 677–683, who discussed the question of the devil's *ontological* reality.

reflects the iciness of their protagonists' souls – and more particularly, the absence of God's love in their hearts.

But in other ways, the characters of the devils diverge. Leverkühn's devil, though also poorly dressed, is something else:

> Is a man of a more spindled figure, not near so tall as Sch. [the violinist, Rudi Schwerdtfeger], yet shorter than I – a sporting cap tugging cap tugged at one ear, and on the other side reddish hair extending from the temple upward; reddish lashes round likewise reddened eyes, a face pale as cheese, with the tip of the nose bent slightly askew; a stocking-knit shirt stripped crosswise with sleeves too short and fat-fingered hands stuck out; trousers that sit untowardly tight, and yellow, overworn shoes ne'er to be clean again. A *strizzi*. A pimp-master.[84]

He is much like the pimp who had initially guided Leverkühn to the brothel; his red hair is, here, a leitmotif for the devil (Mann, especially in this novel, has a code for various hair and eye colors). The devil first recalls "good old German air from *anno* fifteen hundred or so, shortly before the arrival of Dr. Martinus" – Luther, of course, who was on such intimate terms with devils. Then after recalling several of prodigies of this era – "children's crusades, and bleeding hosts, famine, insurrection, war, and pest in Cologne, meteors, comets, and great signs, stigmatical nuns, crosses that appear upon men's garments,"[85] the devil gets to Leverkühn's own, particular case:

> Good times, devilish German times! Does your mind not take warm comfort in the thought? The proper planets met together in the house of the Scorpion, just as a most well-instructed Master Durer drew it for his medicinal broadsheet, and there arrived in German lands the small delicate folk, living corkscrews, our dear guests from the Indies, the flagellants – you prick up your ears, do you not? As if I spoke of the vagabonding guild of penitents, scourging their backs for their own and all mankind's sins. But I mean the flagellates, the tiny imperceptible sort, which have flails, like our pale Venus-the *spirochaeta pallida*, that is the true sort. But right you are, it sounds so snugly like the high Middle Age and its *flagellum haereticorum fascinariorum*. Ah, yes, they may well prove to be fascinarii, our revellers – in better cases, such as yours.[86]

In contrast to the Europeanized-cosmopolitan speech of Ivan's devil, Adrian's devil speaks the hearty and homely German dialect of Adrian's native Kaisersachern – and that of Professor Kumpf and, in all probability, Martin Luther. This is, in fact, a trope that Mann had used before. In Hans Castorp's vision in the chapter, "Snow," in *The Magic Mountain*, the witches in the inner sanctum of the otherwise idyllic civilization that Hans envisions mock Hans and curse him out (while dismembering and eating a small child) in the native dialect of his hometown of Hamburg. In these tropes, the curses and obscenities are entirely personal, unique to the individual involved – like

84 Mann 1997:240.
85 Mann 1997: 247.
86 Mann 1997: 247.

the door in Kafka's story, "Before the Law," that is intended for only one person. Hell is all about isolation: the absence of God's love, the absence of community; Adrian, like other Fausts, faces his damnation alone. But to the extent that Adrian is emblematic of Germany, Germany becomes a nation of isolated, damned souls.

An additional difference of Ivan's to Adrian's devil is that the former is depressingly constant, while the latter undergoes a series of metamorphoses. Adrian's devil starts as a pimp, then becomes a university scholar – resembling Theodore Adorno – with whom Adrian discusses music, and then into the familiar and traditional form of the chapbook devil. It is in this guise that he conducts the contractual business – but here, with a refinement. Mann, in a droll touch, makes this traditional devil a lawyer, who legalistically demands that Leverkühn sign the contract with his own blood. Venereal disease is often referred to, in German as in English, as "bad blood." The devil does not *exactly* demand Leverkühn's "soul"; what, exactly, would the devil do with such a thing, and who still fully believes in them anyway? No, not that; after a learned disquisition on the distinction between *attritio* and *contritio*, this devil exacts a different price – though it amounts to much the same thing. When Leverkühn asks what this clause in the contract would say, the devil answers:

> He: "It would say renounce. What else? Do you think jealousy is at home only in the heights and not in the deeps as well? You, fine creature well-created, are promised and betrothed to us. You may not love."
> I (must truly laugh): "Not love! Poor Devil!"[87]

Adrian objects that it is impossible to prohibit love, since our universe was created and is sustained by God's love, and there is, besides, both human lust and human *caritas*. The devil will not stand for such cavils. He retorts, quite plainly: "My proviso was clear and upright, ordained by hell's legitimate zeal. Love is forbidden you insofar as it warms. Your life shall be cold-hence you may love no human."[88]

This is what horrifies Zeitblom most, writing in retrospect, knowing the manner of Adrian's death. The central concern then becomes whether Adrian and, by implication, all of Germany can ever be redeemed. This is played out in the story of Leverkühn's nephew, and in Leverkühn's culminating composition, "*The Lamentation of Dr. Faustus*." The composition of his last piece was the long-promised breakthrough, beyond dry intellectuality, to barbarism. Although of course we cannot hear this piece, and have only Zeitblom's descriptions, we are meant to take Zeitblom at his word that this piece has both the intellectuality and emotional impact that we have, for example, in Beethoven's late masterpieces. But what we *can* know

87 Mann 1997: 264. H.T. Lowe-Porter's earlier translation has two nice features for this passage: she translated the devil's "Du" by "Thou"; and in the manner of Kumpf, she rendered "Armer Teufel!" as "Poor divvel!".
88 Mann 1997: 264.

is the words of the Cantata, which consist of the repetition, in 12-tone scale, of the sentence: "Denn ich sterbe als ein böser und guter Christ" ("For I die as a bad and as a good Christian"; in German, the word "Christ" carries the double meaning as either "Christ" or "Christian").

This piece was inspired by the death, from meningitis, of Leverkühn's cherubic little nephew, Nepomuk, whom everyone called "Echo." His death, it is implied, results from the fact that Adrian had dared to love him, in violation of his pact. Even Zeitblom expresses his distress in terms that recall Ivan Karamazov's rebellion against God's creation: "Ah, My God, why do I seek gentle words for the most inconceivable cruelty I have ever witnessed, that even today goads my heart to bitter complaint, indeed, even to revolt."[89] Soon after the child's death, Adrian gathers his friends in his home, for a first performance at the piano of his completed *Lamentation*. It is more than a performance: it is a confession of all of the sins of his life. Zeitblom and the others all watch Adrian perform the piece, and then witness in horror as Adrian collapses at its end, for his contractual 24 years of heightened creativity have elapsed. He sinks into madness and incapacity; after a 10-year period, similar to the length of Nietzsche's madness (and possibly from the same cause), Leverkühn dies.

As he listens to the last moments of the *Lamentation*, Zeitblom muses that there might be hope for Adrian, coming out of the depths of hopelessness. This would be the hope for redemption that "is born out of . . . a religious paradox, which says that out of the profoundest despair, if only as the softest of questions, hope may germinate".[90] Whether this "paradox" is real or only Zeitblom's wishful thinking for his friend, as some critics have claimed, remains a matter of debate, but one can imagine that Mann might have hoped for even such a glimmer of hope for Germany. Such a hope rests in the Christian's plaint in Leverkühn's last piece: to die as a "good Christian" is to die in the recognition of the sins of one's life, that one has been a "bad Christian" all along: this is true contrition.

The term "breakthrough" is ambiguous in itself, and when applied to Leverkühn, it can be taken to refer to two distinct events. On one hand, as described, the breakthrough can refer to Leverkühn's breaking out of his intellectual sterility by engaging in the demonic and barbaric. There is, however, a second breakthrough in Leverkühn's life, and that was his genuine love for his nephew Echo, his rage at Echo's cruel death, his true contrition for the sins of his life, and finally, the genuine emotional content of his last piece of music. In writing this piece, probably for the first time, Leverkühn did indeed *break through* to the emotional and spiritual reality of love.

89 Mann 1997: 497.
90 Mann 1997: 515.

The opposite of *evil* is not only *good*, but also what is the essence of the good, and that is *love*. In this sense, too, Mann portrays disease as the privation of nature in a subject. God's creation of the universe was an act of love, and the falling away of human beings from God is also, in both Mann's *Dr. Faustus*, as in *The Brothers Karamazov*, a falling away from love: the iciness of Adrian's soul, like that of Ivan Karamazov, is the iciness of being without God's love.

Concluding Remarks

In the works discussed above, and others, Mann focuses on disease not only as a topic important in its own right, but also as a prime example of evil. This is not an uncommon way to frame a discussion of disease, but aside from his pre-occupation with this topic, there are also problems with this emphasis. Some moral philosophers (Kant comes to mind) reserve the word evil for immoral actions, i.e., those requiring *volitional* acts against the good. Others (e.g., Thomas Aquinas) divide evil in *poena* and *culpa* – pain[91] and fault, respectively.[92] Brian Davies, in discussing Thomas's conception of evil, distinguished between "evil suffered" and "evil done,"[93] corresponding approximately to *poena* and *culpa*. The question about Mann, then, is this: in what way is disease an evil?

In Mann's fiction, the sufferer from a disease is not entirely a passive recipient of misfortune: there is also a volitional aspect to the acquisition of a disease. As a (nearly) constant ironist, it is difficult to decide, in the end, whether, or to what extent, Hans Castorp's acquisition of tuberculosis is a volitional act. Does he get tuberculosis because he has inherited a disposition to it – the weak lungs to which his family seems prone? Or is this weakness more akin to the stain of original sin, affecting the entire human species? Or is he enticed by the allure of the mountain – and with it, the allure of Clavdia Chauchat, and her predecessor, Pribislav Hippe? Or is he, perhaps, a mere victim of historical forces beyond his control; for we see him, at the end of the novel, descending from the mountain, like countless others, to fight and perhaps die in the Great War?

It is difficult, or rather, impossible to say simply that disease is or is not a volitional act in Mann's characters. It is also difficult or impossible to say whether disease is an evil for them. For, although Mann ironizes the German romantic (Novalis, among others) tendency to view disease as spiritualizing, he also partakes of it.

91 *Poena*, of course, can be translated as either *pain* or *punishment*. This is an important distinction in many contexts, but for the present, the focus is on pain. Nevertheless, to keep this ambiguity of translation in mind, I use the Latin words.

92 See, for example, *Summa Theologica*, I, Q48, a6.

93 Davies 2011.

Adrian Leverkühn's art, for example, might not have been made had he not voluntarily, purposefully contracted syphilis.

There is, however, one thread that I believe runs through all of Mann's fiction, and puts forth a (relatively) straightforward view about disease and the problem of evil. Ultimately, Mann's position is an Augustinian (and Thomistic) one. For Augustine and Thomas Aquinas, and contra the Manicheans, evil is not a *thing*, i.e., a substance, but, again, is the privation of the good in a subject. Against their position is one, still quite prominent in popular culture and elsewhere: to view evil as *something* – that is, as a substance with its own essence or nature. As an example, consider the Star War series of movies, in which there is a battle throughout the universe between the dark and light sides of The Force. Whether or not the filmmakers intended to portray a Manichean view of evil,[94] this is one common interpretation among viewers of the film, even if they do not use the word "Manichean." The power of evil to do harm is so blatant that it not only seems intuitive, but unavoidable to view evil as *something*, i.e., an entity (being) with a nature or essence. Such a view can be called the *reification of evil*, and it is understandable, in a way. For what could be more natural and commonsensical than the idea that good and evil are forces warring against each other? The view – that of Augustine and Thomas Aquinas – that evil is both overwhelmingly powerful but not a substance seems, in contrast, both counterintuitive and even, for some people, repulsive. And yet, Augustine and Thomas Aquinas, neither one of them a fool, to put it mildly, argued that this is the case: that evil is both powerful and not a substance, and that it has no nature or essence.

The argument, initiated by Augustine and developed by Thomas Aquinas, concerns the language by which we discuss "being." In Latin, as in English, a being can refer to an entity (*ente*) or being-in-itself (*esse*), which can also be called by a separate word, "existence." In what was probably Thomas's first book, *De Ente et Essentia*, he made the critical distinction that was central to his philosophy: what we call a "state of being verb" in English (or Latin) is, in reality, two words in one. First, a state of being verb can "signify the truth of a proposition,"[95] as when one says, "the rose is red"; for one is not saying that the rose is the color red, itself, but only that it has the quality of red color. In this sense, the word "is" is a grammatical copulative. Second, however, a state of being verb "posits something in reality,"[96]

94 Some observers contend that the philosophical outlook of the film series is Augustinian.See, for example, George Dunn, George. 2015. "Why the Force Must Have a Dark Side." In J. T. Eberl and K. S. Decker, Ultimate Star Wars and Philosophy, Wiley-Blackwell, 195–207, 2015. See also, Terrance MacMullan, "Balance through Struggle: Understanding the Novel Cosmology of the Force in The Last Jedi", The Journal of Religion and Popular Culture 31:1, 101–112, 2019; Eberl and Decker 2015: 195–207. See also, MacMullan, 2019: 31:1, 101–112.

95 Aquinas 2010: *De ente et essentia*, Ch.1. For the original Latin text see: *Documenta Catholica Omnia*, http://www.documentacatholicaomnia.eu/04z/z_1225-1274__Thomas_Aquinas__De_Ente_et_Essentia__LT.pdf.html.

96 Aquinas *De ente et essentia*, Ch. 1.

as when one says, "God is the Creator," or "Quarks are the fundamental units of matter." Thomas wrote that in this second sense, being (*ens per se*) refers to that which can be sorted into the ten categories of Aristotle (in Aristotle, *Metaphysics*, Bk. 5, c. 7, 1017a22–3, and in *Categories*).

In saying that evil has no nature or essence, one can nevertheless say that is real, bloody real: but then, to use the problematical word, what "is" it? The privative view of evil is that it exists only parasitically and destructively in something good, diminishing and corrupting its goodness. To use a simple, physical example, Thomas Buddenbrooks would not have died of a bad tooth if he did not first have teeth. One can never state often enough that Augustine and Thomas Aquinas were *not* saying that evil is illusory or only a state of mind; it is as real as can be to us; yet it is not a created entity with an essence or nature of its own, i.e., outside of the context of the good that it diminishes.

Mann's fictional treatment of disease is a recognition of both the temptation to *reify* evil, and the dangers of doing so. Hans Castorp is *attracted* to disease and to death. The same is true for Thomas and Hanno Buddenbrook, though they are attracted to death more than disease, and perhaps view disease mainly as an avenue towards death. Adrian Leverkühn, most volitionally of Mann's characters, seeks his disease, and uses it as a pathway towards artistic creativity. In these cases, and others in Mann's fiction, disease has this attractive aspect – and as it attracts, it becomes for them a thing, it is reified as if it were a good. But this attraction is also morally ambiguous. For Thomas Buddenbrook, the attraction of death is also attraction to the type of life his brother has lived – the attraction of all that he resists in the name of family tradition and propriety. The "spiritualizing" or deepening aspect of disease and death were noted earlier in this essay in the quotations from Novalis. There are, of course, good reasons for Mann to cast an ironic eye on this tradition, but in *Buddenbrooks*, succumbing to the attractiveness of death did lead to the fall of the family line. For Hans Castorp, the attractions of the mountain are much the same – and Settembrini warns against it accordingly, when he tells Hans that disease and death should never become "*actu* – do you understand?" For Leverkühn, the dangers are made most explicit: for not only did he die a horrible death but, it is implied, his angelic nephew Nepomuk also died as a result of his transgressions.[97] More broadly, Adrian's crisis stands, albeit in a complex and ambiguous way, for that of Germany in World War II.

Mann tended to sort and oppose his characters by *types* or, more disturbingly, by "race," especially in his earlier fiction. But as Todd Kontje put it, "When Mann writes of racial difference he is not only talking about 'racialized Others,'" to use today's parlance, "but also about himself."[98] A similar statement can be made

97 There is a whole set of additional problems, here, which space does not allow me to go into. For example, exactly which of Adrian's "transgressions" – his lovelessness or his pact with the devil or others – "leads to" Nepomuk's death, and how, exactly, does it do so?
98 Kontje 2006: 495–514.

about Mann's tendency to medicalize his characters, which, indeed, often overlaps with the racialization. Like his racial dichotomies, the poles must be taken with a large dose ironic salt. For Hans's affair with Clavdia is not only a "fall" into the morally lax East, but also a fall into disease. Thomas Buddenbrook likewise feels the allure of death, and Adrian Leverkühn the allure of disease.

If, however, one tries to identify a single thread of steadfast truth running through all of the irony and ambivalence, it might be this: increasingly in his career, Mann's view of evil became the Augustinian one, on in which he refused to reify evil. This, as stated earlier, was a position towards which his fiction *evolved.* In the early short fiction and still somewhat in *Buddenbrooks*, the moral perspective derives mainly from his portrayal of the grotesque, which appeals to the reader's decency or taste. We feel, for example, the cruelty of Frau von Rinnlingen rejection of Herr Friedemann. In *Buddenbrooks*, there is more subtlety, but the moral terms are terribly murky. The primary value for most of the characters, it would seem, in upholding the family tradition, is that doing so is congruent with upholding the business for financial gain and social position. Doing so exacts a price: the maintenance of the burgher's propriety requires tense vigilance, and in each generation, the strain increases until Thomas Buddenbrook succumbs to the allure of death. In any case, as Christian points out, and as becomes obvious from the plot in any case, businessmen are not necessarily moral pillars, and quite of few of them in the novel really are swindlers. Thus, neither the business nor its accompanying social conservatism is so good after all, irrespective of the strains it causes. On the other side of the coin, the exoticism towards which Thomas Buddenbrook is attracted and to which Christian Buddenbrook fully yields is personified by physically healthy characters – as attested by the healthy white teeth of Aline, Anna, and Gerda, their tokens of health. When one comes to *The Magic Mountain*, the terms have changed considerably. Hans Castorp, like Thomas Buddenbrook, is attracted to exotic, "morally lax" beauty, is enticed by Naphta's murky irrationality and above all, yields to the attractions of disease on the mountain, which frees him from the tedium of ship-building in the flatlands. But several things clearly undercut the pervasive irony. First, Settembrini transcends his slightly clownish limitations and becomes a prescient voice against Naphta's suicidal darkness. Second, the most affecting death in the novel – *truly* affecting – is that of Hans's cousin, Joachim Ziemssen, and this dispels the perverse grotesqueries of many previous deaths. Third, Hans's true sense of repulsion at moral wrongdoing during the séance, in which he had summoned the spirit of Joachim, anachronistically wearing the soldier's gear of the Great War, sets some limits and puts an end to his seemingly endless experimentation. And finally, the Great War itself, towards which Hans Castorp descends from the mountain, is the prime reminder of why, as Hans foretold during his vision in the snow, death must have no dominion over life. In these and other instances, the novel points to the fundamental flaw in Manicheism: that it tends to level the moral landscape, such that the definitions of good and evil become somewhat arbitrary.

That is, without reference to an absolute good, which, by definition, a Manichean philosophy lacks, it can become merely a matter of subjective bias which of the two warring forces in the universe gets labeled as "good" or "bad." All-pervasive irony also tends towards moral leveling, but in *The Magic Mountain*, instances such as the ones just cited, put limits on what can be ironized. Finally, in *Doctor Faustus* the moral perspective is almost the opposite of that of *Buddenbrooks*. Adrian Leverkühn is willing to make a pact with the devil, whether literally or figuratively, for the sake of his art, with disastrous results. The most morally ambiguous character in the novel might be Serenus Zeitblom, the fictional narrator of his friend's biography. As a member of the Catholic minority, and forcibly retired by the Third Reich, he becomes something of an outside witness to the descent of Germany, but it remains an open question whether or how much he shares in his country's guilt in spite of that. But by this time, the moral perspective has gained a startling degree of clarity. The devil can be called by many names, but in this novel he is evil reified: he is evil made concrete, physical, real – *actu*, as Settembrini put it. For Mann, too, the war was a reckoning, and he had come a long way from those early days in which he wrote *Reflections of a Non-Political Man*.

Finally, to revisit and add a few words about the four features that define Mann's concept of disease:

1. Diseases are hidden, secret and they reveal the state of the soul; but one must add to this that in Mann's novels, the truth will out: diseases cannot remain hidden forever.
2. Diseases are sought after: they do, like neurotic symptoms or dream manifestation in Freud's writings, fulfill a wish – of a kind. But as paths towards the spirit or enlightenment or heightening or art (as exemplified by Novalis's speculations), diseases are false paths in Mann's writings, and in general, taking such a path ends badly. This was especially the case in his later writings.
3. Disease is a type of decay: to clarify, this means a type of diminishment of the individual who has the disease, even if the disease, like the neurotic symptom in Freud, once seemed to offer a solution to an impasse.
4. Disease is something for which the character feels guilty: this is most apparent in Dr. Faustus, but one need to add to this that when Leverkühn shows contrition, Mann was referring mainly to the metaphorical sense of the disease, not the physical disease itself.

In all of these features of disease in Mann's novels, there is a uniting principle: that disease, as an evil, is conceived not the "Manichean" sense as having an essence or nature of its own, but rather, in the Augustinian / Thomistic sense of the privation of good in a subject.

Pablo Maurette

Monstrosity and the Monstrous Revisited: Fortunio Liceti's Medical Imagination

Almost seventy years after its publication, Georges Canguilhem's 1952 essay "Monstrosity and the Monstrous" continues to offer a template that helps us imagine new ways of bridging the natural sciences, in particular medicine, and the humanities. By "the monstrous," Canguilhem understands marvelous and oftentimes horrifying, repulsive or ominous figures born from the imagination. "Monstrosity," instead, refers to exceptionally deformed individuals of any given species. By means of this pair of concepts, the French thinker ponders different ways in which scientific research and the religious, artistic, and literary imagination collaborate in order to shed light on one of the most disconcerting natural phenomena, genetic malformations. Rare and refreshing though approaches such as these may seem to us, who are trapped in an ever-expanding bubble of hyperspecialization, they were by no means exceptional in times past. In fact, going back to Aristotle (384–322), western science has traditionally developed alongside and in close relation with forms of knowledge anchored in the imagination, as well as influenced by the belief in the supernatural. In early modernity, and especially within the context of the so-called scientific revolution, the richness of the connection between experimental science and the artistic imagination comes to the forefront with unprecedented force.

With Canguilhem's distinction acting as a theoretical backdrop, this essay will focus on the work of physician and proto-teratologist Fortunio Liceti (1577–1657) whose *De monstruoroum caussis, natura et differentiis libri duo* (Padua, 1616) constitutes a historical threshold that at once separates and connects two very different approaches to the natural world. The first one is steeped in the ancient view that monstrous creatures can be the consequence of parental sins or of the hyperactive imagination of the mother. The second one is grounded in the conviction that the supernatural must be removed from the equation when studying nature. Liceti's work is a rare example of how the imagination that fills itself with wonder in the face of nature's seemingly infinite variety and ingenuity, coexists with the modern impulse to taxonomize and explain the exceptional circumscribing itself exclusively to what is observable and commensurable. By looking at Liceti's analysis of the causes of monstrosity through the lens of Canguilhem's dichotomy, I hope to bring forth the rich complexity of an approach that successfully combines areas of the human intellect today commonly perceived as disconnected from one another. And in doing so, I trust to ultimately show its theoretical potential for current research in the medical humanities.

https://doi.org/10.1515/9783110788501-007

Canguilhem's Monsters

Georges Canguilhem (1904–1995) was a physician, an epistemologist, and an omnivorous reader who engaged systematically with biology, medicine, psychology, history, literature, art history, and philosophy in his writings. In short, he was someone about whose professional endeavor the term "multidisciplinary," too often used loosely, actually provides an accurate description. In 1943, while an active member of the French Resistance, Canguilhem published his most groundbreaking essay, "The Normal and the Pathological," where he puts into question the notions of "normalcy" and "abnormality" within the context of medicine. The main issue for Canguilhem is whether to consider the living being "a system of laws or . . . an organization of properties."[1] At the center of the argument we find examples from embryology and teratology which provide extremely visible cases of "abnormal" physiology. By distancing himself from the Aristotelian notion of a fixed set of ontological laws, and entertaining the idea that the natural world is a constant process of "hierarchization of possible forms," the epistemologist concludes that "nothing can be lacking to a living being once we accept that there are a thousand and one different ways of living."[2] Through two quotations, Canguilhem begins to sketch out a conclusion: "All living forms are, to use Louis Roule's expression in *Les poisons*, 'normalized monsters.' Or, as Gabriel Tarde puts it in *L'opposition universelle*, 'the normal is the zero of monstrosity,' with zero here meaning the vanishing point."[3] Canguilhem's conclusion is that "the term *normal* has no properly absolute or essential meaning."[4] Consequently, "pathological," commonly understood as a state during which there are no norms, is not its opposite but merely a descriptor of a state of being ruled by a *different* set of norms. More importantly, such norms are not only a matter of medicine and biology, Canguilhem adds, because both these disciplines are "necessary parts of an anthropology."[5] By understanding the medical and natural sciences as subdisciplines of a much larger epistemic endeavor whose center is the human being, not only does one check specific methodological blind spots proper to particular disciplines, but also one is forced to reexamine the critical foundations on which science in general rests. In other words, as they grow from and refer to the human, all forms of knowledge are branches of the humanities.

A similar argument informs "Monstrosity and the Monstrous," where Canguilhem explores once again the notion of "order" in nature, this time focusing exclusively on the figure of the monster. By "monstrosity," the epistemologist understands "the accidental and conditional threat of incompleteness or distortion in the formation of the form . . . a limitation from within," that is a phenomenon which pertains to the

---

1 Canguilhem (1952) 2008: 123.
2 Canguilhem 2008: 126.
3 Canguilhem 2008: 126.
4 Canguilhem 2008: 127.
5 Canguilhem 2008: 133.

realms of medicine and biology.[6] The monstrous, on the other hand, is originally a juridical term. It refers to grave religious violations, or to appalling immoral behavior. In the pre-modern world, Canguilhem notes, monstrosity tended to be understood as an *effect* of the monstrous; monstrous births were commonly perceived as either a punishment for the sins of the parents, or admonitions to a community for collective faults. As an interest in the preternatural arises and grows among natural philosophers from the late sixteenth- and into the seventeenth centuries, Canguilhem continues, the monstrous is progressively relegated to the realm of the artistic imagination. In this way, while science begins searching for the natural causes of extreme deformity, malformation, and genetic deviation what Canguilhem calls "the monstrous" recedes into the realms of literature, painting, sculpture, and, eventually, film.

However, even as they grow apart these two notions continue to be connected and feeding off of each other. We find examples in which mythology inspires medical science (the sirens give their name to sirenomelia, a rare congenital condition of the fetus; the shapeshifting sea god Proteus lends his to the Proteus syndrome, an extremely rare condition characterized by the overgrowth of bone and tissue); as well as of natural science injecting new ideas into the literary imagination. Mary Shelley's *Frankenstein* is the most obvious example, but there are many and they shape to a large extent the science-fiction genre. More significantly, even after teratology becomes an established branch of the medical science (the science of monstrosity) in the nineteenth century, Canguilhem continues to see the intrusion of the monstrous in the realm of the strictly scientific through the discipline known as teratogeny, ie. the scientific manufacture of abnormalities and deviations. Experiments aimed at engineering malformations, like those carried out by pioneers such as John Hunter (1728–1793), whose name Canguilhem strangely omits, Étienne Geoffroy Saint-Hilaire (1772–1844), and Camille Dareste (1822–1899), and that act as preface to the still dawning age of genetic manipulation, are deeply influenced by the roving imagination that finds nourishment in the unlikeliest of places. In Canguilhem's words, "the ignorance of the ancients held monsters to be games of nature, while the science of our contemporaries turns them into the games of scientists."[7] And since the imagination is a function without an organ, he goes on, it feeds on itself and has a never-ending capacity to produce new forms out of old ones.[8] Considering how rare monstrosities are in nature and how prolific the imagination is when it comes to conjuring the monstrous, Canguilhem concludes that "life is poor in monsters [while] the fantastic is a world."[9] This world of the fantastic has no borders. In it, the lines that separate disciplines like history, biology, mythology, religion, medicine, literature, and the visual arts are blurred. Finally, in this world all roads lead to that vortex where

6 Canguilhem 2008: 136.
7 Canguilhem 2008: 144.
8 Canguilhem 2008: 145.
9 Canguilhem 2008: 136.

everything is an exception, an "anticosmos,"[10] as Canguilhem calls it, that acts as both counterweight to and source of inspiration for scientific ingenuity.

With these considerations in mind, we must now direct our attention to those crucial decades between the end of the sixteenth and the beginning of the seventeenth centuries, when the science of the preternatural began taking shape and consolidating itself as an intellectual enterprise on and of its own. As we shall see, while the natural sciences timidly begin their slow divorce from religious dogma becoming ever more secular, and monstrosity is driven further and further away from the monstrous for the sake of reason, the figure of this "anticosmos" in the form of a cluster of sources of inspiration, many of them superstitious, literary, or religious, never disappears completely. Its enduring presence in the very origins of modern science is a reminder of the inability to fully divide the waters of reason and the irrational, experience and imagination, corroboration and belief. Thanks to its ever-effective power to shock and awe, the figure of the monster constitutes a particularly illustrative example of the close-knit tie between disciplines.

Early Modern Monsters

In what remains to this day the most comprehensive study of monstrosity in early modern European thought, *La nature et les prodiges* (1977), Jean Céard distinguishes three main approaches to the phenomenon in the period:[11]

1. Monsters are natural phenomena and must be examined, classified, and understood as such. The natural philosopher must collect specimens and examples in order to account for their etiology and typologies. This approach stretches back to Aristotle, the first to systematically engage with monstrosity in his writings on the generation of animals. From his pen come two *dicta* that would guide many in the understanding of monstrosity for the following two millennia. "Nature does nothing which is superfluous" (translated as *natura nihil agit frustra*, from *Generation of Animals*, II.4.739b); and "nature takes no leaps" (translated as *natura non facit saltus* and cited *ad nauseam* throughout the Middle Ages and the Renaissance).
2. Monsters are effects from transcendent causes and they function as signs and warnings from the divine – signs that we must decode through means such as divination. Cicero is perhaps the first notable proponent of this approach. In it, the Latin etymology of the word comes into light. *Monstrum* is, in fact, derived from the verb *monere*, which means "to warn" and "to admonish." Although it could also derive from *monstrare*, "to show," as monsters both show us something, they point in a specific semantic direction, but they also are worthy of being shown and exhibited.

10 Canguilhem 2008: 146.

11 Céard 1996: 4.

3. Monsters are no more than errors of our limited perception. When a creature seems monstrous to us is because we have not yet understood it, or are not yet used to it. The monster, in short, reveals more about our own capacity to rationally access the natural world than about the world in itself. Céard calls this the "rationalist approach," and signals Michel de Montaigne as the most influential of its proponents.[12] This notion has classical antecedents like Varro, whom Isidore of Seville, one of the widest read authors through the Middle Ages and into the Renaissance, cites in his *Etymologies* under the entry for *portent*: "A portent is therefore not created contrary to nature, but contrary to what is known nature."[13]

We can add a fourth approach that takes elements from the previous three: Monsters are wonders that prove the ingenuity and even the playfulness of Nature. This notwithstanding, the natural philosopher must strive to classify them and investigate their causes. Among such causes, he will find God's will to admonish an individual for private faults, or signal to community future ills. As we will see in the next section, Fortunio Liceti is a prime example of this fourth approach. He follows, however, in the footsteps of pioneer teratologist Ambroise Paré.

Few works are more representative of the growing interest for the preternatural in the early modern period than *On Monsters and Marvels* (1573). Ambroise Paré (1510–1590) was trained as a surgeon, cut his teeth on the battlefield where he introduced new techniques to treat the wounded (he is reputed to have perfected the ancient technique of the tourniquet), and came to be Chief Surgeon to Charles IX and to Henri III. The fact that he was a surgeon and not a physician is of the utmost relevance within the context of a growing concern with the importance of first-hand experience as a necessary complement to book learning. Paré bases his considerations regarding monsters and marvels on three main sources. First, Pierre Boaistuau's popular *Histoires prodigieuses* (1560), a collection of reports on monstrous births and other wonders that range from the outrageous to the testimonial.[14] Second, Jacob Rüff's *On the concept and generation of man* (1554), a classic on generation that

12 In the essay "Of a Monstrous Child," Montaigne describes a monstrous birth with scientific precision and concludes: "What we call monsters are not so to God who sees in the immensity of his work the infinity of forms that he has comprised in it; and it is for us to believe that this figure that astonishes us is related and linked to some other figure of the same kind unknown to man. . . . We call contrary to nature whatever is contrary to custom, nothing is anything but according to nature, whatever it may be. Let this universal and natural reason drive out of us the error and astonishment that novelty brings us." See Montaigne (1580) 2003b: 654.

13 Isidore of Seville, *Etymologiae* IX.3.2 (ca. 625) 2006: 243.

14 Although his work is mostly a collection of sensational examples, Boaistuau does entertain the notion that the causes of monstrous births are varied and range from the natural to the divine. Montaigne too was an avid reader of his work, which due to its success was translated into English soon after its publication. See Boaistuau 1569, Chap. 5: "Of the bringing forth of monsters, and the cause of their generation."

includes a section on monstrosity.[15] Whereas for Rüff monstrosity was merely a chapter in the study of generation, for Paré it opens up to problems that necessitate careful treatment and require a treatise of their own. Third, and more importantly, Paré draws from his own experience of which he provides eyewitness account with a compelling use of the first-person narration. Two elements give away the incipient modernity of Paré's approach. On the one hand, he is more interested in material and efficient causes than in finality. On the other hand, he is not deeply invested in the difference between monster and marvel, and even though he lists God's will as one of the causes for monstrous births he quickly moves on to other types of causes. In fact, while he does not deny that the birth of a deformed child might mean something, or that its ultimate cause is divine, for Paré these are just some among many features of the phenomenon.

Monster is anything born with a peculiarity that transcends the ordinary, Paré begins, and in Chapter one he lists the causes of monstrosity. Among them, there is the glory and wrath of God, but also an excess of seed, or a scarcity thereof, smallness of the womb, hereditary ailments and conditions, the agency of demons, etc.[16] It is precisely the sheer variety of causes for monstrosity what fascinates Paré. In the list of possible origins of a monstrous birth, theology, demonology, and divination are paired with embryology, medicine, and biology. Connecting as it were the supernatural and the natural causes like a hinge, the fifth cause for monstrous births, says Paré, is the imagination,[17] and in Chapter nine he provides examples of how this can occur. At the very instant of conception, the mother or the father can have a "nocturnal vision," or "receive an ardent and obstinate imagination," which induce the monstrosity.[18] For instance, "Damascene, a serious author (sic), attests to having seen a girl as furry as a bear whom the mother had bred thus deformed and hideous, for having looked too intensely at the image of Saint John [the Baptist] dressed in skins along with his [own] body hair and beard, which picture was attached to the foot of her bed while she was conceiving."[19] Such a cause listed along with "the smallness of the womb" or "the quality of the seed" is one of the reasons why Paré's book is such a crucial turning point in the history of preternatural science. In his analysis we can already see how the notions of monstrosity and the monstrous simultaneously converge and differentiate themselves from one another. Discussing medical approaches to monstrous births in the period, Huet suggests that "one of the reasons why monsters fascinated surgeons may lie in the fact that monstrous creatures were directly related to a field of knowledge deeply involved

---

15 See Rüff 1580.
16 Paré 1982: 3–4.
17 Paré 1982: 4.
18 Paré 1982: 38.
19 Paré 1982: 38.

with the effects of a prodigious world . . . [they were a] link between the art of diagnosis and deciphering secrets and riddles."[20]

When Francis Bacon (1561–1626) codifies and advocates for the pursuit of knowledge of the preternatural in *The Advancement of Learning* (1605), however, his focus is set almost exclusively on observational and quantifiable phenomena. In book two, Bacon argues that the history of nature is divided into three branches: Nature running its course, Nature erring or varying, and Nature altered or wrought. In terms of objects of study, Bacon clarifies, this amounts to the study of creatures, the study of marvels, and the study of artificial beings. According to the philosopher, the first one is in good shape and has received enough attention. The other two have been neglected and Bacon considers it is high time to consolidate them as legitimate branches of science. The most consequential point of this division of science is that it is not conceived as a vertical hierarchy, but as a triangle. No branch of science is more important than another; they are all equally crucial to the understanding of nature as a whole. This demolition of long-established hierarchies, which constitutes Bacon's most enduring legacy, is what allows for a new understanding of otherness, and in particular of what is perceived as exceptional otherness, ie. monsters. For Bacon, the way to study abnormalities is by collecting them and classifying them.[21] This spirit of collecting rarities is the one that is producing at this precise time the first cabinets of curiosities, which are the origins of the modern museum.[22]

In their groundbreaking study of wonder, Lorraine Daston and Katharine Park claim that the early modern concern with "wonder and wonders in the study of nature marked a unique moment in the history of European natural philosophy, unprecedented and unrepeated."[23] The exceptionality of this period, the authors assess, is due to a fleeting alignment in the positions of two emotions that were

20 Huet 2004: 128. For more on Paré's groundbreaking contribution to teratology, see Wilson 1993; and Berriot-Salvadore and Mironneau 2003. See too, Ciavolella's introductory study to the Italian edition of Paré 1996: 10–18; as well as Céard's critical edition of 1971. Of great value is Baille 2017. On the topic of Paré's notions of procreation and embryology, see Gilles-Chikhaoui 2010. Kathleen Long discusses at length Paré's interest in hermaphrodites. Touba Ghadessi looks at certain fascinating correspondences between Paré's ideas and late sixteenth-century iconography, see Ghadessi 2013.

21 See Bacon (1605) 2008: 176–177. In the *Novum Organum*, Bacon says: "For a compilation, or particular natural history, must be made of all monsters and prodigious births of nature; of everything, in short, which is rare, new, and unusual in nature." See Bacon, *Novum Organum* (1620) II.29, 1841: 392.

22 One of the most famous early modern collectors, Ulisse Aldrovandi, also wrote a greatly influential treatise on monsters, *Monstrorum historia* (1642, published posthumously). For more on the birth of the modern museum out of early modern *Wunderkammern*, see two classic studies: Impey and MacGregor 1985; and Findlen 1996. More recently, also Findlen 2019, as well as Genoways and Andrei 2009. Other noteworthy studies are Bennett 2005: 602–608; and Carrier, 2002: 44–50.

23 Daston-Park 2001: 13.

traditionally separated and that parted ways again as the scientific revolution took steam: Wonder and curiosity.[24] Traditionally deemed as either sinful or naive, curiosity acquires during the early days of modernity a positive value and is considered a fundamental drive towards the advancement of knowledge. Concurrently, natural philosophers "reclaim wonder as a philosophical emotion," Daston and Park add, and locate it as the basis of a new form of inquiry: Preternatural philosophy.[25] While late Medieval thinkers like Albertus Magnus (d. 1280) and Thomas Aquinas (1225–1274) largely dismissed the possibility of the particular, let alone the exceptional, being a proper object of philosophical enquiry, Marsilio Ficino (1433–1499) in the fifteenth century, and Pietro Pomponazzi (1462–1525), Girolamo Cardano (1501–1576) and others in the sixteenth, rehabilitate the philosophical worth of the individual and build the basis for preternatural philosophy, whose birth is announced in the dawn of the seventeenth century with Bacon's manifesto.[26]

This peculiar alignment of wonder, curiosity, and the scientific imagination that Daston and Park stress in their book, takes place against the backdrop of an event that would prove to be enormously significant to the development of preternatural philosophy, the discovery of the new world. Beginning with the first journeys of Columbus and Vespucci, perspicuously recorded in their own words, two very different positions in relation to monsters begin to take shape. On the one hand, to some travelers these voyages do nothing but disprove ancient myths and superstitions that located monstrous races at the farthest ends of the world.[27]

24 Daston-Park 2001: 15.

25 Daston-Park 2001: 137.

26 For Ficino 2002, see especially the third book of *De vita* (1489), where he discusses the manifest and hidden properties of stones, plants, and animals. Pomponazzi 2011 deals with these matters in *De incantationibus*, his one work to make it to the Index. As for Cardano, see his *De natura* as well as his *De subtilitate* in *Opera omnia* (1663). Talking about Cardano and this "new type of wonder" drawn from the particular, Daston and Park argue that "the wonder of the connoisseur, so familiar with a multiplicity of extraordinary phenomena that he knew which truly deserved his amazement. This wonder was a finely graduated register of response that only the best informed and the most philosophically sophisticated could deploy" (Daston-Park 2001: 167).

27 Daston-Park 2001: 25 cite Ranulph Higden, a fourteenth century English monk, who wrote that "at the farthest reaches of the world often occur new marvels, as though Nature plays with greater freedom secretly at the edges of the world than she does openly and nearer us in the middle of it." See Higden 1865–86: 361. Lindquist and Mittman also stress this aspect of transatlantic exploration and discuss early maps with the example of Abraham Ortelius, friend and competitor of Mercator, who published Europe's first atlas of maps in 1570, *Theatrum orbis terrarium* (1570). In it, off the coast of *Hispania nova,* a ship in full sail is menaced by a massive sea serpent. The texts accompanying the map hint at the presence of other monsters and monstrous peoples, such as the Amazones in Brazil; and the Caribbean was labeled "Mar de caníbales". These new monstrous races, the authors argue, were inspired by ancient monsters from Pliny, Augustine, Isidore. See Lindquist and Mittman 2018: 20–21. The fear of monstrous races lasts well into the nineteenth century, notes White, when "all the blank spaces or black holes in our maps of the world have been filled in with both physical and demographic data." See White 1991: 208.

A good example is Maximilian Transylvanus (d. 1538), secretary of Charles V, who wrote an account of Magellan and Elcano's first circumnavigation of the world (1523) after interviewing some of the crew of the Victoria. In it, he concludes that circumnavigating the world refutes the belief in monsters. In his own words: "Who will believe now in the existence of *stipadas* and *pygmees* and other monsters that ancient writers told us about, when Spaniards and Portuguese going around the world never found anything of the sort?"[28] On the other hand, as Anthony Grafton points out, other chroniclers understood the New World and its peoples using categories from classical literature. These writers knew from reading Herodotus, Pliny, Mandeville, and others that faraway lands are inhabited by monsters and are full of wonders, and this is precisely what they see in the newfound land, the giants and cannibals they had learned about in classical sources, the fantastic beasts they had read about in bestiaries, and the barbaric habits that they had studied are exactly what they find in America.[29] Natives of southern Patagonia, for example, were identified with giants due to their height. Certain tribes living in the Caribbean who practiced ritual anthropophagy were deemed to be the cannibals Herodotus and Pliny described in their works. Furthermore, native people from the New World and Africa were commonly understood to be the monstrous inhabitants of the Antipodes that writers like Saint Augustine and Pope Zachary I had mentioned in their works. Such misconceptions derived in catastrophic consequences when, for example, during the famous debate with Bartolomé de Las Casas in Valladolid concerning the problem of "natural slavery" in regards with the natives of the New World, Luis de Sepúlveda argued that the dominion and enslavement of the Indians was a moral duty due to their inherently monstrous nature, which he demonstrated by appealing to Aristotelian concepts.[30]

Faced with a plethora of never-before seen human types, animals, plants, and minerals coming from across the ocean, early modern Europeans who were not dedicated full-time to annihilating and exploiting, set out to collect, to taxonomize, and to understand natural and cultural variety. This coincides with an ever more explicit insistence upon rationalism – a rationalism that understood itself not yet diametrically opposed, but certainly in conflict with religious dogma. Physicians, perhaps due to their constant engagement with the body which is mortal and corruptible,

---

28 Cited Fernández de Navarrete 1964: 2.558. See also Bénat 2006: 235–254.

29 Grafton 1992: 55. In an article that is already a classic, Wittkower discusses the history of the survival of the Greek "conception of ethnographical monsters" and their "power of survival"; monsters, the author concluded, "did not die altogether with the geographical discoveries and a better knowledge of the centuries but lived on in pseudo-scientific dress right into the 17th and 18th centuries." See Wittkower 1942: 159.

30 According to Greenblatt, Jean de Léry, one of the first travelers to arrive in Brazil, wrote after seeing the New World with his own eyes that, "he now believes Pliny when he describes the marvels and prodigies of distant lands" (Greenblatt 1991: 22). For more on this see Flint 1984: 65–80; Hanke 1959: 6; and O'Gorman 1961: 55.

were on the front lines of this battle to the point of arising suspicions of atheism. In the mid-1630s, a fledgling English physician, Thomas Browne (1605–1682), writes a first-person narrative of his profession (published in 1642–43 with the title *Religio Medici*) and feels the need to defend himself against such suspicions evoking a common saying at the time according to which out of three physicians, two of them are atheists.[31] This conflict was particularly visible in the works of those who investigated monstrosity. Few expressed it more clearly than Martin Weinrich, who in his *De ortu monstrorum commentaries* (1595), says: "Theologians and physicians do not follow the same goal nor do they base their conclusions on the same principles. The former base their knowledge on the holy scriptures . . . the latter ground themselves solely on reason and look for the causes of things in their vicinity."[32] This growing intellectual clash and the encounter with the New World combined make the period especially fascinating to appreciate how science and the imagination (or, in Canguilhem's terms, *monstrosity and the monstrous*), drew from one another as they slowly started to part ways, one to consolidate modern experimental science, the other one to nourish the artistic and poetic imagination. By 1616, the year of the publication of Fortunio Liceti's first edition of the treatise on the nature and causes of monsters, the ground was ripe for one of the most remarkably hybrid approaches to the problem of monstrous births – one that combined a strictly Aristotelian theoretical foundation, a methodology centered on direct observation, and a surprising number of inherited beliefs from a quasi-mythical tradition.

Liceti's Monsters

In "Monstrosity and the Monstrous," as he presents a rather sweeping historical survey of approaches to the study of monstrosity, Canguilhem says that "the first teratological works of etiological intent, by surgeons or physicians like Paré or Fortunio Liceti, can hardly be distinguished from the chronicles of prodigies by Julius Obsequens (4th century) and Conrad Lycosthenes (1557)."[33] In these works, Canguilhem adds, the iconography brings monstrosity and the monstrous together, and reason merges with the imagination as one seamless surface. Whereas the latter statement holds water against the backdrop of Canguilhem's theory, the former is questionable at best. Liceti's *On the Causes and Nature of Monsters* (*De monstruorum caussis, natura et differentiis libri duo*, published in Padua by G. Crivellari in 1616, with subsequent editions in 1634, 1665, and 1668) might very well be the most influential and innovative work on monsters from the first half of the seventeenth century. To quote

31 See Browne 1977: 61.
32 Weinrich 1595, part 1, p. 6a, cited by Céard 1996: 438.
33 Canguilhem 2008: 139–140.

Daston and Park, Liceti and other preternatural philosophers "shifted the marvels of nature from the periphery to the center [and] in the process, they reclaimed for natural philosophy not only wonderful phenomena but also the emotion of wonder itself."[34] Far from being a collection of sensational chronicles, like those by Obsequens, Lycosthenes, and Boaistuau, and by following in the footsteps of Paré, Liceti asks the crucial question about causality and searches for answers grounded on robust philosophical foundations as well as on particular case-studies.

In order to do so, Liceti first needed to clear up the ground, distinguish literal and metaphorical meanings of the word "monster" (*monstrum*), and circumscribe his object of study. A monster is not a man who lives or behaves like an animal, it is not a woman of indescribable beauty, and it is not a creature with a slight malformation (a baby born with six fingers, for example), he says and adds, "we call monster any animal that is born with a disposition and an arrangement of limbs completely different and totally contrary to nature . . . like, for example, a child with no feet, a girl with two heads, a boy with a dog's head, a Centaur, and other similar beings."[35] As Céard points out, Liceti's Aristotelianism shines through this definition which will be the main working premise in this treatise. After all, the man held the chair of Aristotelian Philosophy first at Pisa and then at Padua. Following closely the research made by the Greek philosopher in his *Generation of Animals*, Liceti argues that, on the one hand, monsters are only animals, living beings, while on the other, monstrosity is determined by a certain arrangement of the limbs that marks an astonishing difference between the individual and its parents.[36] The first two examples (i.e., a child with no feet and a girl with two heads) are cases of genetic malformations that derive in deformities. The third one, a boy with a dog's head, could also be a case of the latter that has not been yet fully understood by science, although the appeal to the long-standing legend of the cynocephali, a monstrous race of men with dog's heads that were a common feature in travel narratives and bestiaries since antiquity, does bespeak a lingering attachment to pre-modern beliefs and legends. With the fourth example, the centaur, we are completely in mythological terrain. In Chapter three, Liceti dedicates a chapter to such monsters and includes testimonies both drawn from tradition (Aristotle, Plutarch, and others) and his own experience, like the story of a woman in Padua who gave birth to a feline with a pair of human legs coming out of its rear.[37] As we can see from early on, Liceti's notion of the natural world is one that includes impossible, indeed mythical creatures. In other words, monstrosity and the monstrous intertwine from early on creating the theoretical fabric of the text. This in turn is connected to the fact that, for Liceti, the very notion of

34 Daston-Park 2001: 160.

35 Liceti, *De monstruorum natura* I.1, 1616: 4. Translations from Liceti are my own.

36 See Céard 1996: 443. For more on Liceti's Aristotelianism, and especially its development throughout the physician's intellectual career, see Del Soldato 2017: 531–547.

37 Liceti, *De monstruorum natura* I.3, 1616: 13.

monstrosity is inextricably linked to the emotion of wonder that these creatures elicit in others. Towards the end of book one, he reformulates his definition and argues that a monster is "a subcelestial being the very sight of whose enormous constitution and members inspires horror and admiration."[38]

The effect that monsters produce among people, or rather Liceti's concern with that effect, is what leaves the door open for the monstrous, that dimension informed by legends, beliefs, superstition, and the imagination to creep in. Liceti insists time and again on the *admiratio* that one is seized by when face to face with a monster. Monsters "attract us" and "capture us" due to their rarity and novelty, producing "extreme admiration" in us, he says at the beginning of the preface.[39] However, there is no doubt for Liceti that monsters are natural phenomena and not divine signs nor punishments for sins, either individual or collective. In a strictly Aristotelian vein, the physician argues that there are four causes of monstrosity: material, formal, efficient, and final.[40] By affirming this, Liceti hopes to establish that monsters are not against nature since nature, according to Aristotle's famous dictum, "does nothing in vain."[41] The fact that they have not only a formal but also a final cause means that monsters occupy a purposeful place in the overarching plan of nature. They have a formal cause because they are animals, and all animals have a formal cause in the soul.[42] Even transmitting the soul guarantees the perpetuity of the species' substance. As for the final cause, Liceti does not deny that it is God, who is the final cause of everything. But, like Roger sharply points out, Liceti insists on separating the physician from the theologian by reminding the reader that God is a faraway cause that does not help us come to a better understanding of the physiology and taxonomy of the monster; and that after acknowledging that he is in fact the ultimate cause, one must get to work on deciphering and describing the more immediate causes.[43] Galileo Galilei (1564–1642), with whom Liceti over the years corresponded actively, had recently expressed a similar idea in his famous 1615 letter to Duchess Christina, where he writes that the Holy Scriptures help us get to heaven, but do not help us much in figuring out how the heavens work.[44] In his 1616 treatise, Liceti explains the purpose of monsters in strictly pragmatic terms that seem to resonate with the spirit of this scientific and intellectual secular

38 Liceti, *De monstruorum natura* I.11, 1616: 41. He repeats this verbatim in I.12, 1616: 42, and in I.12, 1616: 45.

39 Liceti, *De monstruorum natura* 1616: 1.

40 See Liceti, *De monstruorum natura* I.6, 1616: 26–27.

41 Cited by Liceti in *De monstruorum natura* I.6, 1616: 27: . . . *quod natura ex Aristotelis decreto nihil agit frustra: sed omnia, quaecumque producit, propter determinatum procreat finem.*

42 See *De monstruorum natura* I.8, 1616: 30–32.

43 See Rogers 1963: 90–91. Monti argues that Liceti proposes a preformistic embryology by which "monstrosity is pre-formed in the seed" so the direct responsibility lies within the parents "who were the carriers of the malformation". See Monti 2000: 7.

44 The letters between Liceti and Galileo are included in Galilei, *Opere* (1640) 1909: vols. 10 to 18.

renewal. Contrary to vulgar opinion, Liceti says, nature does not fail when it produces monsters. Instead, the finality of the monsters is "to preserve the species of [the monster]'s progenitors when the matter at hand is organized differently."[45]

The mystery of monstrosity is thus relegated to the sphere of matter and to the logic of material causes. According to Liceti, it is only when faced with materials that are defective, insufficient or unfavorable, that nature produces such creatures. In his own words, upon encountering impediments to reach its ultimate goal, nature rather than relinquishing its creative drive, generates a creature that albeit different from its progenitors in the disposition of its limbs, resembles them in essence and form.[46] In *De perfecta constitutione hominis in utero*, a treatise on embryology, published the same year as *De monstrorum caussis*, Liceti once again excuses nature for monstrous births and praises its capability for infinite variety (*varietas*) of expression.[47] The physician compares nature to an artisan who deals with scarcity and imperfection in the raw material at hand with awe-inspiring ingenuity. "In fact, I see a common element between nature and art here: unable to make what it wants to make, it instead produces what it can."[48] The ingenuity demonstrated by nature in the production of monstrosities is what, in Daston and Park's view, qualifies Liceti as a prime example of what they characterize as the attitude of "wondering pleasure" vis-à-vis monsters. In "Monsters: A Case Study," the fifth chapter of *Wonder and the Order of Things*, the authors distinguish three main emotions that monsters provoke in the early modern mind: Horror, pleasure, and repugnance. These categories, at once "cognitive and emotional," overlap and coexist sweeping through different disciplines and at times inhabiting the very same individuals.[49] By horror, the authors refer to the emotion produced by the belief that monsters were signs of divine wrath – a belief that persists throughout period and well into the Enlightenment among certain circles. Pleasure, on the other hand, is the affect inspired by the notion that monsters are benign beings designed to produce wonder in us – "ornaments of a benevolent creator."[50] Finally, repugnance is the product of a complete dismissal of the problem of monstrosity as one worthy of legitimate scientific interest. To an increasing number of natural philosophers, monsters are mere errors that bear absolutely no relevance to the study of nature.

As one of the most notable examples of this second attitude, Fortunio Liceti even argues for a tweak in the etymology of the word "monster." Since antiquity, the word had been understood to derive from the verb *monstrare*, "to show," and taken as referring to something that the gods (or, upon the advent of Christianity, God) use to

45 Liceti, *De monstruorum natura* I.7, 1616: 30.
46 See Liceti, *De monstruorum natura* I.12, 1616: 45.
47 See Liceti, *De perfecta constitutione hominis in utero* I.11, 1616: 42. Cited in Luppi 2010: 55.
48 Liceti, *De monstruorum natura* I.11, 1616: 40–41.
49 Daston-Park 2001: 176.
50 Daston-Park 2001: 177.

communicate signs of his will to humans. In Liceti's secular reading, instead, monsters are called thus not because they show humans that God is angry at them, but because "their novelty and enormity provoking much wonder as surprise, everyone shows them to one another."[51] The remarkable thing about Liceti's attitude, and the reason why Daston and Park associate it with what they call "wondering pleasure," is that, like Girolamo Cardano before him, he considers that certain wonders of nature continue to cause wonder even after science has studied and understood them.[52] It is the persistence of wonder what stimulates the imagination and allows for what Canguilhem calls the monstrous to coexist and collaborate with the strictly scientific pursuit of understanding of monsters. Furthermore, this sense of wonder and surprise in the face of the naturally exceptional is shared by the layman and the man of science alike. These characteristics set Liceti apart both from contemporaries of his who also investigated the natural causes of monstrosity, like Martin Weinrich (1548–1609) and Jean Riolan (d. 1657), but who continued to stress the religious and divinatory aspect of the phenomenon, as well as from Baconian scientists who dismissed wonder as incompatible with reason and experimental science.[53] Standing in the threshold between two worlds, Liceti positions himself as a Janus-like creature with two faces, one looking at the past, the other glancing into the future.

This amphibian approach to natural philosophy comes through in his assessment of the causes of monsters. Liceti writes:

> In regards to the generation of monsters we must first consider how many causes there are and what their common origin is. On this issue, I think we can name three main causes for their generation. The first one is supernatural since God allows for and procures them in order to punish men for the sins and crimes they have committed, as the Scriptures state that happened to Nebuchadnezzar. The second one is under nature as the devil deceives humans through their senses, as we can see in those who were transformed into something else by means of witchcraft. And the third one is purely natural and comes from a certain defect or impairment found within the principles that are naturally destined to formed the bodies of men. We will not be dealing here with the miraculous origin of monsters, nor with those created by the devil through illusion. Here we will only discuss the natural origin of monsters.[54]

It was due precisely to the emphasis placed on natural causes that nineteenth-century physician Saint-Hilaire, who coined the word "teratology" and formally inaugurated the discipline, identified Liceti as his most direct predecessor.[55] When Lazzarini argues that, "Liceti's originality lies in his attempt to identify the physiological causes of malformations for the first time," rather than neglecting the influence of Paré, Rüff, and

51 Liceti, *De monstruorum natura* I.2, 1616: 7.
52 See Daston-Park 2001: 228.
53 See Martin Weinrich, *De ortu monstrorum commentaries* (Leipzig, 1595), Chaps. 9–12. For more on Weinrich, see Céard 1996: 438. Also, Riolan 1605: Chap. 8. For more on Riolan see Bates 2005: 85–87.
54 Liceti, *De monstruorum natura* II.1, 1616: 51–52.
55 Fulcheri 2002.

others, she is ascertaining that Liceti is among the first to go in depth into the natural causes for monstrosity.[56] As he surveys the natural causes of monstrosity, Liceti follows Aristotle closely. There are two main sets of causes that explain a monstrous birth, excess of matter and scarcity thereof. Most of book two of *De monstrorum caussis* is dedicated to enumerating specific causes in both instances. Liceti lists superfetation (an excess of semen), inherited conditions, narrowness of the uterus, dysfunctional nourishment of the fetus, trauma on the pregnant body of the mother, diseases of the embryo, etc. More importantly, Liceti understands that there could be other causes and that one must continue investigating. In his own words, *nobis ex rei natura nunc est indaganda*.[57] In fundamental agreement with Bacon's call to investigate the preternatural by collecting samples, Liceti spends the remainder of the treatise giving examples that he draws from three sources: Ancient writers, early modern accounts, and first-hand experience. His conviction that science is an ever-ongoing process and his preference for first-hand experience over book learning positions him as typical early seventeenth-century natural philosopher. And yet, Liceti believes that monsters can also be caused by god and the devil.

Whereas Liceti does not engage long with the first cause of monstrous births (God), he does address the second one, the belief that the devil is capable of manufacturing monsters by means of intervening in natural processes. This does not mean, however, that monsters produced by demons are not also natural. Their efficient cause for monstrosity *cannot but* be natural. The devil simply uses nature in his advantage to disseminate fear and chaos. He can debilitate the seed, or act upon the parent's imagination and induce a monstrous conception, Liceti adds. In any case, "the efficient cause of the monster is strictly natural, but the demon is the minister, or the artisan who places what is active on what is passive, piling features upon the creature's regular arrangement, removing or adding impediments, but only interfering with the causes of monstrosity through localized motion."[58] Two elements here are crucial to understand the phenomenon of the monstrous birth. First, and this goes back to the etymology proposed by Liceti, a creature is a monster if and only if it elicits awe and wonder among people. A pygmy born in the Congo basin, for example, is not a monster because he or she is born of pygmies and lives among pygmies who will not feel the urge to "show" him or her off to others seeking to induce awe. One born in Scandinavia, on the other hand, can be considered monstrous due to the wonder and surprise it is bound to arouse among locals. A white girl born in Ethiopia is also monstrous, says Liceti, concluding that

56 Lazzarini 2011: 430. In the mid sixteenth century, in Sicily, physician and anatomist Giovanni Filippo Ingrassia describes a two-headed baby born in Regalbuto. This, as Cusumano points out, is one of the earliest medico-physiological descriptions of a monster that we know of. See Cusumano 2012: 860.

57 Liceti, *De monstruorum natura* II.19, 1616: 108.

58 Liceti, *De monstruorum natura* II.73, 1616: 231–232.

it is resemblance to one's kin and one's ethnic group what determines the phenomenon.[59] What Liceti is suggesting here is a certain relativism, or rather, contextualism of monstrosity. Here too we can appreciate Liceti's dutiful Aristotelianism. For Aristotle, monstrosity was a phenomenon of non-resemblance.[60] Resemblance to one's progenitors happens within a spectrum. The farther one is from one's parents (from one's father, really), the less one resembles him, the more monstrous one is.[61]

The second element that is key to Liceti's understanding of the monster is craftsmanship and ingenuity. It is here that Canguilhem's notion of the monstrous again comes into play. Nature is a resourceful artisan that makes do with what it has in order to fulfill its plan, which consists on the preservation of the species through their form (i.e., their soul). In book two, Chapter fifty-eight, Liceti provides a long list of outlandish examples taken from classical and contemporary sources, both treatises and pamphlets, but also from his first-hand experience. He talks about a man who was born with the head of a dog, a woman who gave birth to twins conjoined by their back, one of whom was human and the other, a dog, or a pig with a human head, or a cat with a set of human legs coming out of its bottom, as well as children born with horns, a cow that gave birth to a boy, and a boy born with the head of an elephant.[62] This ingenuity of nature, however, has its counterpart on human imagination that also plays a role, albeit limited, in the production of monsters. This is how monstrosity and the monstrous once again become entangled as the mother's imagination can be the cause of certain changes in the fetus that derive in monstrous births. Liceti lists the imagination (*phantasia*) as the first possible cause for the type of monstrosity based on excess of matter, but also for other types of monstrosity: amphibian, or dual-natured monsters (2.28), formless monsters (2.42), enormous monsters (2.47), multiform monsters (2.56). Fantasy cannot really add or subtract matter from a fetus, and it only operates once the abnormal multiplication of limbs has already taken place in the womb; it can, however, and only when it acts "vehemently" (*vehementer*), function as a principle of order that shapes the fetus, say, in the shape of a two-faced Janus after seeing an image thereof.[63]

---

59 Liceti, *De monstruorum natura* I.12, 1616: 42–45. See also II.46, 1616: 152. In spite of this argument, which indirectly debunks the belief in monstrous races, in the illustrated editions of *De monstrorum caussis* (Padua, 1634), images of traditional specimens belonging to monstrous races are included. Among them, cyclops (1634: 192), cynocephali (1634: 182), Blemmyes (1634: 10, 84, 99), and men completely covered in hair (1643: 149, 158).

60 In his work on embryology, *De perfecta constitutione hominis in utero*, 1616: 37 and 49, Liceti reiterates this notion.

61 Liceti wrote a treatise on generation, and specifically on spontaneous generation, *De spontaneo viventium ortu* (1618), where he follows up on a number of these issues stressing in particular the importance of the soul of the parents during generation. For more, see Hirai 2006: 451–469.

62 Liceti, *De monstruorum natura* II.58, 1616: 180–194.

63 Liceti, *De monstruorum natura* II.11, 1616: 93–94.

This can be understood also taking into consideration Paula Findlen's argument concerning the intersections between the notion of *lusus naturae* and *serio ludere*. In the seventeenth century, monsters can be understood as products of nature's infinite ingenuity because scientists approach their exceptional morphology in a way that is at once serious and ludic, stressing the paradoxical in the natural world.[64] In Liceti's work, as Leeming, points out, "*lusus naturae* calls upon us to capture Nature's *ingenio* that both underpins and is the object of the *serio ludere* of medico scientific investigation."[65] This is, in Canguilhem's terms, the monstrous at the service of understanding monstrosity. But whereas in the pre-modern period, the monstrous was commonly understood to be the cause of monstrosity, in the context of the scientific revolution, and contained by the power of the imagination, it becomes one among many causes for it as well as a tool for its understanding.

Conclusion

In 1548, art critic and theorist Benedetto Varchi (d. 1565) gave a lecture at the Florence Academy (originally in Latin, and later published in Italian), entitled "On the Generation of Monsters." In it, Varchi anticipating Liceti by over half a century, focuses on the exhibitionist aspect of monstrosity appealing to the etymology. Varchi believes that some monsters are signs of God's wrath, whereas others have merely natural causes. Among the many possible natural causes for monstrosity, Varchi includes the imagination. This, he explains, can happen in two ways. First, he mentions the traditional belief according to which the mother's fantasies can induce malformations, birthmarks, or other exceptional features in the fetus. Second, he adds, the imagination can become an actor due to its immense creative potential, the clearest example of which would be the artistic imagination that possesses an unbridled capability to envision and portray monstrosities.[66]

Liceti's insistence on the multifarious importance of the imagination in the study of monsters resonates even more with Varchi's ideas when we consider that *De monstruorum caussis* was perhaps more influential due to its remarkable illustrations than to its actual content. It is worth noting that the first edition did not include the famous images. Liceti must have understood the appeal of visual aids and the need for them after the 1616 publication, because the second edition almost two

64 Findlen 1990: 294.
65 Leeming 2019: 22.
66 Varchi 1858: 664. Varchi goes further and claims that this does not just refer to the work of art, also to the artist himself. Michelangelo, for example, is a monster of nature: "Chiamansi ancore mostri dell'animo tutti coloro i quali sopravanzano gli altri nelle opere o di mano o d'ingegno che vincono quasi la natura, ciò è fanno quello che non è solito farsi ordinariamente dagli altri. Ed in questo significato diciamo che . . . Michelangelo è un mostro della natura."

decades later (1634), appeared with the daunting and now famous engravings by Paduan artist Giovanni Battista Bissoni.[67] Aside from the fabulous frontispiece – an unlikely assembly of incredibly athletic monsters standing on each other's shoulders like a group of cheerleaders at a football game – and what might very well be the most notorious depiction of the Ravenna monster, the treatise includes one illustration that evinces with particular clarity Liceti's profound intellectual investment in the importance of the imagination. It portrays a trifecta of monstrosities (See Figure 1). The first monster has a human female body covered in scales, the head of a donkey, and a tail in the form of a dragon. The second one has an elephant's head, eagle's claws and a human body that includes a face on its stomach and several tiny animal heads protruding from his breast, elbows, and knees. The third one is covered in hair, has horns, and a long umbilical cord that falls onto its feet. The trio is surrounded by small flying creatures that urinate and defecate over them. We see here a combination of recognizable genetic malformations and pathologies (such as hypertrichosis, elephantiasis, keratinous skin tumors – also known as cutaneous horns – and hermaphroditism) and imaginary deformities. The picture thus portrays with great perspicuity that assembly formed by first-hand experience, book learning, and the imagination that forms the base of Liceti's scientific enterprise.

Giovanni Imperiale, a colleague of Liceti at the University of Padua, once argued that physicians "find their humoral likeness in poets for both vocations must mediate between theory and practice, the contemplative soul and the body in action."[68] The humoral likeness also extends to artists, we might add. That liminal area between theory and practice, and between action and contemplation, is the arena of the imagination, a space in which distinctions between disciplines become increasingly hazy. It is in this space – a space leading to a vortex which Canguilhem associates with an "anticosmos" where the exceptional is the norm –, that the scientist, the artist, the poet, the humanist converge into one as they allow the imagination to take the reins and guide the way.

67 In discussing the influence that new anatomical works, including early treatises on monsters like Liceti's, had on the work of plastic artists of the sixteenth and seventeenth centuries and in particular the stress on humor and on *lusus naturae*, Sandra Cheng mentions Guercino's caricature drawings and stresses that "his ability to construct creatures out of such disparate parts is a display of artistic *invenzione* similar to nature's ingenuity in her ability to create monsters." See Cheng 2012: 198.

68 Cited by Barbour 2013: 152.

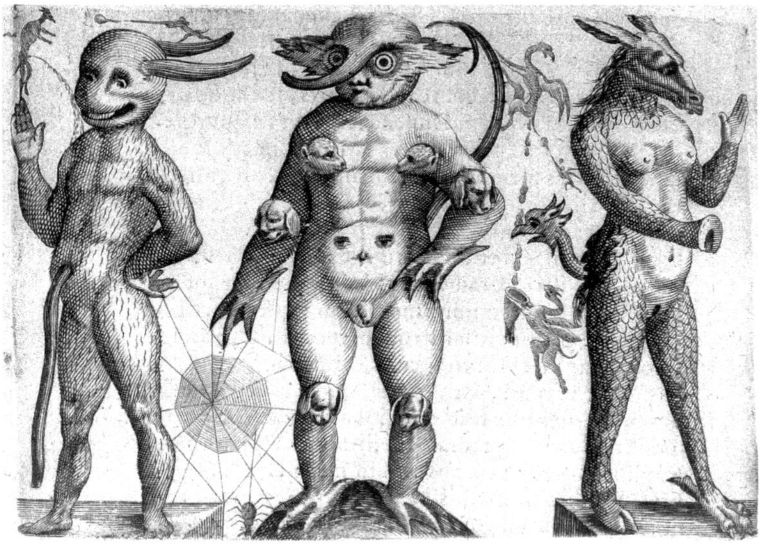

Figure 1: Liceti Fortunio 1577–1657 *De Monstrorum caussis, natura, et differentiis libri duo.* Credit: *De monstrorum caussis, natura, et differentiis libri duo*. Wellcome Collection. Public Domain Mark Wellcome Collection. Public Domain Mark.

Roberta Ballestriero

The Flesh of Wax: The Use of Scientific Collection in Medical Humanities

A Brief History of Wax Modelling

For its particular physical and chemical characteristics, wax was known from ancient times and used for a variety of different purposes by the Egyptians, Greeks and Romans. The first known use of wax for modelling was the sculpting of bronze and jewellery with the lost-wax casting process (cire perdue). In this case it was used as an intermediate medium to obtain a final result in other materials: metals such as bronze, gold or silver. Beginning in the 4th century BCE it was also used in the elaborate encaustic painting technique where, to bond the pigments, paint and bees' wax were heated together. This "burning in" of the colors with bracers (the ancient *cauterium*) was an essential element of the true encaustic method. Described by the 1st century CE Roman scholar Pliny the Elder in his Natural History, this technique was used to obtain bright, stable colours that we can still see in the oldest surviving examples of some panel paintings such as the Romano-Egyptian Fayum mummy portraits (middle Egypt), dating from the early centuries of our era. Encaustic painting was used up to the 9th century CE when it was replaced by other techniques such as eggs tempera and eventually oil paint.[1]

The advantageous characteristics of wax such as malleability, resistance to atmospheric factors and the possibility of accepting paint, explain why it was used widely for the creation of a variety of figures and portraits. This tradition began in pre-Christian Rome, where the ancestral portraits of the patriciate were set up in the atrium or in niches and carried in processions at funerals accompanying the deceased.[2]

From pre-Christian times to the present, wax has and continues to be used for votive and ex-voto images. The offering of objects to a divinity, or to a saint, to ask for a grace (propitiatory ex-voto), or to give thanks for a received grace (gratulatory ex-voto) is a very ancient custom. Votive offerings could be of any kind but often

Acknowledgments: Photographs by Owen Burke and Roberta Ballestriero. I would like to express my thanks to the Gordon Museum, Kings College London, and the University of Cagliari with the Collection of Clemente Susini's Wax Anatomical Models, for kindly permitting the publication of the photographs of this manuscript.

1 Piva 1991: 177–182.
2 Pliny the Elder 1997: 19–20. See also Panzanelli 2008.

https://doi.org/10.1515/9783110788501-008

reproduced parts of the human body, representing healthy or diseased organs that were generally placed in places of pilgrimage, churches, chapels and shrines.[3]

The mimetic potential of wax models led to a development of this art from the beginning of the 16th century, reaching its artistic peak particularly in Florence, the cradle of the Renaissance. From the 13th to the 17th century, the habit of donating votive offerings was very common in Florence, so common that it created a real industry of *ex-votos* in different materials but mainly in wax. Votive offerings of all kinds were present in different Florentine churches but especially in the church of SS. Annunziata. Initially these 'bóti,' as they were known in the Florentine vernacular, represented limbs or parts of the human body. Eventually, these anatomical votive offerings progressed to proper statues. Nobles, Florentine as well as foreigners, commissioned life-sized figures of themselves in coloured wax. These were dressed in their own clothes and then offered, as an act of devotion, to the Santissima Annunziata where they became a major feature, turning the sanctuary into an enormous museum of wax figures of all types. In 1786 Leopold II, Holy Roman Emperor from 1790 to 1792, introduced several ecclesiastical reforms amongst which he ordered the clergy to free the churches of all votive offerings, thus the practice of donating wax votive offerings ended, and the existing ones were melted to make candles.[4]

As noted, wax as a medium has several unique characteristics and it is capable of a remarkable mimetic likeness superior to other materials. It is flexible, easy to work, can be coloured, and can be adorned with organic materials such as body-hair, hair, teeth and nails. However, these same characteristics led the practice of this art to a slow decline and wax objects to be considered as a mere craft of questionable taste. From an artistic point of view, it virtually disappeared, continuing to survive only in minor areas, such as votive uses or in waxworks during the 19th century such as Madame Tussauds in London or the Musée Grévin in Paris. On the other hand, the use of this art for didactic purposes increased considerably in the scientific world, in the study of normal and pathological anatomy, in obstetrics, in zoology and in botany.

The Renaissance and the Rediscovery of the Human Body

From ancient times physicians and artists had tried to illustrate anatomy, but it is only during the Renaissance that the human body was rediscovered to become the

3 Antoine 2001, Chilvers 2003: 197, 204.
4 Lanza et al. 1979: 19.

centre of attention and to be reproduced through major and minor arts. As is well known, the Renaissance was not a purely artistic phenomenon but an extremely innovative period that embraced the sciences, the liberal arts, literature and philosophy. Regarding the representation of the human figure, the art during Renaissance was based on the rapid expansion of anatomical knowledge resulting from the direct investigation of the body's morphology through the dissection of cadavers. This practice was limited during the 14th century due to the Papal Bull issued by Pope Boniface VIII (1294–1303), "De sepolturis," which prohibited the manipulation of corpses. Although the Papal Bull did not refer specifically to dissections it affected the practice. However, these ambiguous interpretations ceased with the permissive Papal Bull issued on 4 April 1482 by Pope Sixtus IV (1471–84) for the University of Tübingen, Germany. The bull explicitly authorised the performance of autopsies, with the consent of the local bishops, indicating the Church's approval of the practice. This permission that was later reiterated by Clement VII (1523–34) leading to the flourishing of anatomical studies, the scientific basis for medical advancement.

This renewed scientific interest in anatomy motivated not only doctors but also Renaissance artists to go further to study and dissect the cadaver. The new discovery of perspective helped the representation of the body in foreshortening and the human figure returned to those of ancient schemes, positions and proportions with the adoption of classical models. The need to further research to obtain knowledge, especially in anatomy, drove artists to surpass the limitations of the two-dimensional nature of drawing by using materials like wax to perfect morphological research. And it is precisely in the Renaissance, during the initial phase of research, that wax played an essential role, especially in the work of artists.

The use of wax during the 16th century is mainly related to the making of models and sketches. Wax was found to be one of the most suitable materials for artists for research, surpassing the limitations of the two-dimensional nature of the drawing. Numerous artists such as Donatello, Benvenuto Cellini, and Michelangelo Buonarroti used preparatory wax models on a reduced scale. Ludovico Cardi (1559–1613), known as "Cigoli," created a small statue of an "écorché," which was of particular importance being the first known anatomical representation modelled in wax. This model, exhibited in the Bargello National Museum in Florence, is of extreme interest in view of the careful rendering of the external musculature and because, being of considerably reduced dimensions, it was clearly not created from a cast of muscles taken from a cadaver.

Ceroplastics and Anatomy: The Birth of the First Collections of Wax Models

From the outset, the study of anatomy was hampered by difficulties in obtaining cadavers. In the 17th century, as anatomical studies became more scientific, means of preserving the cadaver for investigative and teaching purposes were sought. The 17th to the beginning of the 18th century was also the golden period of anatomical specimens obtained by injection. Even if dissected anatomical preparations were known since antiquity, methods were tested for the preservation parts of the human body subject to rapid deterioration, such as viscera, and the circulatory and lymphatic systems. Air, colored liquids, preserving fluids containing alcohol, mercury and other metals such as lead, tin or bismuth, and wax were injected into the vascular system. The Dutch naturalist Jan Swammerdam (1637–1680) and the Dutch anatomist Frederik Ruysch (1638–1731), whose works are still well known today, developed new techniques for the preservation of specimens by injection. The results were reasonably successful but not completely satisfactory, because over time the specimens deteriorated and could no longer be considered completely accurate (be it in terms of color or morphology). An alternative method of providing an accurate reproduction of the various organs of the human body was clearly needed.

Towards the end of the 17th century a collaboration between Gaetano Giulio Zumbo (1956–1701), a Sicilian wax artist, and the French surgeon Guillaume Desnoues (1650–1735) resulted in the creation of the first realistic anatomical models made from colored wax, representing a valid alternative to dissected human specimens. Wax gradually took its place as a material capable of allowing the creation of extremely realistic (as to form and color) and long-lasting artefacts.

Gaetano Giulio Zumbo (or Zummo) was born in Syracuse in 1656.[5] The scarce evidence available refers to the last 10 years of his life, following the official recognition of his art in Florence, when Cosimo III de' Medici, Grand Duke of Tuscany, took him into his service in 1691.[6] It is thought that due to his cultural and spiritual preparation (Zumbo was an abbot) he initially undertook works of a predominantly religious nature. Subsequently, his work turned towards the macabre: death and disease. He produced four gruesome compositions known as 'Theatres of Death,' which convey a general sense of decay and the precariousness of life that seemed to please him: "The Plague," "The Triumph of Time," "The Vanity of Human Glory," and "The Syphilis" (or "The French Plague"). These tableaux, still on display at "La Specola" museum in Florence, portray destruction in minute detail with a morbid attention even to the most trivial aspects. Death is illustrated as an event that suffocates and eliminates any vestige of beauty. Zumbo depicts, with a relish typical of his time, the process of

5 Cagnetta 1988: 62.
6 Cagnetta 1988: 62.

deterioration of the flesh with the meticulousness of an anatomist and a taste for the macabre absorbed from the culture of Mannerism and early Baroque. This was clearly illustrated in paintings of 17th century artists such as Mattia Preti (1613–1699) or Luca Giordano (1634–1705), where the *Plague in Naples* was depicted. Thanks to Zumbo, wax anatomical modelling evolved and expanded from Italy and France to spread throughout Western Europe.

Zumbo was the first wax modeler working in collaboration with men of science, and the invention of anatomical ceroplastics must be attributed to him. However, the first workshops and first collections of wax anatomical models were created in Bologna by Ercole Lelli (1702–1766) and Giovanni (1700–1755) and Anna Manzolini (1716–1774). During the first half of the 18th century Cardinal Prospero Lambertini, Archbishop of Bologna, expanded the scientific studies in the city after becoming Pope Benedict XIV. He commissioned from Ercole Lelli a series of wax models representing the anatomy of the human body to be used for teaching. For a brief period, Giovanni Manzolini was one of Lelli's collaborators, but following an argument they went their own separate ways. Manzolini continued to work as a wax modeler assisted by his wife Anna Morandi, who after his death continued their trade to become an expert sculptress and anatomist. She became so famous that her services were requested by numerous Italian and foreign universities and by the Royal Society of London and the Empress Catherine II of Russia who repeatedly and unsuccessfully invited her to leave Bologna. Her wax self-portrait is on display in the Science and Art Museum of Palazzo Poggi alongside the collection of anatomical waxes. The Osteology and Myology waxes created by Lelli, and those of the sensory organs, intestinal tract and female organs of reproduction produced by the Manzolinis, housed in the Bolognese museum, probably constitute the oldest known collection of anatomical models in wax.

The art of anatomical ceroplastics spread from Bologna to Florence, where the second great wax modelling workshop was created, probably towards the end of 1771, at the Natural History Museum "La Specola." It was created by the physicist Felice Fontana (1730–1805) a great scientist in the court of the Grand Duke of Florence, Peter Leopold (1747–1792). When the workshop was started it had only one modeler, Giuseppe Ferrini (18th century) and no dissector, with Fontana initially performing all of the dissections. In 1773, Clemente Susini (1754–1814), who was later to take the Florentine art of wax modelling to the highest peak of artistic perfection, was admitted to the workshop as second modeler.

Towards the end of the century the fame of La Specola's workshop had surpassed that of Bologna's. Between 1771 and 1893 entire collections were created for the museum and for Italian and foreign universities including models of normal and pathological anatomical parts and botanical specimens.

The creation of large collections of wax anatomical models was intended to facilitate the education of students of medicine and, in fact, their scientific and didactic value has proved indisputable. The possibility of studying anatomy by means of

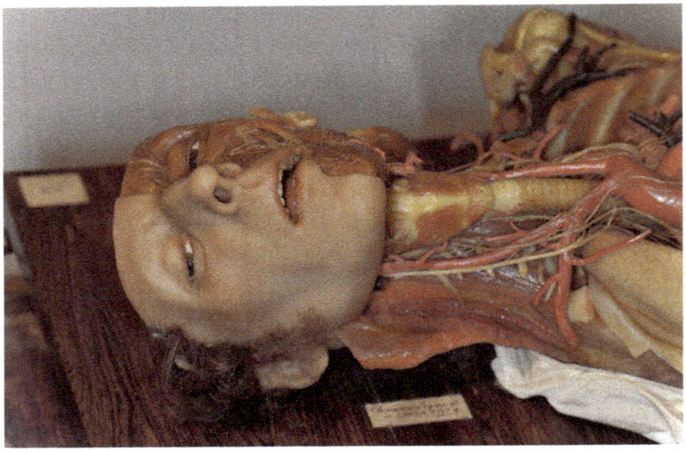

Figure 1: Clemente Susini, *Table XII Head, trunk and left upper limb of an adult male with vessels and nerves, detail.*
1803–1805 Clemente Susini's Anatomical Waxes Museum, University of Cagliari.
Photo: Roberta Ballestriero and Owen Burke.

realistic life-sized wax models was undoubtedly of considerable benefit as an adjunct to the anatomy manuals, whose two-dimensional illustrations were of limited use.[7]

Felice Fontana had the idea of making molds to produce anatomical waxes to instruct students, surgeons and medical staff. As Prof Alessandro Riva explained in his "Notes from the lessons on the History of Medicine," it was a useful idea for different reasons: Colour printing was, at the time, technically inadequate and exceedingly expensive and, because most of the surgeons did not know Latin,[8] they could not read anatomical and medical textbooks, generally written in that language. Also, since there were no freezing facilities, it was very difficult to keep cadavers for teaching.[9]

Acclaim for the Florentine wax modelling workshop soon spread throughout Europe. Emperor Joseph II, on arriving in Florence with the eminent surgeon Giovanni Alessandro Brambilla (1728–1800) was so impressed by the collection of wax anatomical models that he commissioned an even larger collection for the Military Academy of Medicine and Surgery he had founded in Vienna. In 1786, the collection was delivered by mule and is still today preserved in the Josephinum Palace.

7 Ballestriero 2010: 225.
8 For about two millennia surgery was considered a secondary discipline without any practical use: in fact, it was considered science only at the end of the 18th century.
9 Riva 2020: 42.

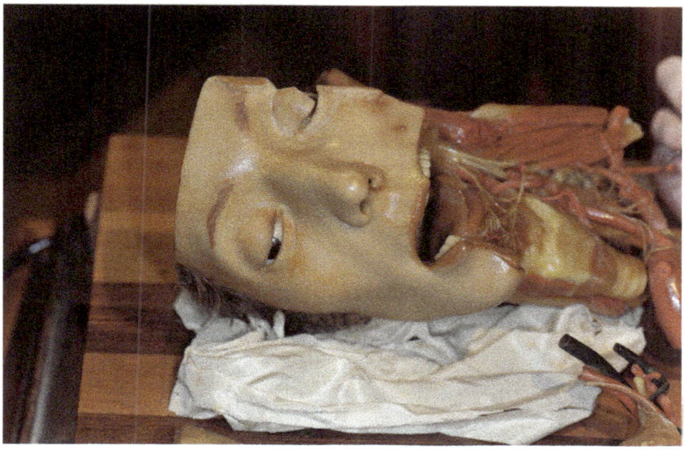

Figure 2: Clemente Susini, *Table XVI Organ of taste, detail.*
1803–1805 Clemente Susini's Anatomical Waxes Museum, University of Cagliari.
Photo: Roberta Ballestriero and Owen Burke.

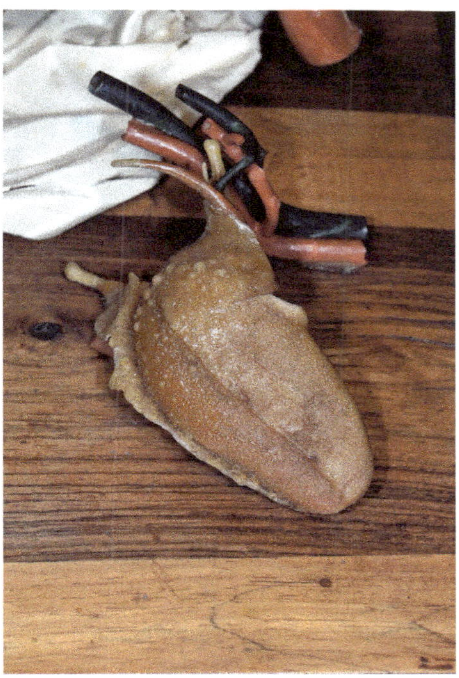

Figure 3: Clemente Susini, *Table XVI Organ of taste, detail of the tongue.*
1803–1805 Clemente Susini's Anatomical Waxes Museum, University of Cagliari.
Photo: Roberta Ballestriero and Owen Burke.

Between 1801 and 1805 the Sardinian anatomist Francesco Antonio Boi (1767–1860) travelled to Florence to further his studies. Thanks to Boi's sojourn in Florence we have the Susini collection of anatomical waxes, made when he was at the height of his artistic skill; they are still today on display in the Museum Complex in Cagliari.[10] These waxes were produced later than the Florentine or Viennese collections, between 1803 and 1805; they are not mere copies of previous works, but are unique pieces, each signed and dated, for which a considerable sum was paid at the time.

Figure 4: Clemente Susini, *Table XII Head, trunk and left upper limb of an adult male with vessels and nerves, detail.*
1803–1805 Clemente Susini's Anatomical Waxes Museum, University of Cagliari.
Photo: Roberta Ballestriero and Owen Burke.

At 'La Specola' museum, with the help of several anatomists, Fontana equipped a workshop in which molds of parts of the human body were made of plaster to produce wax models by skilled artisans. His main aim was to create anatomical models of scientific value for teaching purposes whilst removing the sense of repulsion produced by cadavers (Fig. 1–4). In fact, referring to the invention and the use of artificial anatomy, he stated in 1786: " . . . it will be of infinite use in providing a perfect knowledge of all organs of the human body, allowing [all and sundry] to learn without any feeling of disgust or hesitation the more intricate details of anatomy . . . ."[11]

Like paintings, sculptures, architecture and other forms of art, anatomical collections often differ in style. Even in anatomical drawings the representation of

10 Riva et al. 2010: 218.
11 Fontana 1786, quoted by Castaldi 1947: 41.

organs, muscles and bones changed over the centuries. Collections of anatomical models made in coloured wax were usually very accurate from the scientific point of view, but their artistic result often changes depending on the country of provenance. The artistic style presented in the Italian waxes can probably be explained by the rich artistic environment that surrounded physicians, anatomists and wax modelers. It is not then a surprise that the anatomical models were recalling previous famous masterpieces in their pose and attitude.

Michelangelo Buonarroti, with Leonardo da Vinci and other artists of the Renaissance, investigated the human body, personally dissecting cadavers. The results of Michelangelo's discoveries are clear in some sketches and especially in masterpieces such as the Sistine Chapel. Apparently, one of the most beautiful écorché of Clemente Susini kept in La Specola was sometimes nicknamed '*The Superficial lymphatic system in Michelangiolesc pose.*' It shows the superficial lymphatic vessels in a male subject but the elegant pose of this statue, and others similar to it, recall some of Michelangelo's masterpieces of the Higher Renaissance and Mannerism such as *The Creation of Adam* in the Vatican's Sistine Chapel or the sculpture of the *Twilight* kept in the New Sacristy of the Basilica of Saint Lawrence in Florence. The inspiration for the pose of these figures was clearly taken from the classical past, a key point of the Italian Renaissance. The pose of Adam and Noah, in the Sistine Chapel, recalls the prototype of the ancient river god represented by Phidias' (c. 480–430 BCE) statue in the west pediment of the Parthenon.

It is an acknowledged fact that the arts tend to exert an influence on each other. In fact, the Florentine anatomical collection was not only appreciated by physicians and scientists but even Antonio Canova, a contemporary of Susini, is reported to have bestowed particular praise on his écorché of the superficial lymphatic system.[12]

On the other hand, as an artist Susini could not avoid being also influenced by the Neoclassical taste where the predominant perception of beauty was the smooth, refined and sensual expression portrayed by those such as Canova who, ironically, was apt to cover his statues with a fine layer of wax in order to soften the coldness of marble.

Pathology, the Representation of Deformity

The versatility, flexibility and above all its extraordinary mimetic qualities and the fact that it so remarkably resembles human flesh, explains why wax was used to create funeral masks, effigies and especially anatomical models. Collections of human remains and artificial models have contributed considerably towards enhancing progress in

12 Martini, 1895, cited by Castaldi 1947: 48.

many scientific fields. But, are medical collections of these wax models still relevant today? Or are we keeping them, as has often been suggested, only for historical and artistic reasons?

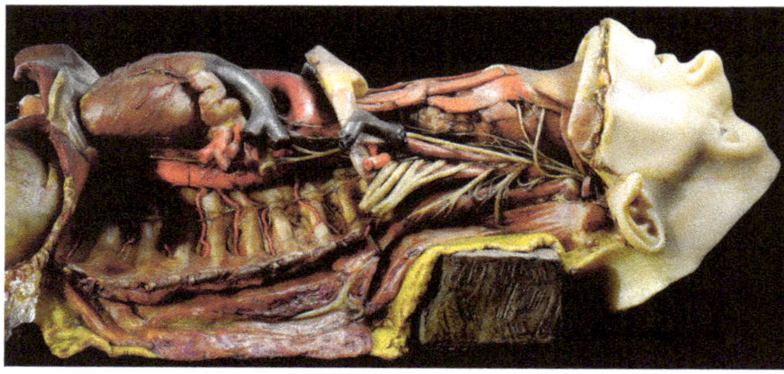

Figure 5: Joseph Towne, *Dissection of head and neck.* 1851, Gordon Museum, Kings College, London. Photo: Roberta Ballestriero and Owen Burke.

One of the fields in which the use of wax has been indispensable and possibly unrivalled is the representation of pathological anatomy. Even today there is probably no other material that can be used to reproduce the pathologies of the human body in such a realistic manner. As disease began to be studied scientifically it was described by means of drawings, paintings and sketches. However, as with normal anatomy, wax appeared to be the most suitable material to faithfully record the observations. This was a new representation of pathology. These models were incredibly realistic and life-like and anticipated the role of photography in recording pathologies belonging to the past centuries. It was, in fact, only in 1839–40 that photography (Daguerreotype) was applied to medicine thanks to the French physician and cytologist of the Charité Hospital in Paris, Alfred François Donné (1801–1878).

Although Zumbo was the first to represent the decay of the body caused by the plague and syphilis in his four gruesome compositions (see above), it was in Bologna and then in Florence where wax models were employed for the demonstration of pathological anatomy and dermatology, often through the technique of moulages.[13] Currently, more than 100 of these models are kept in the Museum of Pathology at Careggi Hospital, University of Florence. They are the wonderful works of artist Giuseppe Ricci and two other modellers from "La Specola" workshop, Luigi Calamai (1800–1851) and Egisto Tortori (1829–1893) and are of special interest

13 According to the Oxford Dictionary 'Moulage' is a cast or impression, especially of a person or a part of the body; the process of making a cast or taking an impression.

because from an artistic point of view the results are different from La Specola's previous works.

Prof. Gabriella Nesi has pointed out that a wax modeler reproducing a pathological condition develops a special relationship with the patient studied. Most of the workshop artist's time was spent modelling hospital inpatients, each a unique pathological case requiring their full attention with each such intimate contact resulting in a special rapport between the patient and the modeler.[14] Creating such replicas was incredibly useful for recording the improvement or deterioration of the condition. This process was very different from those used in the representations of normal anatomy that were accomplished using molds and several different cadavers to create an accurate model, but generating an impersonal general anatomy.

'Medical Humanities' at the Gordon Museum of Pathology in London

The first serious attempt at introducing the art of modelling anatomical works from coloured wax in England was made by the sculptor Joseph Towne (1808–1879). Previous attempts had been made, although all noteworthy anatomical wax models that had been exhibited had been brought from abroad.[15] In England cadavers were more readily available than in Southern Europe and were habitually used in anatomical studies. The Murder Act of 1752 introduced the punishment, post-mortem for murder, of anatomization and dissection allowing the College of Surgeons to obtain six criminal bodies a year for dissection. During the 18th and 19th centuries, the continuous stealing of corpses and also murders – *Body snatchers* or *Resurrectionists* – led to the promulgation of the Anatomy Act in Britain in 1832 and similar measures in the United States in subsequent years. These decrees, which were amended and refined over the years, recognized the necessity of corpses for medical education and research and sought to control theft by making more bodies, both the unclaimed ones of the poor and sick and those donated by family members, available for the study of anatomy. Following the regulation of cadavers' use an attempt was made to create anatomical models accurate enough to replace them for teaching purposes.

Joseph Towne, the first British wax anatomical modeler, worked at Guy's Hospital Medical School in London from 1826 until his death. He is said to have created

14 Personal communication with Prof. G. Nesi, Careggi, Florence, May 2017.
15 Wilks and Bettany 1892: 415.

Figure 6: Joseph Towne, *Lateral view of dissected male body.* Mid-19th century, Gordon Museum, Kings College, London. Photo: Roberta Ballestriero and Owen Burke.

approximately 1000 anatomical and pathological models from coloured wax, some of which were sent to India and Russia, amongst other countries (Fig. 5–6).[16]

The Gordon Museum houses the largest collection of Towne's models of the internal structures of the human body and moulages of the skin conditions of patients sent to him by Thomas Addison (1795–1860), the foremost authority on skin diseases of his day. The museum still preserves the information of the individual patients and each of the specimens is fully documented with all the relevant details of his/her condition and complete medical history. All the dermatological moulages are derived from patients treated at Guy's Hospital (Fig. 7–8).[17]

The dermatological moulages were considered and treated as if they were real specimens and for this reason they were exhibited in glass containers similar to those used for human specimens.[18]

16 Ballestriero and Richardson 2014: 1.

17 Founded in 1721 by Thomas Guy, it was originally established as a hospital to treat the "incurables" that St Thomas's Hospital refused to accept. The hospital was known for the care of "the incurably ill and the hopelessly insane."

18 In practical terms the models are heavy, solid, thick and therefore particularly suited to frequent handling by physicians during teaching demonstrations.

After 190 years since their creation Towne's models are still used for didactic purposes and before the nineteen seventies dermatological moulages were frequently used in classes. Since then, dermatologists at the Gordon Museum began to use the models more systematically and, nowadays, on Mondays some dermatologists teach classes at the museum using Towne's waxes. At times, some students seem to prefer using the wax preparations as opposed to studying the actual patients as in this way they are able to spend more time to freely observe the features of the disease (and the models are not contagious). At the Gordon Museum students interact with the wax moulages as if they were real people and only afterwards they begin to see patients in clinic. As Jonathan White, Honorary Senior Lecturer at Kings College London suggested, students enjoy role-playing clinical cases in a friendly environment and it has been observed that, unlike those who use only books and two-dimensional images, the ones who also study the models subsequently show greater empathy towards patients.[19] The aim of this program is to enhance specific skills in health-care students to improve social cognitive abilities, especially empathy. Using artworks as 'patient surrogates' is an ethical way to practise these skills, with the added benefit of not disturbing sick patients.

The convenience of having accurate and realistic pathological models to readily allow the detailed study of a disease throughout its various stages resulted in the rapid spread of moulages throughout Europe. The English collection is not only interesting from a historical point of view but it is still relevant to third world countries where some of the diseases, represented by moulages, are still observed. Today, thanks to them we are still able to see, in three dimensions, diseases such as smallpox, a disease that was officially declared eradicated on May 8, 1980, by the 33rd World Health Assembly.[20]

These models are a useful tool that remind us of the dangers of past diseases. In fact, in recent years, in Western countries we have seen the return of some long-since eradicated, or nearly eradicated, diseases.

For example, reports by Public Health England on the rise of syphilis were of concern. In 2015, nearly 3,000 cases were diagnosed in the capital, accounting for 56% of all cases in England (5,042).[21] Furthermore, in 2017 the UK witnessed a 20% increase in cases of syphilis compared to 2016 with more than 7,000 cases reported to Public Health England in 2017.[22] Even if most cases of syphilis can be cured with

---

19 Ballestriero and Edwards 2017: 16.

20 Smallpox was the first disease to have been fought on a global scale. This disease caused repeated large-scale epidemics and in the 18th century smallpox was the cause of death of more than 400,000 people in Europe each year Eradication of smallpox is considered the biggest achievement in international public health to date. In www.cdc.gov/smallpox/history/history.html.

21 www.gov.uk/government/news/phe-publishes-report-on-syphilis.

22 BBC News, 5th August 2018, in www.bbc.com/news/health-44368741.

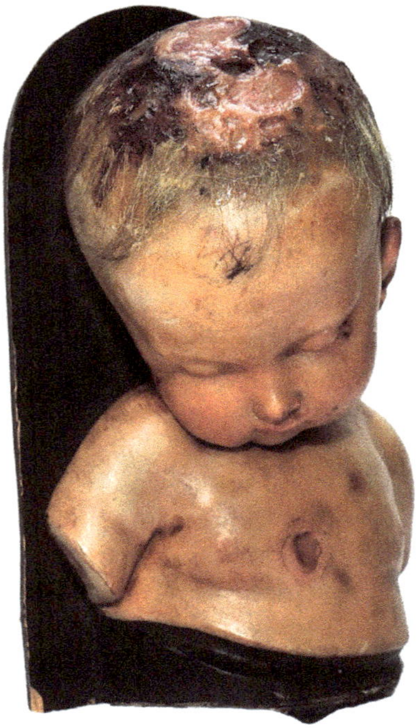

Figure 7: Joseph Towne, *Child with congenital syphilis.*
Mid-19th century, Gordon Museum, Kings College, London.
Photo: Roberta Ballestriero and Owen Burke.

antibiotics, the moulages are a dire warning of how serious the disease can become if left untreated.

Other infectious diseases that appeared to have been under control are also becoming more prevalent, due in part to the continued weakening of vaccination coverage perhaps caused by the surge of Anti-Vax movements. The latest news on the spread of measles in first world countries is troubling. According to the World Health Organization, over 82,500 children and adults in the WHO European Region were infected in 2018 and 72 died. To put this into context, this number of cases far exceeds the annual totals for every year of this decade; the number of cases in 2016 was the lowest with 5273 while by 2017 had already risen to 23,927.[23]

Even more recently, the number of measles cases in the United States in 2019 has surpassed the highest number on record since the year 2000 when the disease was declared eliminated.[24] According to the CNN *Heath* website analysis of data

23 Copenhagen, Denmark, 20 August 2018.
24 CNN, https://edition.cnn.com/2019/04/24/health/measles-outbreak-record-us-bn/index.html.

from state and local health departments, there have been 681 cases across 22 states.[25] It is easy to underestimate the seriousness of measles but, in fact, of every 1000 cases it will kill 3.[26]

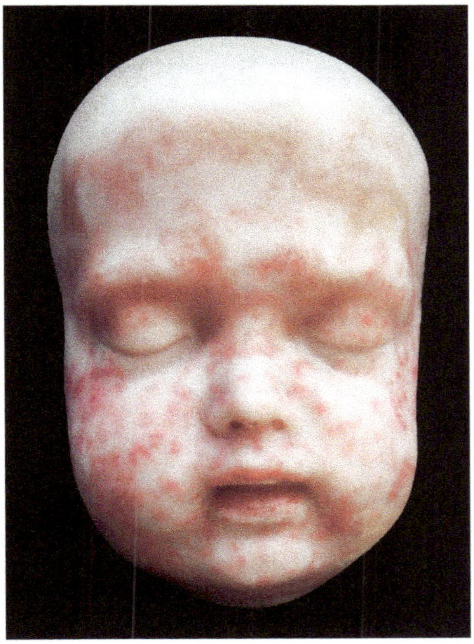

Figure 8: Joseph Towne, *Child with measles. Mid-19th* century, Gordon Museum, Kings College, London. Photo: Roberta Ballestriero and Owen Burke.

In the broad subject of 'Medical Humanities', a relatively new and not very well-defined field of knowledge, there are numerous situations where the arts and medicine work in unison to improve learning, research and medical care. While 17th century scientists explored methods for the preservation of cadavers to be used in teaching and research, different materials were used to create beautiful models, thus leading to the birth of "artificial anatomy." Among them, wax, for its resemblance to human flesh, allowed for the creation of extremely realistic and long-lasting models of normal and pathological anatomy. Objects such as dermatological wax moulages have often been kept more for historical or artistic reasons than for their didactic value. However, they are still valid teaching aids that allow students to study, in close proximity and with no pressure that might arise from

25 CNN, https://edition.cnn.com/2019/04/24/health/measles-outbreak-record-us-bn/index.html.
26 Galassi 2017: 74.

an examination of a real patient, a variety of conditions, both contemporary and from the past. It is important to appreciate that these scientific collections are still relevant today and that their educational value from a medical teaching perspective is undiminished. The Gordon Museum in London recognises this and uses the important educational potential of its antique and modern scientific collections to enhance the observation skills and empathy of health-care students.

Ariella Minden and Paolo Savoia

The Body between Life and Death: Berengario da Carpi and the Anatomical Image of the Sixteenth Century

Introduction

Jacopo Berengario da Carpi is among the most famous anatomists of the so-called "Pre-Vesalian" era. His rise to fame tells the tale of a careful cultivation of powerful patrons accompanied by a lifelong curiosity surrounding the inner workings of the human body. As a protagonist of the 'anatomical Renaissance' that unfolded over the course of the sixteenth century, Berengario advocated for the primacy of touch and sight in medical education and clinical practice as described extensively in his printed works. It is impossible to understand the novelty of sixteenth-century anatomy in any other way than looking at the diversity of visual, cultural, and intellectual stimuli that leant themselves to a new, varied, dynamic, and innovative approach to the human body. Berengario's figure and work stand precisely at the crossroads of the history of medicine, the history of art, and a visual and material history of death and religious ritual. This essay looks at Berengario's life to understand the social, political, cultural, and visual contexts of Renaissance anatomy. It is precisely for these reasons that this figure serves as an important case study for understanding the medical humanities.

This chapter explores how the son of a barber-surgeon entered one of the most prestigious universities of the Renaissance, rising up the ranks to become a professor of surgery at Bologna, a position that he would go on to hold for twenty-five years. Straddling these two diverse paths of surgical training, we show how Berengario absorbed and thought critically about medical authorities of the recent and ancient past by translating the knowledge he gained through private dissections and clinical practice into commentaries and manuals that were widely circulated throughout Europe. After providing a sketch of the surgeon's prolific career in order to situate Berengario in the broader landscape of medical education in and out of the university as well as in the vast world of private dissections performed by the teacher with select students, we will go on to consider Berengario's texts and how he utilized personal anecdotes and artisanal analogies to style himself as an expert. Finally, we will analyze the woodcuts across the texts to reconsider the role and development of

---

This work is the outcome of interdisciplinary collaboration, a shared approach, and fruitful discussions between the two authors. Ariella Minden authored sections 1, 3 (partim: pp. 178–179), 4, 5; Paolo Savoia authored sections 2, 3 (partim: pp. 180–181), 6, 7.

https://doi.org/10.1515/9783110788501-009

medical illustration two decades before the publication of Vesalius's *Fabrica* (1543) while also looking at the myriad technical, cultural, and religious practices that were used to depict life, death, and the human body. We argue that the opening of human bodies in a medical context – as well as the visual representation of thereof – must be seen in continuity with a series of religious, ritual, and legal practices that account for a certain ease in handling dead bodies. Such ease is not equal to indifference, but on the contrary signals a complex system of honor that concerned the criminal and the saint, the illustrious and the vile body. The existence of a specific skillset tied to medical centers on the Italian peninsula and ways of handling dead bodies in their crude materiality allowed for particular changes to occur and subsequently be disseminated across Europe over the course of the sixteenth century through the proliferation of printed manuals.

Through an exploration of cultures of medical practice and empirical discernment surrounding one of the most celebrated moments in the history of Western medicine, this chapter presents a rich case study the humanities.

Surgeon, Teacher, and Celebrity Doctor

Jacopo Barigazzi, later to be known as Jacopo Berengario da Carpi, was born around 1460 in the small, but culturally vibrant court city of Carpi, near Modena.[1] His father Faustino, a relatively well-known and well-esteemed barber-surgeon, was Jacopo's first teacher, apprenticed to his father from the time he was a young boy. Faustino had a successful practice and was known to intervene in difficult cases including that of one Bernardino of Vicenza, who was impaled in the forehead by a billhook.[2] The barber-surgeon's career was not confined to just Carpi, but Faustino was seconded throughout the Emilia to perform similarly technically difficult and delicate operations. Berengario helped him treat patients from a formative age.[3] His early education was entirely practical, and it was in his father's workshop that Berengario began to familiarize himself with tools, flesh, and bones.

This was a common for surgeons in the late-fifteenth century. Training could follow one of two paths: either that of a university education with a strictly mandated curriculum or the apprenticeship system which eventually qualified the prac-

1 For a discussion of Berengario's name see Putti 1937: 7–11. This early work on Berengario written by the famous Bolognese surgeon and collector Vittorio Putti is itself a monument to the medical humanities.

2 Savoia 2018 and 2019: 27–54.

3 Berengario da Carpi 1522: 2r (English trans. Lind 1959: 35); Putti 1937: 12–13.

titioner as a barber-surgeon.[4] It was only from the late Middle Ages onwards that surgery became an academic discipline taught in the most important medical schools on the Italian peninsula, namely: Bologna and Padua. In these two cities, university educated surgeons could sit on the elite Colleges of Medicine which had, among other mandates, the task of examining and granting licenses to barber-surgeons to perform certain surgical procedures such as bloodletting.

In Bologna, the first written record of a professor of surgery appeared in 1388, but it was not until 1405 that there was a consistent holder of the position. Even at that, it is clear that the barber-surgeon trajectory remained far more common, where, of the 65 medical degrees granted between 1419 and 1435 only one was in surgery. This persisted into the sixteenth century as indicated by an annotation on the Bolognese *rotulus* of 1512 where one of the *Riformatori dello Studio*, the body responsible for appointing professors, notes that despite surgery not being an honorable discipline, due to its vast popularity especially among foreign students they had to find a replacement for Berengario while he was on secondment.[5]

This constitutes the background necessary to understand what was at play in the development of a distinct branch of Italian anatomy conducted somewhere between the "workshop" and classroom. Renaissance surgeons formed a complex and composite spectrum of professions ranging from the barber-surgeon to the licensed practitioner, from the itinerant seller of remedies and the bonesetter or lithotomist to the university-graduate physician specialized in surgery. In European cities the typical sixteenth-century institutional arrangement could take three forms: the first was a division between a College of Physicians, a College of Surgeons (graduate or otherwise Latin-reading surgeons) and a guild of barber-surgeons (this was the case of Venice and some northern European cities); in the second one, learned surgeons were part of the College of Physicians, and non-graduate surgical practitioners were part of the barbers' guild or independently licensed by the College (the case of Bologna and Padua); and the third model, most widespread north of the Alps, was a division between a guild of barber-surgeons and a College of Physicians, with surgeons sharing their practice with barbers.[6]

Berengario himself, towed this line between action and learning, the hand and the mind. The surgeon fondly recalled his friendship with Alberto Pio, *signore* of Carpi, to whom he dedicated his *Isagoge Breves* of 1522. The court of Carpi was the site of great erudition and intellectual exchange with perhaps the most famous

4 For an ample discussion of these two paths to becoming a surgeon in Italy see: Siraisi 1990: 48–77 and 153–186; Pesenti 1978: 1–38; Palmer 1979: 451–60; Gentilcore 2006: 182–87; Conforti 2008: 323–340; Bartolini 2015: 83–100.

5 Siraisi 1990: 63; Ferrari 1987: 50–106.

6 On the institutional settings of surgery in England see Pelling 1998: 203–229; Chamberland 2009: 300–332. For Edinburgh, see Dingwall 1995: 34–98. For Paris, see Gelfand 1980: 21–27; Guerrini 2015: 25–30. For the Netherlands, see de Moulin 1988: 46–94.

courtier being Aldus Manutius, the prolific printer, who served as a tutor to the young prince in 1479. Berengario recalls the Muses he and Alberto pursued together under the tutelage of Aldus and notes an episode where the two dissected a pig.[7] However, his style, command of Latin (or lack thereof), and that neither Aldus nor Alberto ever mention Berengario in their writings betrays this aspect of his biography. Nevertheless, Berengario's inclusion of these details in his dedicatory letter is noteworthy as part of his program of self-fashioning aimed at endearing himself to, and legitimizing himself within humanistic circles at Bologna.[8]

In 1480 Berengario moved to Bologna to pursue a degree in arts and medicine at the University, from which he graduated on August 4, 1489. The medical curriculum there, as in other major universities at the time, was not exactly forward looking, and medical teaching was still based on a few texts, including a collection of Galenic writings and Avicenna's *Canon.*[9] From the early fourteenth century, however, human dissections were becoming more commonplace in medical pedagogy as Mondino de' Liuzzi describes in his *Anatomia* of 1316. However, it was not until almost a century later that dissections actually came to be institutionalized at Bologna. The statutes of 1405 and 1442 detail the acquisition of cadavers and stipulate that at least one anatomical demonstration was to take place each year. Initially, the professor was supposed to obtain the cadaver and the students were to pay for it, but by 1442 civic authorities were charged with procuring corpses and a new rule was added that bodies for dissections had to come from criminal executions and must be foreigners, meaning at least thirty miles from Bologna. After the dissection took place the cadavers had to be given a proper burial at the expense of the professors and their students. Between 1490 and 1543 a new wave of printed texts, often illustrated with detailed images rendered in close collaboration with artists also came into being, pushed forward by the humanistic enterprise of the publication of the complete works of Galen in a new translation.[10] Finally, the importance of dissection was definitively sanctioned by the building of permanent, elaborate anatomical theaters which crystalized the practice of public dissection's intellectual and spectacular importance.[11]

After graduating, Berengario left Bologna and spent the 1490s in Carpi, practicing with his father. Berengario accrued modest fame during this period, above all for the use of mercury as a cure for syphilis, the terribly painful and disfiguring

---

7 Berengario da Carpi 1522: 2r (Engl. transl. Lind 1959: 35).

8 Putti 1937: 14. Putti among others think that the timeline is implausible because Berengario would have been far too old; however, French speculates that given the apprenticeship, the date at which he would have received Latin education could have been later. French 1985: 44–45.

9 On the evolution of the teaching of medicine in Italian universities and the medical curricula of the time see Siraisi 2001: 1–10; Agrimi and Crisciani 1988.

10 Fortuna 2019: 437–452.

11 See Carlino 1999: 170–188; Ferrari 1987: 50–106; Park 1993: 1–33; Siraisi 1990: 78–114.

disease that gripped the Italian peninsula from 1494 onwards.[12] It is in the 1502 *rotulus* that the first surviving reference is made to Berengario as lecturer in surgery, a remarkable achievement for a non-Bolognese graduate given the strict hierarchy and communal control governing teaching appointments.[13] Soon after obtaining his professorship, Berengario married a Bolognese noblewoman, and in 1506 was conferred Bolognese citizenship by Pope Julius II. As already noted, Berengario's teaching attracted a large number of students from across Europe who flocked to the city to watch the professor perform dissections.[14]

Between 1508 and 1512 Berengario was put in charge of a special commission that served to enforce public health measures during a plague outbreak in the city. Berengario's rise to fame continued throughout central Italy thanks to both his popularizing of a treatment for cranial fractures that his father had pioneered called the '*cerotto umano*' – a special powder made with human bones – alongside his continued care for patients stricken with the French pox. His reputation as more or less a celebrity doctor took him to Milan, Florence, and Rome in the service of some of the most important aristocratic families on the Italian peninsula.[15]

In 1517 war broke out between the Medici and the Della Rovere families over possession of the Duchy of Urbino. Lorenzo de' Medici endured a critical head wound at the battle of Fossombrone and was subsequently transported to Ancona for medical treatment. The pope sent an entourage of the best physicians which included Berengario. It is unclear whether Berengario was directly involved in the trepanation of Lorenzo's skull, but he was most certainly took part in the postoperative care, a delicate matter, and clearly effective, with Lorenzo having made a full recovery only one month later.

Berengario once again returned to Rome and was there between the end of 1525 and the beginning of 1526 for a duration of four or five months. It was this stay in the Eternal City that both Vasari in his "Life of Raphael" and Benvenuto Cellini in his autobiography recalled. The surgeon was famously summoned by the pope in order to treat an outbreak of syphilis among the cardinals, which he did, once again, using mercury. Cellini wrote that the treatment did more harm than good, but that Berengario had the good sense to leave before any of the adverse effects started to manifest.[16] Even more bitingly, the 17th-century physician Bernardino Ramazzini stated that: "He most certainly had a much better knowledge of potion making than alchemists, with the real

12 See Arrizabalaga, Henderson and French 1997.

13 On the professional structures of guilds and medical academic associations see Naso 1983; Park 1985. For the make up of the Bolognese professorate see Grendler 1999: 475–485.

14 Ferrari 1987: 50–106.

15 Putti 1937: 39.

16 Cellini 1728: 33 (Cellini 1980: 117–118): "Egli era persona molto astuta, e saviamente fece andarsene di Roma; perché non molti mesi a presso tutti quelli che aveva medicati si condusson tanto male, che l'un cento eran peggio, che prima sarebbe stato amazzato, se fermato si fussi."

transformation being that he turned mercury into gold, with such rare propensity and entirely unseen in our own time."[17]

In 1527, Berengario lost his university post which he had held for 25 years. In all likelihood he did not leave the post voluntarily. The abrupt suspension of his salary points to some kind of condemnation or perhaps even threat of exile.[18] After leaving his chair and shortly before his death in 1530, Berengario became court surgeon to the Este in Ferrara. A document dated November 25, 1530 states: "and for 24 *lire* 1, 14 *soldi* for the funeral rites of master Berengario da Carpi, physician, and he had been buried in San Francesco."[19] On that day the last wishes of Jacopo Berengario da Carpi were observed and he was buried in the Franciscan monastery.[20]

In this rich and varied biography, we have seen how the deft maneuvering among intellectual and courtly circles coupled with a vibrant and successful clinical practice allowed Berengario to cultivate a storied career and garner immense fame in his own lifetime as a Renaissance physician.

The Author

Most of the information about Berengario and his life's work comes directly from the surgeon himself and his printed body of writings. As a prolific author, his medical manuals, commentaries, and treatises are littered with autobiographical details and both personal and familial anecdotes that serve as testimony to his innovative medical practices. Berengario's first printed work dating to 1514 was an edition of Mondino de Liuzzi's *Anatomia* which served as the nucleus for his most ambitious project to be published seven years later in 1521, a comprehensive commentary on the text.[21]

Prior to his magnum opus, however, in 1518, Berengario published a treatise on cranial fractures. Riding on the coattails of the fame he had recently garnered from his high-profile treatment of Lorenzo de' Medici's head wound, the book provides a practical and expansive guide to the treatment of head injuries.[22] The text walks the

---

17 Ramazzini 1745: 28 (first edition 1700).

18 Tiziano Ascari and Mario Crespi reported that some believed that in 1527 Berengario was charged of heresy and exiled from Bologna by the Inquisition on the account of his naturalistic treatment of the watery substance flowing from the crucified body of Jesus Christ (see pages 30–31 below), but they do not provide any source: see Ascari and Crespi 1964. On this point see Arieti 1999: 428.

19 Di Pietro 1971: 41–42: "e de avere adi 24 lire 1, soldi 14 per le esequie de maistro Iacomo da Carpi medicho e fu sepulto a San Francesco la Compagnia."

20 Martinotti 1923: 1–11.

21 Berengario da Carpi 1514.

22 The success of the *Tractatus de fractura calvae sive cranei* is attested to in its seven reprintings between 1518 to 1728; see Lippi 2017: 1–5.

reader through the whole process from initial diagnosis to surgical and non-surgical interventions to postoperative care, which included suggested changes to diet and perspective adjustments to sleep and exercise. There are also eight woodcuts towards the end of the treatise which were integral to his guide to the tools of cranial surgery and their appropriate usage.

As with all of Berengario's writings, he engages with and comments upon ancient sources such as Galen and Avicenna, but ultimately what gave his works their authority and led to their long afterlives was his evocation of dynamic communities of medical practitioners who sought to create proprietary technologies and treatments to elevate their status and renown. We gain an appreciation of the importance of the training he was afforded in his father's workshop. In particular, the knowledge and skill he needed to employ and popularize the "*cerotto umano*," a technology for which his father was offered a large sum of money, but instead chose to leave to his sons as "precious inheritance."[23]

The book also provides fascinating insight into the interfaith dialogue that was possible in the medical discipline in a way that in other parts of daily life was much more limited. There was a strong community of Jewish doctors practicing in the Emilia as we know from the abundance of luxurious Hebrew medical manuscripts produced in the region at this time as well as documentary sources recording the presence of and treatment by Jewish physicians. In Berengario's case he refers to a certain Jacob, a Jewish doctor in Ferrara who treated many noblemen including Ercole d'Este himself. Berengario characterizes this man as a "dear friend" of his father and writes that he created highly effective pharmaceutical cures to treat certain head traumas. He did this, however, under the cover of secrecy and Berengario, in need of these recipes, one day followed the doctor into a field in order to spy on him and see which herbs he would collect.[24] This anecdote not only reveals the interfaith exchange that could occur in medical practice, but also demonstrates the competition and secrecy that surrounded the propriety and inherited knowledge of these practitioners.

After this treatise, having likely met the polymath reformer Ulrich von Hutten during his visits to Bologna between 1512 and 1517, Berengario became intimately involved in the publication of von Hutten's first-person account of the French disease, *De Guaiaci medicina et morbo gallico*. The work, which was first published by Girolamo Benedetti, received Berengario's editorial support through the subsequent editions printed into the 1520s.

23 Berengario da Carpi 1518: 88: "isto cerato vidi patri meo offerri magnam pecuniae quatiatem: et ille totaliter renuit dicens hoc ceratum non esse dandum alicui nisi propriis filiis tamquem si esset haeretditas praeciosa."

24 Berengario da Carpi, *Tractatus* 1518: 58: ". . . Iacob haebraeum cui pater meus erat amiccissimus . . . ."

It was then in 1521 that Berengario published his tour de force *Commentaria super Anatomia Mundini.*[25] Dedicated to Cardinal Giuliano de' Medici, later Pope Clement VII, and printed by Girolamo Benedetti, the towering tome of 1056 pages superficially follows the traditional structure of the scholastic *commentarium* with an exposition of the content of each chapter that then proceeds to introduce the *questiones* and *dubia*. Berengario wrote in this conservative format," but his commentary on Mondino's anatomical treatise, as with many other works written in this genre in this period, are full of digressions, experimental reports, and true departures from the Aristotelian-Galenic orthodoxy that was supposed to have ruled the Medieval and early modern universities.[26] The text is indeed much more than simply an Aristotelian exercise, where in addition to these stalwarts of the genre Berengario included *digressiones*, which allowed the surgeon to depart from certain conventions by way of clinical case studies and experimental reports. In order to resolve age old spats, Berengario performed a series of experiments, among the most famous of which was the one he conducted on fetal bladders in order to better understand secretion in utero.[27]

While this investigational approach to anatomy was revolutionary in many respects, the book was unwieldy in its heft, and likely for that reason was not a commercial success. The following year on December 30, 1522, with a different printer, Benedetto Faelli, Berengario published a much-condensed dissection manual, the *Isagoge Breves.*[28]Although no explicit reasons have been found to account for the change in printer, Faelli's biography would imply that he brought a certain business acumen to the condensing and repackaging of Berengario's work as a book that was consciously and explicitly didactic, an essential for any student of anatomy. Faelli began his career not as a printer, but as a book seller with close ties to the Benedetti family, where he had a formal agreement to sell books printed by Francesco 'Platone' Benedetti, Girolamo's uncle, throughout the 1480s.[29] It was only in the 1490s that Faelli set up a press of his own and surviving documentation demonstrates that commercial success was of paramount concern. For instance, a contract dated May 22, 1499 between Faelli and Filippo Beroaldo the Elder, a professor of rhetoric and poetry at Bologna, stipulates that Beroaldo was to lecture on Apuleius's *Golden Ass* to coincide with the publication of his commentary on the

25 Berengario da Carpi 1521. The book contains 21 figures: 6 figures of the abdominal muscles; 3 of the vessels of the members; 3 of the female genitals; 1 of the vertebrae; 5 of the muscles of the whole body; 2 of the whole skeleton; 1 of the bones of the hand and the foot.

26 Siraisi 2007; Nutton 2019: 472–486.

27 French 1985: 49–52. See also Agrimi and Crisciani 1990; Park 1999: 347–368; Crisciani 2005: 297–324.

28 *Isagoge breves* was the most successful book by Berengario and was reprinted five times: 1522, 1523, 1535, 1660 (published in London and translated into English), 1664 (reprint of the English version).

29 Sorbelli 1929: 47.

text in order to maximize the number of copies sold. The contract also specifies that 1,200 copies of the book were to be printed, a strikingly high number for the time.[30] From this archival evidence, we gain an appreciation of Faelli as a printer concerned with effective marketing strategies to promote sales.

It was this business savvy that the printer likely brought to the production of the *Isagoge* as well. Instead of being a burdensome commentary, the reader is presented with a clear, 144-page dissection manual that moves through the human anatomy in the order in which a dissection would be performed, giving tips on how to cut and best see certain anatomical features along the way. The book was handy and easy to use, something that Berengario saw as central to his task as author in order to compensate for other cumbersome works that were not practically arranged, noting of such precedents that: "The authors seem to have borrowed fables from other volumes instead of writing a genuine anatomy."[31] The success of this approach is evident given that less than one year later a second edition of the *Isagoge* was published on July 15, 1523.

In the year before his death, Berengario edited an important new translation of a collection of anatomical writings by Galen published in Bologna and accorded privileges by both Pope Clement VII and Emperor Charles V.[32] This volume is a testament to Berengario's ability to align himself with some of the most powerful patrons on the Italian peninsula, where beyond the imperial and papal privileges, Ercole Gonzaga, to whom the book is dedicated, also sponsored its printing. In his dedicatory letter, the surgeon recalls a conversation on anatomy that took place in the company of the philosophers and philologists Pietro Pomponazzi and Leonardo Bonamici. This kind of prefatory letter once again confirms Berengario's courtly and intellectual ambitions and his ability to maneuver among a wide range of social circles in his capacity as surgeon and professor.

These printed works, in particular the *Treatise on Cranial Fractures* and the *Isagoge*, enjoyed long afterlives and reprinting in some of the larger printing centers. The *Treatise on Cranial Fractures* was reprinted at least two more times, once in Venice in 1535 and again over a century after its initial publication, in Leiden in 1629. The *Isagoge* enjoyed similarly enduring success, where a smaller and more economic edition was printed in Strasbourg in 1530, and an English translation was made in 1660 in London and reprinted again in 1664.

---

30 Archivio notarile di Bologna, atti del notaio Agostino Landi, 22 Maggio 1499, transcribed by Sorbelli 1929: 61.

31 Berengario da Carpi 1522: 2r (English trans. Lind 1959: 35): "quos eorum authores ad alia transferentes volumina fabulas potius quam Anatomiam scribere videbantur."

32 Galen 1529.

The Artifex

Three questions to follow up on this overview of Berengario's printed output are: where did Berengario accumulate this knowledge? How did he understand his task as an author? And how did he establish his authority as an expert qualified to write on such subjects? The answers to these questions are intimately intertwined. Berengario sees and styles himself as an expert precisely by way of his manual, experiential, and experimental practice of medicine and surgery, not yet couched in the university and sometimes even at odds with the institution. It was through his training and the skills that he acquired by way of his apprenticeship with his father, Faustino, as well as his numerous private dissections that allowed him to write such remarkable, even revolutionary works. The surgeon referred to his practice as "*anatomia sensibilis*", an anatomy guided by the senses, and his role as that of the "*artifex*," craftsman.[33] In this framework, manual dexterity and judgement sit comfortably next to knowledge of ancient authorities. This also meant that the act of writing was an act of translation from embodied knowledge and artisanal epistemologies to words on a page.[34] Berengario makes this explicit in the preface to his *Treatise on Cranial Fractures* where he writes that because of the nature of the discipline most of medicine cannot be translated into writing nor expressed in words, instead practice is cumulative and comes from years of training and first-hand experience.[35] This point is reiterated in the section of the treatise that deals with the detection of symptoms in order to diagnose a range of injuries. In concluding this chapter dedicated to distinguishing one trauma from another when there are overlapping symptoms, Berengario writes:

> I deem, however, that the differentiation of such symptoms is very difficult and is only known by experts. They are symptoms that cannot be described in writing and can only be understood by he who possesses ingenuity, introspective, analytical, and synthetic capabilities as well as lots of experience. There are many things that the doctor knows that are not possible to put into writing as is seen every day.[36]

In concert with his view of writing as translation, Berengario was a strong proponent of sight and touch as the guide of the surgeon, anatomist, and physician, chiding his predecessors for blindly following medical authorities without using their own cultivated skills. He is so adamant about this point that in the introduction of his commentary on Mondino's *Anatomia* he thrice reiterates it over the course of

33 See French 1985: 57.

34 The literature on artisanal epistemologies is ever expanding; however, for the most cogent definition see Smith 2004: 3–30. See also Long 2015: 840–847; Gooday 2008: 783–795; Struhal 2017: 501–513.

35 Berengario da Carpi 1518: 3v: "Magnifico ego in medico lucidem nec calamo scribi nec lingua proferri potest . . . ."

36 Translations are our own unless otherwise stated. Berengario da Carpi 1518: 36r.

three pages. The first time he tells the readers that they should not believe everything that they hear or read, but that they should be verifying older assertions through sight and touch.[37] The second proclamation of the primacy of the senses instead goes on to codify this form of empiricism as its own epistemology whereby understanding human anatomy is something that is to be accomplished through manual acts and demonstrations of individual members with dissection described as a "science of understanding the members."[38] In the final appeal, Berengario argues for the use of sensory perception specifically cultivated through experience, a recurring topos throughout his works.[39]

To accomplish this empirical task, Berengario was a strong proponent of the private anatomy, markedly different from the annual public dissection. While the public dissection was very often a performance that could be misleading and did not fully allow students to see everything that was happening, the private dissections took place in teachers' houses or in hospitals among a small group of students who actively took part in the procedure and would discuss specific points about organs and structures.[40] For Berengario it was important that the students knew how to handle a cadaver themselves rather than leaving it to the *sector* who at this time would be the one performing the public dissection as directed by the *lector*.

Berengario advises that students dissect as many bodies as possible, viewing anatomy as a composite. In the introduction to the *Commentaria* he uses this point to rail against the public anatomy. He also notes that it is important to see a variety of types of bodies, as every body is unique and must be treated accordingly.[41] The surgeon later reminds his reader that anatomy is not only to be performed on the dead, but should also be observed while undertaking clinical observations of patients with a range of ailments.[42] There are also repeated references to cemeteries, especially in the *Isagoge*, as a place where one should go to best see the bones of the body on fully decomposed corpses. For instance, while discussing the cranial plates he wrote that they "can best be seen in cemeteries, as also other parts of the cranium and all the bones of the body may be seen."[43]

---

37 Berengario da Carpi 1521: 6r: "Et non credat aliquis per solam vivam vocem aut per scripturam posse habere hanc disciplinam: quia hic requiritur visus & tactus."

38 Berengario da Carpi 1521: 6v.: "Alio modo capitur anatomia pro scientia cognotionis membrorum ubi etiam traditur modus operandi cum manu actu & demonstrandi ipsa membra."

39 Berengario da Carpi 1521: 7r.: "Non credat ergo aliquis sibi soli sed communicat doctorum auctoritates & sui ipsius opinionem cum peritis in anatomia si potest & simul sit sensus & experientia super eod quo sit ferm ut quae forte non distinguit unus distinguant forte alii." On touch in early modern medical culture see Maurette 218: 105–124; Pogliano 2015.

40 Martinotti 1911: 30–47; Klestinec 2011: 142–166.

41 Berengario da Carpi 1521: 5v.

42 Berengario da Carpi 1521: 5r.:

43 Berengario da Carpi 1522: 52v53r (English trans. Lind 1959: 139): "quod potest optime videri in cimiteriis, sicut & aliae cranii partes: & etiam omnia totius corporis ossa."

This insistence on touch and the emphasis on the hand is consistent with Berengario's repeated appeal to the authority of the senses as the ultimate proof in moments of discord regarding the structure of certain parts of the body. Here, the subculture of private dissection was directly linked to epistemological innovation. Well before Vesalius, Berengario was indicative of a medical and scientific culture that was ready to embrace sensory evidence.

The *Commentaria*'s audience was predominantly academics. In maintaining the conservative language of the commentary, *demonstratio* meant conclusion; however, Berengario's conclusions often spoke directly to the senses rather than in the resolution of a syllogism, thus undermining the formulaic and familiar. In Berengario's anatomy, to prove was to expose the structure of the organ to sight and touch.[44] Experimental dissection had the potential to solve problems: one example of this was when the anatomist worked to resolve how a fetus secretes in utero. He set out to conduct a "particular" anatomy, taking a fetus and filling its bladder with water from a syringe. In doing so, he noticed that the water seemed to flow through the umbilical cord to the point where it reached the embryonic membranes. In another attempt, Berengario compressed the bladder of a nine-month fetus with his own hands to see if any urine emerged. Finally, he filled up the bladder with water using a syringe inserted through the penis, allowing him to conclude that the fetus expelled urine through the penis, not the umbilical cord.[45]

Berengario's texts, in particular the *Isagoge Breves* were also sure to reiterate the importance of the manual component of anatomy and surgery. The surgeon, and in turn, a dissection or surgery could only be as good as the skill of the hand performing it. The surgeon and anatomist were expected to be familiar with and skilled in working with a range of tools each of which selected for their suitability in excavating the human body. In several points throughout the *Isagoge Breves* Berengario is sure to caution his reader that a 'skilled' or 'practiced' hand is required for the sake of precision, so as to not obscure or destroy any element of the body intended for study.[46] Not to mention, the cutting open of bodies was a cumulative skill. The more experienced the hand, the more an anatomist was able to understand about the body. Berengario is explicit on these points with respect to the dissection of the eye, where he concludes the passage on the delicate nature of this component of dissection by noting that: "A skilled hand seeks ever more difficult things."[47]

The skilled hand was accompanied by judgement, another critical component of good practice as characterized by Berengario. It was discernment that allowed a physician to swiftly and accurately assess the symptoms of his patient, make a

---

44 French 1985: 52–53.

45 French 1999: 110–111.

46 Berengario da Carpi 1522: 38v (English trans. Lind 1959: 109): "docta manu."

47 Berengario da Carpi 1522: 59r.

proper diagnosis, and prescribe a suitable course of treatment. The face was the first point of reference, Berengario elucidates this both in the *Treatise on Cranial Fractures* and the *Isagoge Breves,* where he writes in the latter:

> Knowledge of the face is much prized by the physiognomist. It is also prized by the physician, since you will make the first prognostication primarily from the face of the sick man; for this is helpful in the recognition of many diseases, such as leprosy, consumption, yellow jaundice, cachexia, and the time of menstruation in a woman. In the face are also recognized those who pretend illness, but not always.[48]

Already in his *Treatise on Cranial Fractures* Berengario made a similar point when he wrote that: "It is advisable for a physician to train himself to know these colors so that he knows how to distinguish them otherwise he won't be able to make the assessment. It is only when he is experienced and trained in similar situations in the same way that experts are able to distinguish real gems from fakes and those of low quality."[49] This is not the only time he used this simile, but it gets recycled in the *Commentaria* with slight modifications: the first change is an emphasis on the duration involved in the cultivation of such experience and the second is the addition of the verb to judge (*iudicant*) so as to specify the way in which good gems are distinguished from the bad or even fake.[50]

That such discernment was a shared capacity of both the surgeon and the goldsmith reappears from the perspective of the goldsmith in Benvenuto Cellini's autobiography. Cellini recalls that while Berengario was in Rome, he stumbled upon the goldsmith's workshop. Entering the shop, the surgeon goes onto inspect the works and is taken by "several sketches of little fanciful vases which [Cellini] had drawn by way of amusement." Berengario goes on to commission the vases from Cellini since they were "very different from any that he had seen before." In this short exchange, before Cellini goes on to cast his doubts regarding the efficacy of Berengario's mercury treatments, the goldsmith acknowledges the surgeon as having "great intelligence for *disegno*" and as such a unique appreciation for his craft, actively

---

48 Berengario da Carpi 1522: 40r (English trans. Lind 1959:113): "notitia faciei multum consideratur a physionomo. Consideratur etiam a medico: ut primo prognosticorum:in primis aegri faciem considerabis. Iuvat in cognoscendis multis morbis sicut lepram, pleripleumoniam, icteritiam, cacesiam aliam cachexiam, & tempus menstruorum in foemina. In ipsa etiam cognoscuntur simulantes aegritudinem, sed non semper."

49 Berengario da Carpi 1518: 28r: "Et oportet qui delectur medicus in cognoscendo istos colores: quia quidlibet non cognoscit: sed tantum ille qui ex expertus & excercitatus: similbus sicut experti cognoscunt gemmas bonas a fraudatis & a non bonis."

50 Berengario da Carpi 1521: 5v: "Qui color non cognoscitur in mortius nec a quodcunque: sed a bona extimatiua Medici et longa ipsius experientia hoc etiam cognosci potest: sicut experti circa gemmas iudicant bonas a falsis et illas cognoscunt: similiter experti Medici praenarrata cognoscunt."

situating the surgeon's and the artisan's practice of judgement or discernment in the same realm.[51]

That manual dexterity and good judgement are requirements of both surgical and artisanal practice reappears in how Berengario chooses to translate certain aspects of practical knowledge. Since Berengario perceived these writings as acts of translation, he had to find ways of using language in order to represent his lived experience to his reader. In doing so, at points Berengario chooses to employ artisanal analogies to clarify certain techniques and better describe anatomical structures.

The first time Berengario does this is when talking about trepanation and which size drill is most appropriate to drill the hole that will first pierce the cranium. He outlines the vigorous debates surrounding the topic and the various pros and cons of a thinner or thicker drill to do the job. Ultimately, he concludes that it is the former because it is the same as any other occasion that a hard material – be it wood or stone or bone – is punctured for the first time, reasoning that "it is better to start with a thinner tool to penetrate the bone as is evident from the experience of any other mechanical skill whether working in wood or stone or any other solid body: any artisan always uses a finer and smaller drill first and then a larger one . . . because it does a better job . . . ."[52] Berengario realized that surgeons and sculptors needed to rely on similar skillsets and tools, and thus the solution to a centuries long disagreement was solved by looking across such disciplinary boundaries, where bones were just one of many hard materials that required a similar technique.

In the *Commentaria*, Berengario again returns to the artisan's workshop, but this time in order to explain certain phenomena that would have been difficult to see in either living bodies or by dissection alone, and next to impossible to articulate in words. There are two instances where Berengario refers his reader to a carpenter's workshop. The first time is in reference to how cranial plates are conjoined, where he says that if one would like to better understand this feature of the human skull, they would be best to visit a carpenter and observe dove tail joints which are fused accord-

51 Cellini 1728: 32–33 (Cellini 1980: 117–118): "Aveva questo valente uomo molta intelligenzia del disegno; passando un giorno a caso dalla mia bottega, vide a sorta certi disegni che io avevo innanzi, infra' quali era parechi bizzarri vasetti, che per mio piacere avevo disegnati: questi tali vasi erano molto diversi et varj da tutti quelli che mai s'erano veduti insino a quella età; volle il ditto Maestro Jacomo che io gnene facessi di argento; i quali io feci oltra modo volentieri, per essere secondo il mio capriccio."

52 Berengario da Carpi 1518: 97r: "Quod etiam hoc sit verum .f. quod melius sit incipiendum a subtili quem alato ferramento dum totum os intendimus penetrare patet experientia in omni alio artificio mechanico sive operetur in ligno sine in lapide vel in alio corpore solido: quia semper artifices utuntur terebro subtili & parvo prius deinde lato & deinde latiori: quia etiam sic operando melius & citius perficiunt quicquid intendunt."

ing to a similar principal as the aforementioned plates.[53] Later in the text, while still discussing the skeletal system, he once again evokes the carpenter. This time, Berengario is attempting to explain the opening of the pubic bones during labor. While it is impossible to see this happen in real time, it would be possible to go, once again, to a carpenter's workshop and look at window hinges which approximated this opening and closing as a way to a better understanding of the birthing process.[54]

The intertwining and sometimes collision of language and experience forced Berengario in his writings to think about how he wanted to position his expertise in order to both legitimize himself to his reader as well as best explain certain physiological phenomena. In mobilizing language intended to recall artisanal epistemologies and the world of craft, the surgeon was making a profound statement on the nature of expertise and the role it should assume in a formal university education.

The Illustrator

Not only were Berengario's texts rich with information, they were beautiful objects that exploited the possibilities offered to publishers, authors, and audiences by the woodcut. The *Commentary* of 1521 contains 22 woodcuts. The prints are found in groupings related to the text and commentary in the section prior. Each woodcut is accompanied by a caption describing what the viewer was intended to observe in the image. Most of the images depict either écorché or skeletal figures in classicizing poses. The attribution of these woodcuts is the topic of speculation and debate, with suggestions spanning the range of artists working in Bologna during these years. The most compelling of the names that have been put forth are Amico Aspertini and Giacomo Francia, two protagonists of Bolognese art in the sixteenth century, who were the heads of large and varied workshop that each had a printmaking component.[55] Another hypothesis is that Berengario himself was responsible for the design of certain prints, especially those depicting anatomical particulars.[56] This proposal as of yet cannot be substantiated, but should not be ruled out entirely. In any event, all

53 French 1985: 57. Berengario da Carpi 1521: 417r: "ut faciunt capentarii iugendo ut firma maneant licet etiam in capite ossa aliqua sint non coniunctura setatili: sed cum alia iunctura quae dicitur supra apodiata."
54 Berengario da Carpi 1521: 493r: "sed debent ire tales medici ad carpentarios & querere qualiter potest aperiri ostium seu fenestra composita partibus & que de tribus suis iunaturis non aperiatur . . . ."
55 There has been little to substantiate the various attributions; however, Marzia Faietti has made a compelling case for the attribution of at least one of the woodcuts to Amico Aspertini on 520v of the *Commentaria*: Faietti and Scaglietti Kelescian 1995: 339–341.
56 Lind has specifically hypothesized that the spine in its various iterations was based on a design by the author himself. See Lind 1959: 26; Putti 1937: 196.

woodcutting involves a collaborative process that requires the convergence of disparate skillsets. It was exceptionally rare at this time that from start to finish a woodcut was the responsibility of a single individual. First a design had to be conceived of by an artist, then the design was passed to the woodcutter who had to have the strength and control to cut into the dense woodblock, and then it would go to the printer who would ink the matrix and make the impressions running the paper through the printing press so that there was enough pressure to ensure that the ink would properly adhere to a slightly dampened piece of paper. As such, singular authorship when it comes to prints is somewhat misleading. In the case of the woodcuts for the *Commentaria* it was certainly a large, collaborative undertaking between multiple artists, woodcutters, the printers, Girolamo Benedetti, and Berengario himself. Whether or not any woodcut can be fully assigned to Berengario, he certainly had a great deal of oversight in many aspects of their design and is directly implicated in guiding the viewing experience through the captions. In the woodcuts for the *Commentaria*, it is possible to identify at least four different designers of the prints on stylistic grounds.

In the following year with the publication of the *Isagoge Breves*, despite a new printer, Benedetto Faielli, the same plates, with few exceptions, are used: two woodcuts are removed, while illustrations of a man with a walking stick and a uterus are added, and a replacement is made for the 1521 woodcut of the spine. In maintaining the same number of woodcuts in a volume one fifth of the size, the outcome became much more densely visual with the concentration of images increasing from only two percent in the *Commentary* to fifteen percent in the *Isagoge*. In both the *Commentary* and first edition of the *Isagoge* only five and six woodcuts respectively do not feature the entire human form, representing an aesthetic break from the broader illustrative strategy of the text. These full-page woodcuts distill the body into its constituent parts without a compositional framework that relies upon artistic conventions of situating bodies in articulated spaces that give context to the human forms. Instead, labels identify veins in the arms and legs, bones in the hands and feet, and numbering of the vertebrae. The overlay of text and image invites the reader to use the image in a different way.

After the evident success of the first edition, it is likely more capital could be invested in the production of the second edition, published less than one year later in July 1523. As a result, about one third of the woodcuts were entirely replaced and the quality of the prints increased with the employment of finer lines and hatching that lend to more volumetric forms. There is also a shift in the visual landscape of the dissection manual with a newfound prioritization of anatomical particulars over schematic écorchés, where the number of prints taking this approach to medical illustration jumps from six to eleven, comprising half the woodcuts in the book.

Berengario's approach to anatomical illustration must be read in its historical context. At the beginning of the sixteenth century the function and worthiness of anatomical images were not the object of unanimous consensus. Even Berengario himself had his doubts. The field of early-sixteenth-century anatomists was divided between those who praised images, and those who were skeptical. For example, Alessandro Benedetti, author of an important anatomical treatise published in 1502 without any illustrations, believed that images betrayed the senses, that nature could only be represented by words, and that discourse was the unique vehicle for "*evidentia*" and "*vivacitas*."[57] In direct criticism of Berengario's woodcuts, Jacques Dubois, a staunch Galenist at the University of Paris wrote in the introduction to his 1539 *Ordo et ratio in legendis Hippocratis et Galeni* that the prints were 'sumptuous, but useless' and that they would only ever help in the treatment of 'picture-people.'[58]

Confronting such doubts surrounding the utility of images, Berengario had a massive undertaking in determining the role of prints in his medical texts. Given that there were few printed, Berengario and his collaborators had to rely on an array of different visual strategies ranging from the technical illustration to devotional imagery to classical sculpture in order to find compositional solutions to overcome the challenges presented by illustration.[59]

The changes between the two editions of the *Isagoge* reflect a certain rethinking as to how images were to function epistemologically and with the text. The playful manipulation of the human body in the *Commentaria* and the first edition of the *Isagoge* was intended to serve as an aide-de-memoire fusing classical and Christian iconographic conventions with physiological systems to help students with information retention. This use of images is by no means new and pre-dates the advent of printing in the West. In printed works themselves we see this as a popular strategy that was likely borrowed from popular printed devotional texts which made use of similar memory aides in relation to the scriptures.[60]

Such illustrations also cohere with Berengario's dynamic instructions for human dissection presented in the *Isagoge*. Berengario saw the corpse as if it was a living body, thus making clear that this science of dissections was in the service of the living. For instance, while discussing the spleen, Berengario gives detailed instructions on how to move the cadaver:

---

57 Ferrari 1996: 155–156.

58 Dubois 1539: 13: "sumptuosa quidem sed nullam in rem utilis.,,nisi sortu pictos homines curatum?" For a broader discussion surrounding the ongoing debates on the utility of images in medical books see: Kusukawa 2011: 188–196.

59 Prior to Berengario's illustrations the so-called Wound Man and the Zodiac Man were the two most prevalent image types to appear in printed medical texts, which came from a long manuscript tradition. The earliest printed example of this being the *Fasciulus Medicina* published in Venice in 1491 by Giovanni and Gregorio de Gregori. For further discussion of anatomical illustration prior to Berengario see: Laurenza 2003: 50, 75–80.

60 For the relationship between image and memory, Lina Bolzoni and Mary Carruthers provide comprehensive assessments of the medieval culture that allowed such tools to flourish. See Bolzoni 2002 and Carruthers 2008. Also Nutton 1999: 61–80.

> You will raise the cadaver, and, when it is in a sitting position, you may better see the location of the spleen under the diaphragm immediately in the hypochondrium, as in a living body. But in a dead body as it lies, the spleen is seen under the ribs because its weight drives the diaphragm readily to the upper region, the lung easily yielding, since it is empty and of a loose texture.[61]

For students to actually gain an appreciation of what this would look like in a patient, they needed to reposition the cadaver in order to approximate a living body. In terms of the illustrations, we see similar techniques employed especially in the final grouping of images which show different muscle groups. This is apparent on page 69r of the 1522 *Isagoge* where an écorché man is depicted in a ruinous, overgrown landscape with a walking stick [Figure 1], his left arm and right leg fully extended allowing the viewer the clearest view of the lateral muscles made possible by way of an artistic composition.

In the images' function as memory tools, the designer of the prints chose to make visual witticisms in alluding to the cadavers used in public dissections, namely those belonging to condemned criminals. These criminals were mostly executed by hanging, the majority were men charged for crimes against property, and had intact bodies, while others were decapitated. Two images refer precisely two these practices, and are included in each of the three anatomical texts. The first woodcut is indented to represent the anterior muscles.

A virile man standing in contrapposto [Figure 2], gazing into the distance, his face portrayed in profile holds a rope alluding to his condemned end, the rope clearly being a noose. In the second image [Figure 3], depicting the posterior muscles, the figure leans against an axe, indicting the other mode of execution.[62]

However, this witty, memorable role of the image takes a back seat in 1523, where instead greater emphasis is placed on singular organs and bone structures, better reflecting the intense investigation and excavation of the human body that occurred during the process of dissection. In tracing the changes made to the illustration of the spine, some conclusions about why this might have been can be deduced and Berengario's ambivalence towards the former strategy revealed.

In the 1521 *Commentaria* the spine is rendered as a flattened schematic form and conveys certain pertinent information that is supplemented by the caption on the left-hand side. Notably, this text also contains a warning that the image is not a "true likeness" neither in number nor appearance of the vertebrae, as such the surgeon goes so far as to direct the reader away from the image and towards a cemetery if they wish to

61 Berengario da Carpi 1522: 13v (English trans. Lind 1959: 59): "Elevabis tamen cadaver: ac si sederet: ut melius videas eius situm qui est infra diaphragma imediate in hypochondrio: maxime in vivo: in mortuo vero iacente videtur esse sub costis: quia sua gravitas impellit diaphragma de facili ad superiora: quia pulmo est vacuus et Rarus."
62 Park 1993: 23–26.

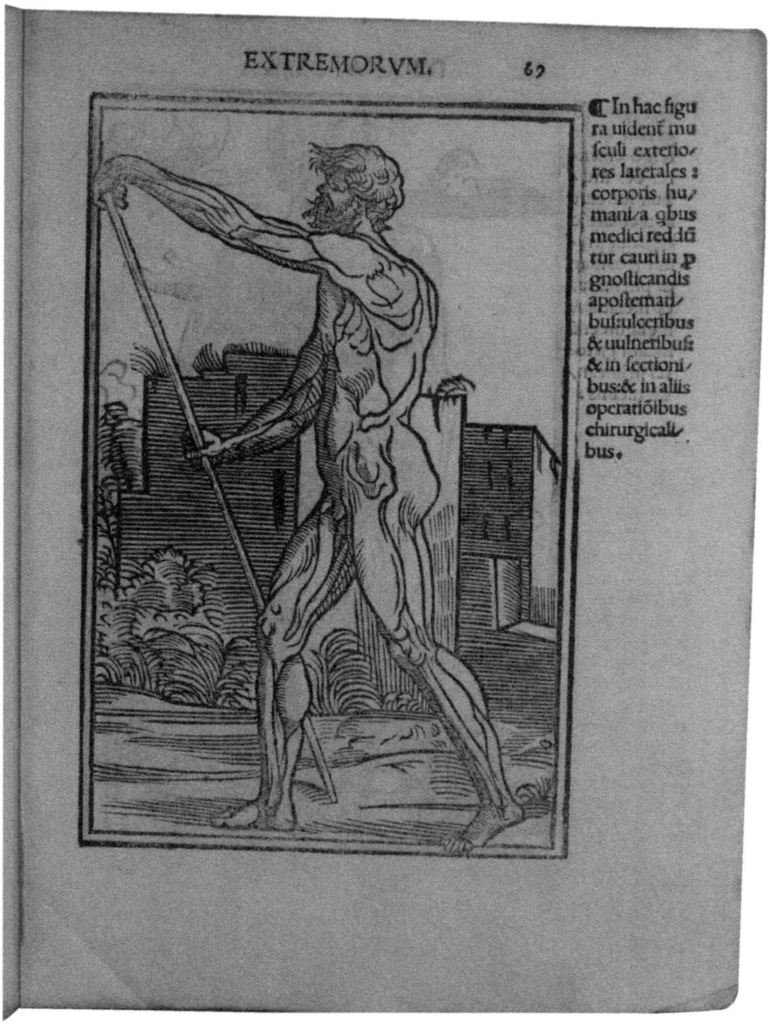

Figure 1: Ecorché showing the lateral muscles. Jacopo Berengario da Carpi, *Isagogae breves perlucide ac uberime in Anatomiam humani Corporis*. Bologna: Benedictus Hectoris, December 30, 1522, fol. 69r. Bologna, 1522.
Photo: Isagoge breves prelucide ac uberime in anatomiam humani corporis. A communi medicorum academia usitatam. / [Jacopo Berengarius da Carpi]. Wellcome Collection. Public Domain Mark, EPB/B/782
https://wellcomecollection.org/works/ujm5ynjj

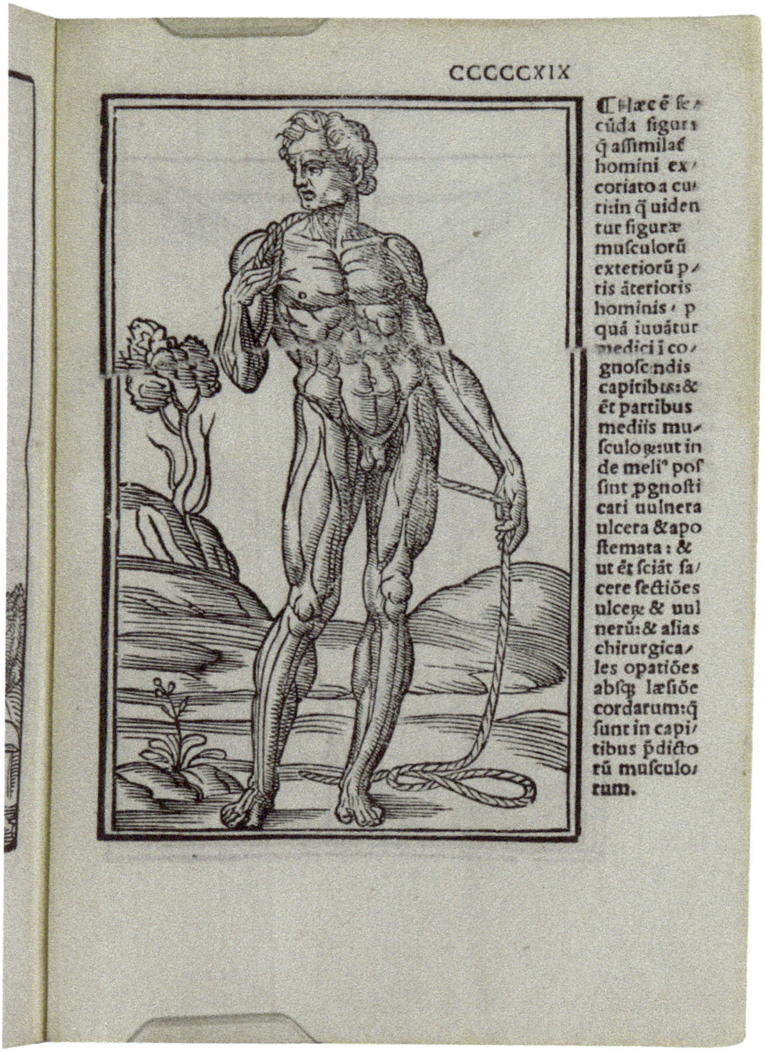

Figure 2: Ecorché showing the exterior muscles of the front. Jacopo Berengario da Carpi, *Commentaria cum amplissimis additionibus super anatomia Mundini una cum textu eiusdem in pristinum et verum nitorem redacto*, fol.591r.
Bologna: Hieronymus de Benedictis, 1521.
Photo: Biblioteca Nazionale Centrale di Firenze, CFMAGL. 1.6.542

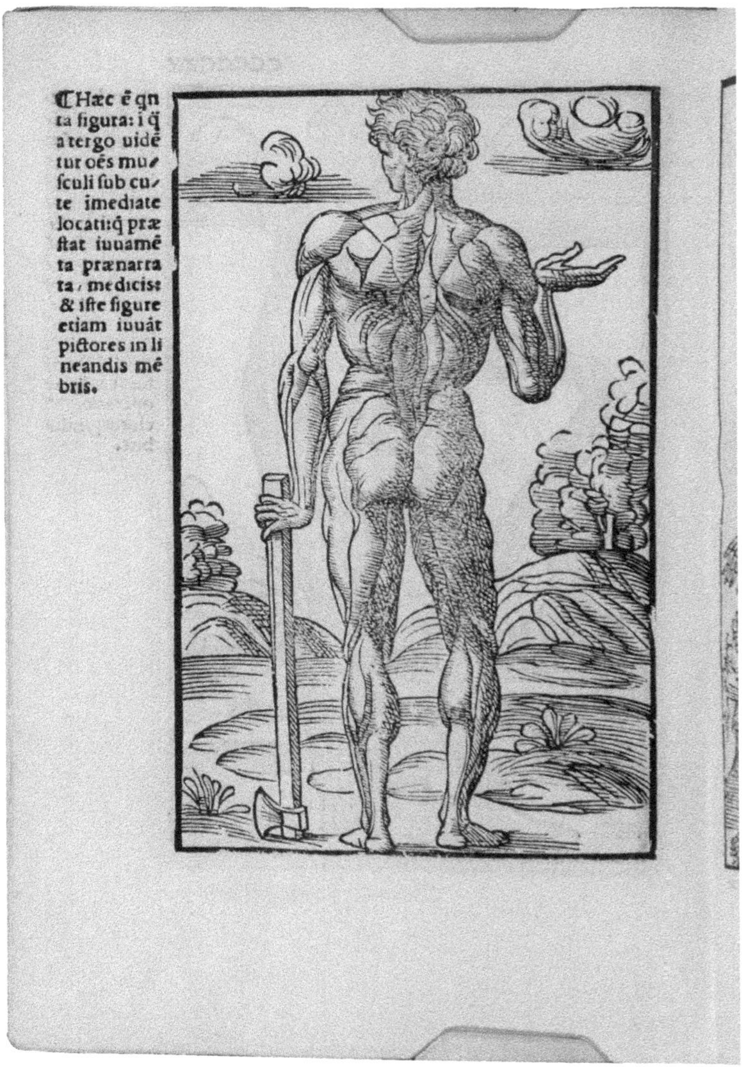

Figure 3: Ecorché showing the exterior muscles of the back. Jacopo Berengario da Carpi, *Commentaria cum amplissimis additionibus super anatomia Mundini una cum textu eiusdem in pristinum et verum nitorem redacto*, fol.520v.
Bologna: Hieronymus de Benedictis, 1521.
Photo: Biblioteca Nazionale Centrale di Firenze, CFMAGL. 1.6.542

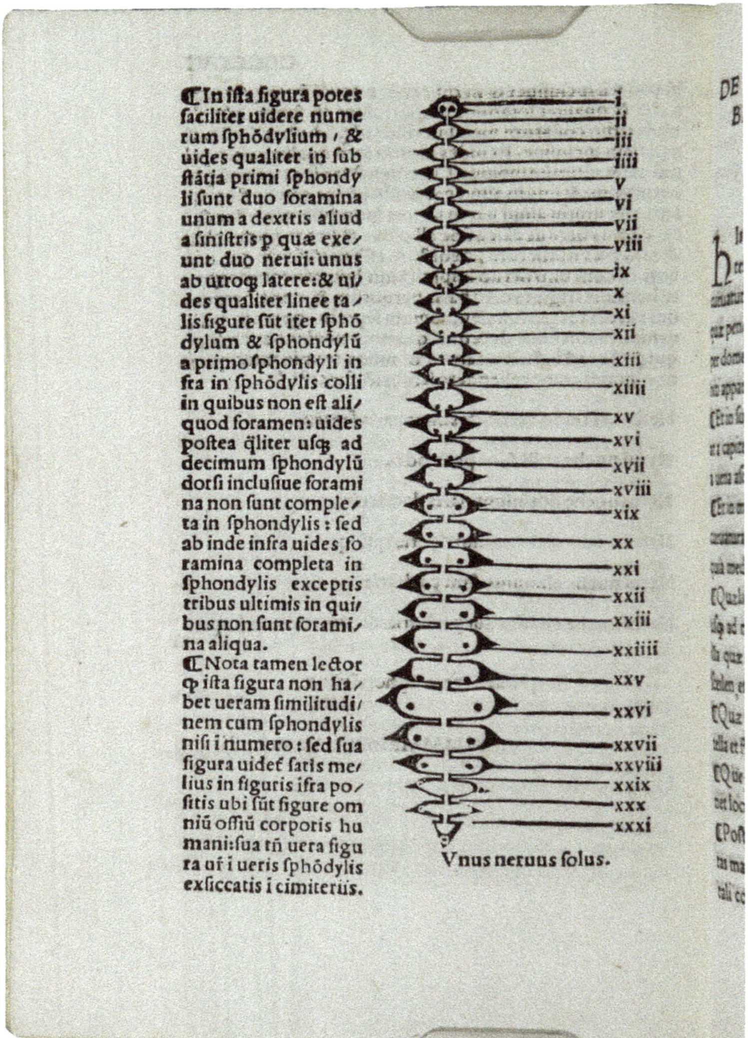

¶ In ista figura potes faciliter uidere numerum sphōdyliū, & uides qualiter in substātia primi sphondyli sunt duo foramina unum a dextris aliud a sinistris p quæ exeunt duo nerui: unus ab utroq latere: & uides qualiter lineę talis figure sūt iter sphōdylum & sphondylū a primo sphondyli infra in sphōdylis colli in quibus non est aliquod foramen: uides postea qliter usq ad decimum sphondylū dorsi inclusiue foramina non sunt completa in sphondylis: sed ab inde infra uides foramina completa in sphondylis exceptis tribus ultimis in quibus non sunt foramina aliqua.
¶ Nota tamen lector ꝙ ista figura non habet ueram similitudinem cum sphondylis nisi i numero: sed sua figura uidet satis melius in figuris infra positis ubi sūt figure omniū ossiū corporis humani: sua tñ uera figura uī i ueris sphōdylis exsiccatis i cimiteriis.

Figure 4: Spine. Jacopo Berengario da Carpi, *Commentaria cum amplissimis additionibus super anatomia Mundini una cum textu eiusdem in pristinum et verum nitorem redacto*, fol.506v. Bologna: Hier. de Benedictis, 1521. Commentaria 1521.
Photo: Biblioteca Nazionale Centrale di Firenze, CFMAGL. 1.6.542

see what the spine actually looks like.[63] Direct observation trumps the consultation of images and in doing so undermines the image's very authority in the text. Such an observation reflects Berengario's aforementioned approach to the body, which in its empiricism is fundamentally skeptical, relying on his own skills of judgement tied to cumulative, practical experience, something that is irreplaceable by a woodcut. It was this dissatisfaction that made the spine the only pre-existing woodcut to be entirely redesigned for the 1522 *Isagoge*. The new woodcut was less schematic with greater delineation of individuated vertebrae and stronger articulation of the transverse process, sacrum, and coccyx [Figure 5]. Despite these changes, the author remained dissatisfied. The illustrator still only depicted thirty-one vertebrae rather than the thirty-three that that actually comprise the spine. Thus, the note to the reader remains, admonishing the reader of the inaccuracy of the image and once again referring them to a graveyard to see the "*vera figura*."[64]

Despite the sustained inaccuracy of the new woodcut of the spine in the first edition of the *Isagoge*, the woodcut remains the same in the second edition [Figure 6], as evidenced by the line on the first vertebra and the open bottom on the sixth, suggesting the same plate. However, the caption changes. There is still the erratum, but rather than pointing the reader towards the cemetery, the author directs them towards another images, saying that the next image allows the reader to better see ("melius videtur") more details of the spine. On the following page [Figure 7], the image goes into much greater depth in its depiction of the spine, providing three different views: the first, a profile view of the entire spine which captures the curvature of the backbone; the second, an overhead view of the second vertebra with an inscription identifying the transverse process; and finally, a frontal view of the sacrum. It is the suggestion of three-dimensionality that affords greater accuracy as well as a proposition of objectivity to the image, with a new way of confronting the transcription of three-dimensional information onto a two-dimensional plane.[65]

One possible source that the illustrator was looking towards was the architectural treatise.[66] Vitruvian architecture and its Renaissance derivatives placed heavy emphasis on the relationship between architecture and the body, so it is not implausible that there would be formal similarities in their representation. Concerns with structures, both macro and micro, as well as the interrelated functioning of systems, be it structural or physiological, could have led to the borrowing of certain representational techniques. One image was not enough to convey all the visual information. Multiple woodcuts were needed to make the image effective from an informational standpoint. The image allows the reader to gain an appreciation of the construction of the form

63 Berengario da Carpi 1521: 526v: "in veris sphondylis exsiccatis in cimiteris."

64 Berengario da Carpi 1522: 62v.

65 For questions of the ontology and objectivity of the scientific image see: Daston 2015: 13–35; Daston and Galison 2007; Daston and Lubeck 2011: 47–80; Elkins 1995; Kemp 2010: 192–208.

66 Long 2011: 50–56.

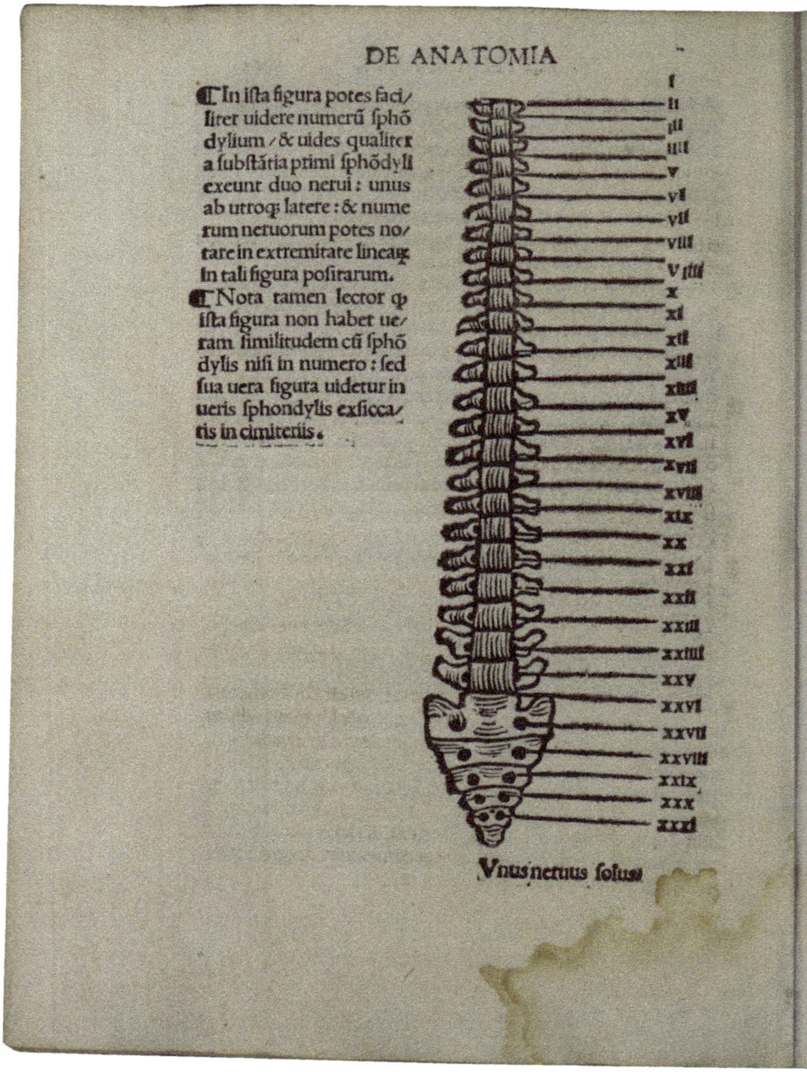

Figure 5: Spine. Jacopo Berengario da Carpi, *Isagogae breves perlucide in Anatomiam humani Corporis*. Bologna: Benedictus Hector, December 30, 1522, fol.62v.
Bologna, 1522.
Photo: Isagoge breves prelucide ac uberime in anatomiam humani corporis. A communi medicorum academia usitatam. *** / [Jacopo Berengarius da Carpi]. Wellcome Collection. Public Domain Mark EPB/B/782
https://wellcomecollection.org/works/ujm5ynjj

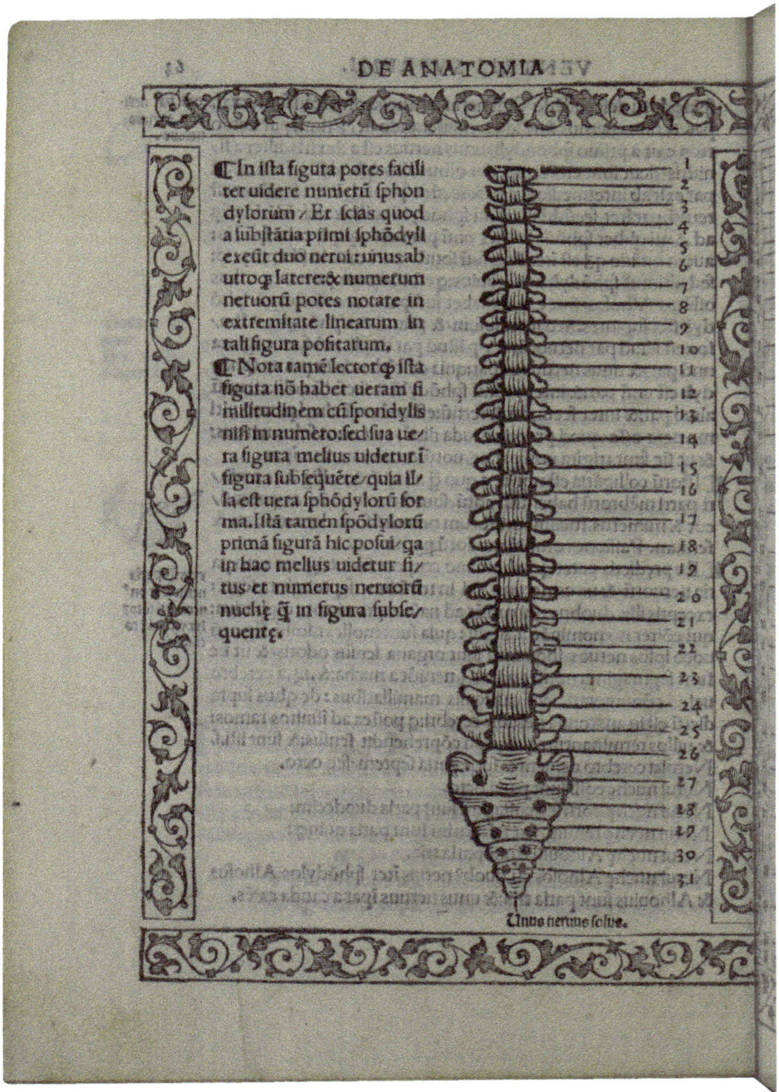
DE ANATOMIA

¶ In ista figura potes facili
ter uidere numerũ sphon
dylorum / Et scias quod
a substãtia primi sphõdyli
exeũt duo nerui: unus ab
utroq latere: & numerum
neruorũ potes notare in
extremitate linearum in
tali figura positarum.
¶ Nota tamẽ lector q ista
figura nõ habet ueram si
militudinem cũ spondylis
nisi in numero: sed sua ue/
ra figura melius uidetur ĩ
figura subsequẽte / quia il/
la est uera sphõdylorũ for
ma. Istã tamen spõdylorũ
primã figurã hic posui / qa
in hac melius uidetur si/
tus et numerus neruorũ
nuchę q̃ in figura subse/
quentę.

Figure 6: Spine. Jacopo Berengario da Carpi, *Isagogae breves perlucide in Anatomiam humani Corporis*. Bologna: Benedictus Hector, December 30, fol.63v.
Bologna, 1523.
Photo: Isagoge breves prelucide ac uberime in anatomiam humani corporis. A communi medicorum academia usitatam. *** / [Jacopo Berengarius da Carpi]. Wellcome Collection. Public Domain Mark
EPB / B 783/B
https://wellcomecollection.org/works/ujm5ynjj

Figure 7: Profile of the spine, coccyx, sacrum, and second vertebra. Jacopo Berengario da Carpi, *Isagogae breves perlucide in Anatomiam humani Corporis.* Bologna: Benedictus Hector, December, fol.64r.
Bologna, 1523.
Photo: Isagoge breves prelucide ac uberime in anatomiam humani corporis. A communi medicorum academia usitatam. *** / [Jacopo Berengarius da Carpi]. Wellcome Collection. Public Domain Mark EPB / B 783/B
https://wellcomecollection.org/works/ujm5ynjj

whether human or architectural and it becomes a study tool; however, because of the three-dimensional source it needs another image to complete it.

Despite these ameliorations, Berengario remained unsatisfied, keeping the note to the reader that sends them to the cemetery to study the original. Berengario makes it clear in this sustained advice that the image should never serve as a substation for the primary object of investigation as it is a mere "figura" the noun repeatedly employed in the caption. Even though images can help the reader they are but one didactic tool. The body is to be handled, poked, and prodded in order to gain the requisite skill and experience. Judgement and experience are gained from looking and touching the corpse, not images.

With the shift from the *Commentaria* to the manual and thereby the increased prominence of the image, it appears that the author and printer saw the need to make the images an integral part of the text. The new woodcuts reflect such changes in their very design and distillation of the human body as well as the overlay of word and image. The image, in capturing three-dimensionality, takes on a new epistemic valence. The changes suggest that there was a desire for an increased reliance on these images as conveyers of anatomical information in line with the text. The captions of the spine represent ongoing concerns not only in medicine, but also theories of art and religion at the time of the relationship between "*similitudinem*" – likeness and "*vera figura*" – true form.[67]

The Theologian

Another sphere of interest is how the body engages or not with the Christian body, in particular the perfect human form, the body of Christ. Included in the *Commentaria* is an image of the crucified Christ used to depict the abdominal muscles. The woodcut does not make an appearance in any of the subsequent editions printed in the 1520s. By looking at religious practice and theology new ways of approaching this image open up [Figure 8].

Late medieval Europe saw the rise of a particular literary genre called *ars moriendi*, handbooks containing instructions for the preparation of a good and Christian death which were widely diffused throughout urban centers. Around the same time, the benefits of confession and absolution came to be extended to criminals condemned to execution. This new concern was reflected by the foundation of confraternities in the early-fourteenth century devoted to ensuring a good death for the condemned. While in northern Europe these duties were administered by clerics, these confraternities of laymen were instead tasked with assisting with religious care from their initial conviction to the very moment they went to the scaffold.

67 See Belting 1994.

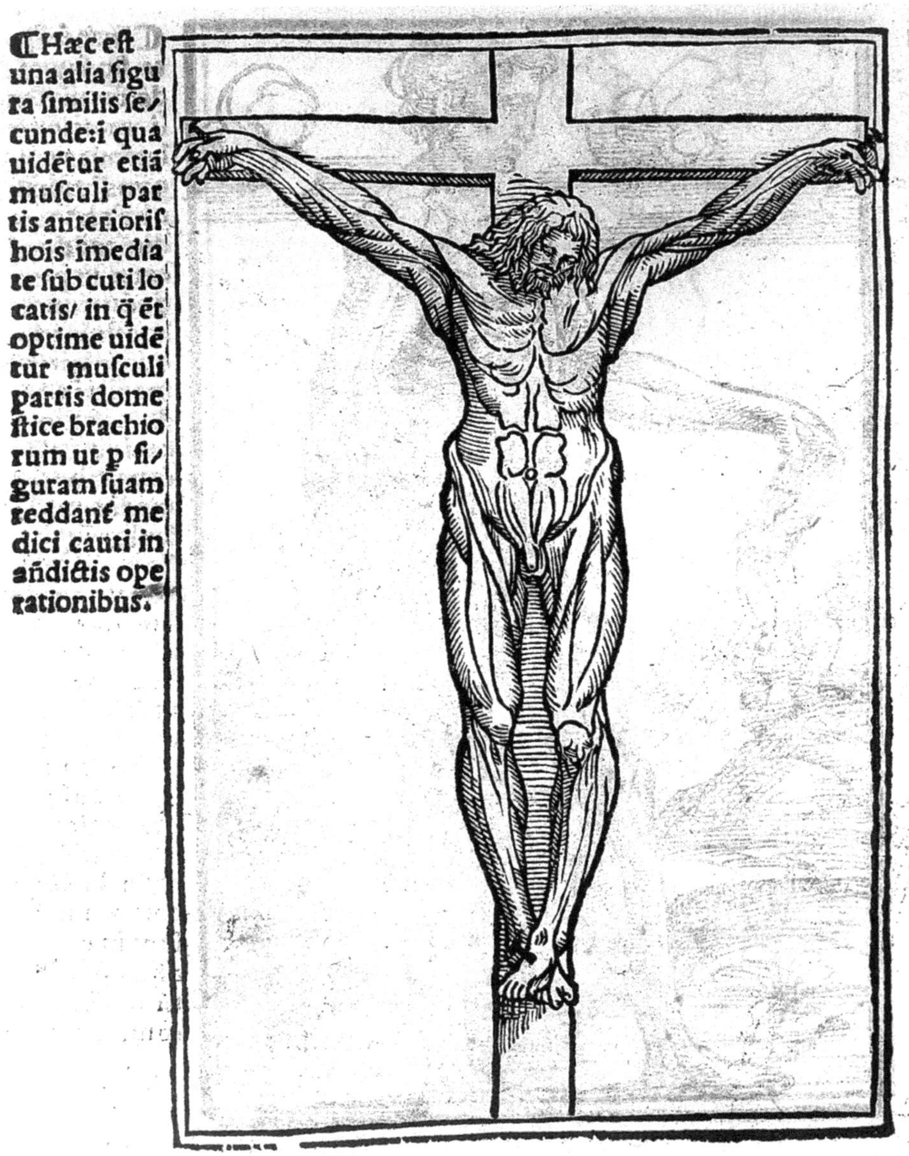

Figure 8: Crucifixion. Jacopo Berengario da Carpi, *Commentaria cum amplissimis additionibus super anatomia Mundini una cum textu eiusdem in pristinum et verum nitorem redacto.* Bologna: Hieronymus de Benedictis, 1521.
Photo: Biblioteca nazionale Centrale di Firenze, CFMAGL. 1.6.542

The Bolognese brotherhood that was committed to this task, the confraternity of *Santa Maria della Morte*, was among the most powerful in the city.[68] Confraternities were responsible both for the comforting of the condemned until the very moment of execution and for the dramatic staging of religious plays, predominantly the Passion and Resurrection of Christ. Audiences flocked to outdoor stages on religious feast days to see these dramatizations of the death of the Saviour and martyrdoms of saints. In these same spaces, audiences would gather to watch the rituals that accompanied a prisoner's final hours. The first "staged" drama of Christ and the saints would clearly have colored their reactions to the second "real-life" drama of the prisoner going to die with piety and dignity. It happened that a single layman could both perform in a Passion play and serve as a comforter at the prison or on the scaffold. This powerfully emphasizes interconnected spiritual realities: "brothers" and prisoners as "brothers penitent" were seeking an intimate union with Christ in his redemptive sufferings where "dramatic and penal forms interpenetrated each other."[69] Dedicated laymen were making strong efforts "to transform a brutal penal event, public execution, into a ritualized and very "real" re-enactment of the death of Christ or one of the martyrs, such as John the Baptist."[70]

The *Comforters' Manual, Santa Maria della Morte's* fifteenth-century "handbook," explicitly told the comforter to incite the condemned to view himself and behave like a martyr. The comforters, besides songs and prayers, presented those about to be executed with a *tavoletta*, a little tablet decorated with images of the instruments of the Passion, the Crucifixion, and/ or a martyrdom. The comforter had to keep these boards as close to the face of the condemned as possible in order to keep his attention fixed on the image while he was on public display, focusing his mind on the virtuous and paradigmatic examples of Christ and the martyrs amidst the jeering and leering of the masses just prior to execution.[71] The image of the crucifix, too, was presented to criminals executed by hanging, and so it may well be the case that Berengario's crucified Christ audaciously alluded to one of these *tavolette* as part of a grouping with the other images that gesture towards the provenance of the dissected bodies.

This idea of sanctity and martyrdom is also conveyed by Berengario's famous flayed man with the radial ornaments, as if the condemned individual, before having been executed and dissected, had obtained the illumination of the spiritual light of redemption after having experienced a revelation. The image also shows something like a tear that runs down his cheek, as if he was sacrificing himself like a martyr for the knowledge of the human body [Figure. 9].

68 On the history of the brotherhoods of lay comforters in medieval and Renaissance Italy see: Edgerton 1985; Fanti 2001; Prosperi 2013.

69 Falvey 2008: 13.

70 Falvey 2008: 13–14.

71 Falvey 2008: 16–17; Fanti 2001: 171–73.

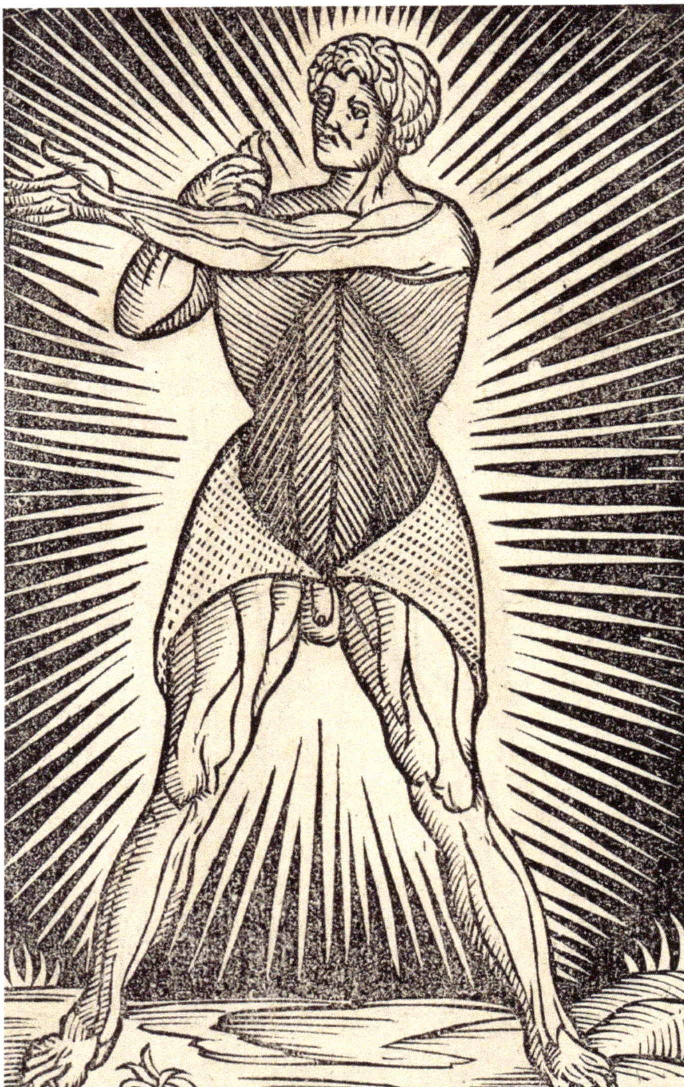

Figure 9: Muscles of the abdomen with rays. Jacopo Berengario da Carpi, *Isagogae breves perlucide in Anatomiam humani Corporis*. Bologna: Benedictus Hector, December 30, fol.6v. Bologna, 1523.
Photo: Isagoge breves prelucide ac uberime in anatomiam humani corporis. A communi medicorum academia usitatam. *** / [Jacopo Berengarius da Carpi]. Wellcome Collection. Public Domain Mark EPB / B 783/B
https://wellcomecollection.org/works/ujm5ynjj

Besides cultural resonances with religious ritual, understanding the nature of and theological debates surrounding the body of Christ might be pertinent to making sense of the crucifixion woodcut. In the section of the *Commentaria* that discusses the heart, Berengario inserted a long discussion on the possible causes of the death of Christ, starting from the Gospel of John, which reported that his heart, when pierced with a sword by a Roman soldier, flowed both blood and water (John 19:32–34). Berengario asked whether this flow of blood and water was natural or supernatural. He finally concluded that this event was miraculous, but he reported that others in the past had made the argument that there was enough water in the veins and in the heart to explain the phenomenon by way of natural causes. Berengario argued against this by claiming that the amount of water could not have been enough to generate a strong flow, and that it would have been impossible for water to spring from the veins separately from blood.[72]

The miracle, therefore, was accounted for by two phenomena: the enormous amount of water in Christ's body, and the fact that blood and water remained separated from each other. In this, Berengario followed fourteenth-century French theologian Nicholas of Lyra's *Postillae litterales* (composed in the 1330s and printed in 1471–72), which insisted on the fact that Christ's body was composed of the Galenic four humors and was therefore completely human.[73] In this way, Berengario proved that observation, and anatomical and surgical experience confirmed the miracle of Christ's body, thus proposing an alliance between anatomical inspection and the verification of miracles which had a long history that was ultimately destined to become epistemologically and institutionally very well established in the early modern period.[74]

Conclusion

As we have seen the dissection manual was a form that was still being experimented with. The place of the image was not yet codified, but images were seen as something with the potential both as a didactic tool and as capable to engage with larger cultural understandings of the body. Berengario and his collaborators brought together an array of visualization strategies borrowed from religious, classical, and technial imagery to find agreeable formal solutions to the representation of the human body. The changes in illustrative strategy present a fascinating pre-Vesalian case study of

72 Berengario da Carpi 1521: 336v. We wish to thank Katharine Park for directing our attention to this passage.
73 Berengario da Carpi 1521: 337r–v. Berengario likely consulted the 1519 Venetian edition of Nicholas of Lyra.
74 Park 1993.

how printed images were indicative of approaches to medical education and a shifting understanding of the body brought about by the greater frequency of dissection.

Looking at Berengario within the larger narrative of the history of medicine, the surgeon formed part of a group of lower-status surgeons who played a significant role in the renewal of anatomy. These men had to present anatomy as a science in the service of the living and with a significant natural philosophical import. This in turn meant that they had to link their science of the human body both to the authorities and to new observational and tactile practices. This is especially true in the case of Berengario, who despite his earlier training with his barber-surgeon father felt the need to receive a university education and a degree. Moreover, Berengario was a man of his times in that he navigated the complex and intricate relationship between medical and religious practice surrounding one of the focal points of Western Christianity: the rituals accompanying the dead in the afterlife.

Berengario was just one of many who were working towards reforming medical education and practice, but the long afterlife of his works and their reception from Italy to England, France, the Low Countries, and beyond are a testament to how his particular approach contributed strongly to a wider anatomical Renaissance.

Jorge A. Lazareff

The Anatomy Lesson of Dr. Jan Deijman and the Social History of the Brain

nor brain alone is worthy for the Muse *Walt Whitman*

During the 16th and 17th centuries, in Holland, professional guilds hired reputed artists to represent their senior members in group portraits. To this practice, we own paintings displaying the likes of syndics, drapers, civic militia guards, and anatomy professors. Of the latter group, Rembrandt van Rijn´s (1606–1669) two anatomy lessons are the better known. The first one, *The Anatomy Lesson of Dr. Nicholas Tulp*, is the most famous. Rembrandt completed it in 1632. The Amsterdam Guild of Surgeons commissioned it to celebrate the appointment of Dr. Nicolas Pieterszoon (1593–1674), nicknamed "Tulp," as the city's "*Praelector Chirurgie et Anatomie*," a position he held for twenty years. When Dr. Tulp announced his intention to retire, the city offered the job to Leiden's Johannes van Horne (1621–1670). The University of Leiden made Dr. van Horne a counteroffer, and that is how Dr. Jan Deijman (1619–1666), the second choice, was nominated and served for thirteen years until his death. In 1656 Rembrandt portrayed him in his first public dissection at the Kleine Veshaal anatomy theater.[1]

In Holland, cities were allowed one event each year held in January so the winter cold would prevent the rapid putrefaction of the corpse. In the protestant European north, the event also had a penal and theological significance. The separation of body and soul is not a rapid process, and the physical mistreatment of the autopsy added a layer of punishment to the criminal.[2] The lectures were held at the anatomy theater of the guild. The public and guild members who wanted to be present paid a fee. Deijman's lecture attracted a large audience. The record shows that his anatomy lesson extended for three days and collected 187 guilds and 6 pennies. Guild members who wanted to be portrayed contributed to the painter's remuneration.[3]

The *Anatomy Lesson* of Dr. Deijman that we see today in the Amsterdam Museum is what is left after a fire in 1723 (Figure 1). The fire consumed the likes of seven guild members but preserved the figure of Dr. Deijman's assistant, Dr. Gijsbert Calkoen (1621–1664) and spared the body of Joris Fonteijn. Of Dr. Deijman, we see the torso and the hands, but not the head. Joris Fonteijn's body stretches on the dissection table at a perspective reminiscent of Andrea Mantegna's (1431–1506) *The Lamentation of Christ*.

---

1 Middelkoop 1998.

2 Aries 2008.

3 Amsterdam Guild of Surgeons 1631–1731.

https://doi.org/10.1515/9783110788501-010

Following the axis of the corpse, we reach the head that is over flexed, perhaps resting on a wooden block for the comfort of the lecturer.

Dr. Deijman hovers his hands over the brain as if waiting for Rembrandt to sketch the moment and then continue dissecting the cerebral membranes. While not known for his advancements in brain science, Dr. Deijman was an accomplished neuroanatomist. He agreed with Rembrandt that while Tulp in his anatomy lesson is forever dissecting the left arm of the corpse of the recently hanged Adriaen Adriaeneszoon, he, Deijman, will forever be dissecting the brain of Joris Fonteijn.

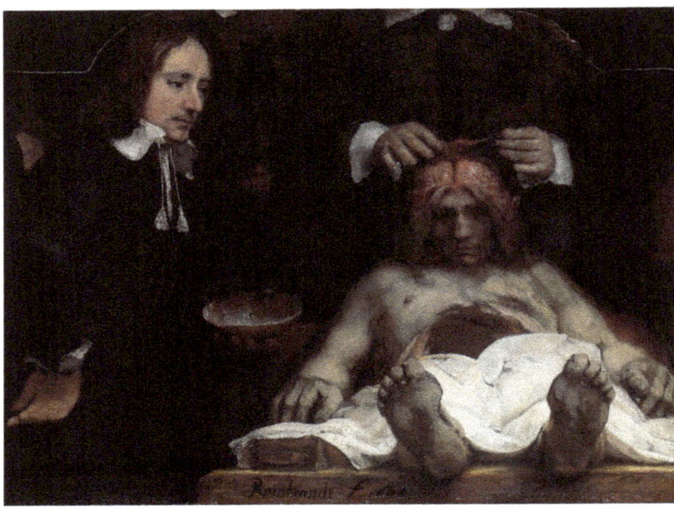

Figure 1: The Anatomy Lesson of Dr. Deijman by Rembrandt van Rijn.
1656. Amsterdam Museum.
Photo by kind courtesy of the Amsterdam Museum, Amsterdam.

Before Rembrandt, the illustrators drew the brain in black ink. The first known print of the brain is found in a book by a Dutch physician, Laurentius Phrysen, published in Strasbourg in 1518. Next, Johan Dryander [Eichmann] (1500–1560) enhanced the details of the cerebral cortex in 1537, followed by Charles Estienne (1504–1564) with his *De dissectione* in 1539. Finally, Jan Stephan von Calcar (1499–1546) set the standard with his work for Andreas Vesalius, (1514–1564) in *De humanis corporis fabrica* in 1543.[4] Andreas Vesalius is also present in Deijman's *Anatomy Lesson*. Rembrandt represents Fonteijn's cerebral hemispheres' exposed dome with the scalp and dura mater dangling at each side of the face. This image resembles Ian von Calcar's illustration of the dissection of the brain in *De humanis corporis fabrica, libri septem.*

4 Clarke and Dewhurst 1972.

The reddish-pink blob at the center of the chiaroscuro captures the attention of even the most disinterested stroller of the corridors of the City of Amsterdam Museum. Next, Rembrandt enhanced the fleshiness of the brain; he painted it red instead of blue-gray, which is the color of the brain of a dead person.

Dr. Deijman, as any neuroscientist, followed the predominant notions about the brain. A *pneuma*, a vital spirit circulated along the cerebral ventricles. These structures, parceled in four communicating chambers, united in a single corridor in the center of the brain. In the anterior portion of the ventricles, the senses gather and generate imagination that flows back shaped reason, and finally, in the most posterior chamber, memory is stored. This theory of ventricular localization of mental faculties has multiple origins. It held the imagination of every historical figure in the medical sciences, from Galen (129–216?) to Islamic scholars and Western Scholars.[5]

It is reasonable to presume that Dr. Deijman planned to slice the dome of the cerebral hemispheres horizontally and expose the lateral ventricles and point towards the sites of superior cerebral functions. In 1656, Dr. Deijman's understanding of the brain was four years short of obsoleteness. Franciscus de la Boë (1614–1672), better known as Sylvius, proposed in 1660 that the human spirit was not a byproduct of whatever happened in the lateral ventricles but a byproduct of the cerebral tissue that surrounded the ventricles. Thomas Willis (1621–1675) forwarded a similar theory four years later. Both anatomists pondered the functional purpose of that immense mass of tissue enclosing the ventricles. For them, the disparity of volume between cerebral tissue and ventricular space determined the location of the cerebral functions.

We should not diminish Dr. Deijman's scholarship and skills. Vesalius acknowledged the difficulties of dissecting the brain and recommended that the head be separated from the torso. Perhaps Deijman wanted to signal to posterity that he was a supreme anatomist, that he could overcome the difficulties that Vesalius predicted would affect lesser talented individuals. In any case, the fire intervened in 1723. Today, Deijman's anatomy lesson survives in the first pages of some neuroanatomy textbooks, in articles cited in this essay, and as the occasional cultural sound bite of PowerPoint presentations delivered at clinical grand rounds. The audience of such displays, or the readers of the textbooks, consider that there is nothing else to garner from the painting.

Michael Foucault (1926–1984), in *The Birth of the Clinic*, defined medicine as the science of the gaze.[6] So, we may ask, where is the gaze of Dr. Jan Deijman? For the design of its frame, Rembrandt sketched Deijman's anatomy lesson. Using the charcoal drawing, the Amsterdam Historisch Museum Curator Norbert Middelkoop and Thijs Wolzak digitally reconstructed the painting. In it, we notice that Dr. Deijman has his eyes yonder the brain as if he was looking at Rembrandt or at the future

---

5 Green 2002: 131–142.
6 Foucault 1994b.

visitor the Surgeon's Guild where the Anatomy Lessons were displayed. We find similar detachment from the *subjectum anatomicum* in the *Anatomy Lesson of Dr. Frederick Ruysch* by Adriaen Backer (1635–1684) and in *The Anatomy Lesson of Dr. Willem van der Meer* by Michiel and Pieter van Mierevelt (1566–1641) and also in Dr. Tulp's. He stands in his anatomy lesson, oblivious of the hand flexors he holds with forceps on his right hand. But Deijman is a pioneer; he is the first who choose the brain. He may be looking at Rembrandt, but his gaze is the gaze of medieval neuroanatomists who were neurophilosophers and theoretical neurophysiologists whose drawings of the brain were a sketch of the likes of a distinguished person with a set of arrows inside their head. Arrows outlining the path of imagination, reason, and memory. Deijman's gaze is the gaze of his master Vesalius who proclaimed that the brain is the holiest of the parts of the body. Deijman's gaze is the gaze of XXI century neuroscientists forwarding through neurocomputing hypothesis about what cerebral structures connect with another cerebral structure to explain the flow from imagination to reason and to memory. Dr. Deijman's and every one of past and present neuroscientists is represented by Emily Dickinson's (1830–1886) poem #632 written about 1860:

> The brain is wider than the sky, /For, put them side by side, /The one the other will include/ With ease, and you beside. /The brain is deeper than the sea, /For, hold them, blue to blue, / The one the other will absorb, /As sponges, buckets do. /The brain is just the weight of God, / For, lift them, pound for pound, /And they will differ, if they do, /As syllable from sound.[7]

And that should settle the subject. But, in the second stanza of "One's Self I Sing in Leaves of Grass," Walt Whitman (1819–1892) questions Emily Dickinson's encomium.[8]

> Of physiology from toe to toe I sing, /Not physiognomy alone nor brain alone is worthy for the Muse, / I say the Form complete is worthier far, /The Female equally with the Male I sing.

With Foucault and Whitman, we look at a painting from 1656 and ask, wherein the *Anatomy Lesson* of Dr. Jan Deijman's the Whitmanian Form? Joris Fonteijn is the Form. An occasional tailor, a petty thief prone to violence, was executed by hanging on January 28, 1656. His corpse was handed by the Court of Justice to the Surgeon's Guild for anatomic dissection and buried in Amsterdam's cemetery on February 2. Dr. Calkoen holds the top of his skull. His abdominal cavity is empty, a white blanket covers his genitals, and the first thing we see is his dirty and ruggedly worn feet. The dead man rarely wore shoes. He was a poor man.

The brain of the poor was for centuries a non-existing concept. Friederichs Engels (1820–1895) does not mention the brain in his *The Condition of the Working Class in England*, published in 1845.[9] The invisibility of the brain of the poor changed in 1919. The medical officers of the Department of Sanitation of the American

7 Dickinson 1961.
8 Whitman 2004.
9 Engels 2009.

Army of Occupation stationed in Trier, a town near the western front, and coincidentally Karl Marx's (1818–1883) birthplace, proceeded to study the physical changes caused by malnutrition in six local thousand children. While the army physicians measured the children's heights and body mass, the schoolteachers pointed that malnourished student exhibited increased "nervousness" and overall slumping in school performance. Lieutenant Smiley Blanton, M.D. (1882–1966) published his observations in 1919 under the title "Mental and nervous changes in the children of Volksschulen of Trier, Germany, caused by malnutrition."[10] It did not escape Blanton that malnutrition among the children of Trier was present even before the war. In 1916 harsher food rationing took place compounded by a failed potato crop. Milk consumption fell from 18,000 liters a day before the war to 3,300 liters in the winter of 1917. The effects of malnutrition on school performance varied according to social class. Blanton points out that "children of rich or well-to-do parents were able to buy much extra food." His report is the first scientific communication about the effects of malnutrition and other environmental stressors on the developing brain.

Inspired by Blanton, two pediatricians, Stoch and Smythe, measured the head circumference of malnourished Cape Town infants. The authors observed that the malnourished group's head circumference and intelligence quotient (I.Q.) were significantly lower than the well-fed children.[11] The effect of malnourishment was further confirmed by a British doctor, Roy Brown, who in 1966 informs about his observations of 1,100 autopsies of children from Uganda and concludes that the brain's weight is decreased.[12]

An abundance of conclusive empirical and experimental data confirmed that the brain is not the immaculate structure that articulates memory from reason and imagination. Brown and Pollitt studied children in the Guatemala Highlands.[13] One group of their study received a nutritious breakfast, the other group, sometimes a sibling, a sugary comfort food. The difference in school performance was so conclusive that, to their credit, the researchers canceled the study.

Understandably, most anatomical evidence on the matter has been gathered from studies on animals. Nonetheless, while scarce, there are robust studies in humans, such as the one reported by Bribiesca and collaborators who studied the brain of children who died at a public hospital in Mexico City. The authors had information about the socio-economic status of the deceased. They observed that the branching of the neurons was deficient in children from a lower economic stratum

10 Blanton 1919: 343–386.
11 Stoch and Smythe 1963: 546–552.
12 Brown 1966: 512–522.
13 Brown and Pollitt 1996: 38–43.

than those from a more fortunate social bracket. This finding implies that children in the first group would have decreased capabilities for associating ideas.[14]

Suppose we trace a world map based on malnutrition indexes. In that case, we can observe that children who are either stunted or wasted inhabit vast areas of Africa and large landmasses in Latin America. The pathological unit that we call the brain of the poor can claim the same territory as any other disease. It has a cause, a mechanism, proven anatomical adaptations, a territorial distribution, a set of symptoms, a differential diagnosis, and a treatment. And still, it is not labeled as such in any medical textbook. Such evidence of vulnerable individuals has not prompted resource mobilization, perhaps because the condition is not contagious.

The zeitgeist swaddles the mind of each observer, be it of art, medicine, or politics. Particularly in what regards the brain, which is, after all, the only organ that studies itself. An organ that is happier with Dickinson than with Whitman, but fully aware that in the Amsterdam Museum is Joris Fonteijn, the Whitmanian form, waiting for us to see him.

14 Benitez-Bribiesca, et al. 1999: e21.

Secrets of the Dead

Valeria Finucci

The Bio-Turn in History Writing: Death, Last Wishes, and Lasting Wishes

Historical analyses, fictional tales, artistic illustrations, philosophical disquisitions, and social practices routinely convey and process non-medical and culturally-sensitive experiences, such as psychological well-being, illness, disability, and death.[1] Every day our cultural memory and commemorative traditions put the "human" and mortality in a dialogic encounter with medicine, for we no longer believe in the objectivity of science and the subjectivity of culture.[2] In this essay, I aim to enter the critical dialogue between cultural studies and the medical arena through narrative histories of death and burial practices, taking my examples mostly but not exclusively from the premodern and early modern period.

To situate the point, I would like to begin with my personal visit a few years ago to a cemetery in the Island of San Michele in Venice, a historical burial ground where a notable number of international literary, musical and politically important figures have chosen to be buried. My goal in engaging in cemetery tourism was to read what epitaphs memorialize as I reflected on the historical significance of burial sites. Passing by the graves of Igor Stravinsky, Ezra Pound, Sergej Diaghilev, Gasparo Gozzi, Carlo Gozzi, Cesco Baseggio, Emilio Vedova, and Christian Doppler, I stopped at the tomb of Josef Brodsky, the Russian poet who was tried in Leningrad for the crime of writing poetry.[3] Brodsky won the Nobel Prize in literature in 1987 and became a United States Poet Laureate in 1991. I met him the year I arrived in the US. At a friendly dinner among Italian graduate students and faculty at Smith College in Northampton in 1976, Brodsky talked about his days in Latina, near Rome, in the refugee camp of Rossi Longhi, where he had

1 Some sections of this essay appeared in a different context and in a much earlier version in a short article I published entitled "Thinking through Death: The Politics of the Corpse." See Finucci 2015b.

2 "There is a need to fundamentally question the cultural distinction between the objectivity of science and the subjectivity of culture," Julia Kristeva wrote recently, "the generality of the natural sciences and the singularity of the humanities." In Kristeva et al. 2018: 55. Or as Ravi Shankar wrote, the field of medical and health humanities "uses subjects which fall within the traditional domain of the humanities in pursuit of goals in medical education." In Shankar 2011: 26.

3 Here are their relative dates: Igor Stravinsky, composer (1882–1971); Ezra Pound, poet (1885–1972); Sergej Diaghilev, dancer (1872–1929); Gasparo Gozzi, dramatist (1713–1786); Carlo Gozzi, playwriter and poet (1720–1806); Cesco Baseggio, actor (1897–1971); Emilio Vedova, painter (1919–2006); Christian Doppler, mathematician and physicist (1803–1853); and Josef Brodski, poet (1940–1996).

https://doi.org/10.1515/9783110788501-011

come against his wishes after being expelled from Russia as a social parasite.[4] He was a chain smoker and in his imaginative Italian told me that he liked big cities and wanted to live in Italy. I had no idea who he was, but the encounter stayed with me.

When I visited his tomb in Venice one wet November day, I noticed a pot of flowers at the foot of his grave. Next to the stele identifying him, I saw an unlikely object, a wet black notebook, with an elastic band around it, which seemed to have been put there recently. Who left it? What was written in it? Did the writer copy Brodsky's poems that were meaningful to him/her? Was the author another expatriate, still mourning the common loss of a beloved native country? Or was he/she a failed writer, who had abandoned in despondency some poems at the grave of a most accomplished poet, a last act of self-erasure? And how about the opposite: the booklet was left as a loving thank you by a successful poet thoroughly inspired by the Russian genius. In the end, what was the relationship of this wet booklet to me? Was I paying respect to the deceased or was I sentimentalizing and estheticizing his burial place? More to the point: is one's biological end also the end of the story? Or is one's death the way to barter mortality for history, since not only the burial ceremony, but also the afterlife of the corpse, time and again, can engage the imagination of the living? As Giorgio Agamben reminds us, a key moment in the history of modernity was not just the execution, but especially the showing of the severed head of Louis XVI during the French Revolution. Corpses and tombs are more than physical objects and are truly important only to those who are alive. They allow us to construct our myths and flesh out our stories, giving us that inheritance of affection ("eredità d'affetti"), to put it in the words of the poet Ugo Foscolo in "Ai Sepolcri," that allows the transmission of our human patrimony, since the tombs of good men, he stated, do spur great souls ("a egregie cose il forte animo accendono l'urne de' forti").[5] Our life, Leonardo da Vinci wrote, is truly "made by the death of others." The turn from flesh to dust, I thus contend, may interest medical professionals and ethicists as much as it interests family members, legal experts, political affiliates, and cultural historians.

4 Josef Brodsky died of a heart attack at the age of 55 in New York, where he eventually chose to reside, but he kept close ties to Italy, especially to Venice, a city which he celebrated through a collection of poems, *Watermark*, and where he chose to be buried because that place constituted, he said, his version of Paradise. The Russian government eventually recognized the importance of his fifth Nobel laureate when President Boris Yeltsin sent a wreath of yellow roses to be placed on his burial site. In 2020 Brodsky former house in St. Petersburg became a museum. See Nechepurenko 2021.

5 Foscolo, *Ai sepolcri*, vv. 151–152 (ed. Pazzaglia 1979). Foscolo was writing against the new Napoleonic law that created cemeteries as discrete spaces enclosed by tall walls in the outskirts of towns. This physical separation between the dead and the living, he wrote, foreclosed all memory of the past.

With this in mind, I would like to go back in time and examine the burial choices, some straightforward and some less so, of some historical figures. I will start with the well-known composer and long-distance traveler, Pietro Della Valle (1586–1652). Journeying through Iraq in the 17th century, Della Valle was thoroughly bereaved when his beautiful young wife, Sitti Maani Gioerida (ca. 1600–1621), whom he had recently married in Baghdad, died an untimely death after giving birth to a stillborn child.[6] Unwilling to bury the Cristian Chaldean Maani in that foreign land and wanting instead to lay her next to the burial place he envisioned for his remains in the Church of Aracoeli, atop the Roman Campidoglio, Della Valle decided to have her corpse heavily embalmed in camphor. There was a body part that he most wanted to preserve in the best condition, her heart, since it was believed that the dead would resurrect where the head or the heart were buried. To safeguard it together with her cadaver he had a suitable casket prepared. The foreign land he was visiting, however, had no iron nails sufficient to provide the airtightness indispensable to preserve the body for the long journey home that he was anticipating, so he commissioned the manufacturing elsewhere of 190 nails (the number is eerily precise) to shut the coffin properly on all sides. He then had it placed at the bottom of a larger leather casket covered with clothes. This was also wrapped in order to make it impermeable and protected with animal hide. And thus, the dead and the living traveled together for the next five years until the musician reached Rome in 1626 and was able to properly bury his wife in his chosen site.[7]

Or here is another case turning again on the cultural politics of place: Gabriele Falloppio (or Falloppia, 1523–1562), professor at the University of Padua and the first anatomist to describe in its entirety the female reproductive system, was rumored to have developed a close, in fact for many a much too close relationship with his Prussian pupil, the botanist Melchior Wieland (1520–1589), better known as Guilandinus. In time, the foreign acolyte was denounced as a "sordid hermaphrodite" by the herbalist Pietro Andrea Mattioli, who maliciously insinuated that Falloppio "loves perhaps the vices of his Guilandinus and the gallantry of so sweet a hermaphrodite, more than truth and my reputation."[8] Mattioli was not the only one to satirize the unmasculine reputation of the botanist. In an anonymous poem written partly in Latin and partly in Italian and entitled "The Lament of the Bo for the Departure of Students from Padua on January 15" (the Bo being the building that housed the medical school in Padua at the time), Antonio the porter is described as observing landladies and prostitutes lining up the streets in tears to say good-by to

6 On Maani, see Baskins 2012.

7 Della Valle later celebrated his wife in "Funeral Oration on His Wife Maani" (1627), most recently reprinted in Spila 2013: 172–85. See also Giazotto 1988; Bianconi 1942; Perocco 1997.

8 Palmer 1985: 154. The vituperative exchanges by Mattioli on Guilandinus are recorded at length in Raimondi 1906 and 1903.

the students leaving the town after the university had to be briefly shut down for the disturbances they had caused. The women were lamenting the rents they lost, although the prostitutes among them, he offered, would be able to make up the difference somehow. The only man named within the group of women mourning the exodus is "the excellent herbalist Guilandinus" ("l'eccellente simplicista Guillandini").[9] We know that Falloppio and his pupil shared a house – a choice that fueled Falloppio's fears that sooner or later inquisitors would take an uncomfortably close look into their arrangement. Thus, he backed Guilandinus's desire to travel to Egypt and Syria to document new plants, and eventually he himself traveled all the way across the Mediterranean in 1588 to ransom for two hundred *scudi* his student/lover who had fallen into slavery for nine months, following capture by Saracen pirates off the coast of Africa.[10] When Falloppio died, the inconsolable Guilandinus, who later in 1564 became the celebrated prefect of the newly established Botanical Garden in Padua (he was the first to import to Europe the seeds of the lilac and of the sunflower), had this lament inscribed on his companion's tomb: "Falloppio, in this tomb you will not be buried alone / With you will also be buried our home."[11] The intimate relationship between the two apparently became in later years a part of the local university folklore. When Falloppio's crypt in the Basilica of Saint Anthony in Padua had to be moved in the eighteenth century to make room for a northern door, an anonymous "pious gentleman" made the decision *ipso facto* to bury Falloppio's bones together with those of Guilandinus, thus reuniting forever the two friends who feared they were too close for comfort when alive.[12] And so they remain reunited even today, their bones mingled within a tomb situated in the ornate Cloister of the Magnolia [Figure 1].

Unusual? Not really. When Petrarch's huge marble tomb was reopened one last time in Arquà Petrarca in 2004, on the occasion of the seven-hundredth anniversary of his birth, a DNA study of the cranium found in the casket revealed it to be not that of a man, but of a woman, which paleopathologists think could be dated as early as the 13th century [Figure 2]. Yet no female cranium was present when the tomb had been officially opened last in 1873.[13] In fact, it has been argued that the anthropologist Giovanni Canestrini, who had been commissioned then by the authorities to conduct this archaeological inspection, accidentally dropped Petrarch's

---

9 See *Lamento del Bo* 1582.

10 Guilandinus, who did not know how to swim, was taken prisoner and became a galley slave. See Grafton 1979: 169.

11 Information on the pair is in Favaro 1928. For more on Guilandinus, see Mieli 1921.

12 As the inscription now reads (my translation), "Here were buried the bones of Gabriele Falloppio and Melchiorre Guilandino." See Favaro 1928; Visentin 2007: 45.

13 On the competition between Florence (to be interred "among his own people," as Boccaccio wrote) and Padua (the place where Petrarch died), regarding where the body of the crowned poet should be buried permanently, see Lummus 2017.

Figure 1: Tomb of Gabriele Falloppio and Melchiorre Guilandinus in the cloister of the Basilica of Sant Anthony in Padua.
Photo: Valeria Finucci.

Figure 2: Tomb of Francesco Petrarch being lifted in 2004 at Arquà Petrarca on the 700th anniversary of the poet's birth.
Photo: Courtesy of Maurizio Rippa Bonati

head when examining it, given his suspicious description of how the cranium disintegrated in his hands when he was manipulating it.[14] Thus, it was unlikely that any

14 See Drusini and Rippa Bonati 2006.

skull should be in Petrarch's thoroughly sealed casket one hundred thirty-one years later. Was this new finding a prank, or was it another surprise concocted by an unnamed "pious gentleman" to reunite the poet with the ethereal, and yet so present, "Laura" figure of his sonnets?

Likewise, when Galileo Galilei's tomb was opened in 1737, almost a century after the astronomer's death, in order to move his body to the Basilica of Santa Croce in Florence – the Pantheon of many male Italian poets and scientists whose graves are staged for cultural and patriotic purposes – archeologists found two craniums in the grave, one belonging to a man of an age similar to that of Galileo at the time of his passing and one belonging to a woman. Who was the woman? It has been hypothesized that she could be Galileo's daughter, the nun Maria Celeste, who had died of dysentery at the age of thirty-three, eight years before her father and was buried on the grounds of her convent. And who put them together? Another hypothesis is that they could have been laid next to each other by Galileo's assistant, Vincenzo Viviani, who after his patron's death often tried to promote his memory, funded the construction of his marble tomb, and even chose to be buried next to him (Galileo could not be buried in a consecrated space when he died, given his problems with the Inquisition). Whether he had previously discussed the idea of this uncommon burial with Galileo is not known, but Viviani seems to have interpreted the desire of his lonely and by now blind maestro (Galileo suffered from reactive arthritis which could have also been responsible for his bilateral blindness) to reunite with the only person who loved him unconditionally, his daughter, as her letters to him recently published reveal.[15]

Not that the option to be buried next to one's daughter worked all the time for desiring fathers. The Bolognese surgeon Gaspare Tagliacozzi (1545–1599), the inventor of a method of reconstructing noses and lips that makes him the father of aesthetic surgery, chose to be buried in the convent of San Giovanni Battista in Bologna, where two of his daughters were cloistered. But he had been interred not even two months when some nuns in the nunnery became convinced that a voice was telling them that the surgeon was a damned iconoclast for remaking with his surgical instruments men's faces as God had originally made them, thus substituting himself for the Almighty. They consequently removed his body from the burial site and threw it on the other side of the convent wall.[16] The Inquisitors were immediately called to address the dispute and after a few months of inquiry they

---

15 See Sobel 1999.

16 "Essendo stato sepelllito Gasparo Tagliacozzo Dottore di Medicina nella Chiesa delle Monache di S. Gio. Battista fù d'indi alcune settimane udita una voce in quel Monastero notificando la sua dannatione, per il che fù il suo corpo d'indi disoterrato, et portato alle mura, per la qual voce s'inspiritarono alcune Monache." The information is in Rinieri 1600: Valerio Rinieri *Diarii delle cose più notabili seguite nella città di Bologna 1520 fino al 1613,* Biblioteca Universitaria di Bologna, MS 434, vol. 4 (reproduced partim in Finucci 2015a: 92 and 201).

absolved Tagliacozzi, condemned the nuns, and buried the surgeon again in his chosen blessed site.[17]

All in all, Tagliacozzi had it easy, but the same could not be said for Leonardo da Vinci, whose tomb in the Chapel de Saint Florentin within the Chateau d'Amboise, where he was buried in 1519, was destroyed a few decades later when the royal castle encountered the ire of religious fanatics during the wars of religion. In 1868 a stele with the words "Leo Dus Vinc" was found by chance close to a coffin containing a skeleton with a large skull. The culturally validated assumption that male geniuses must have large skulls linked this finding to the stone. The presumed corpse of da Vinci was then enclosed in a casket that however went lost for ten years. When it was found, the contents were reburied in the Chapel de Saint Hubert. But is the real artist buried there? A team of microbiologists and paleopathologists would like to offer an answer to the question now by studying the painting "Adoration of the Magi" (1481–1482), which is currently being restored in Florence. Their aim is to see whether it is possible to find in it some trace of the artist's DNA (da Vinci famously used his fingers to spread colors) that would match that of the refound bones.[18] At the same time, another team, led by Agnese Sabato and Alessandro Vezzosi, is trying to find descendants in Italy of the painter's four stepmothers and twenty-four siblings and half siblings, fifteen generations later, confirming one more time that there is no narrative dead-end to dead bodies, whatever the owners may have wished for their flesh while alive.[19]

Such is the case of Grazia Deledda (1871–1936), the only Italian woman to win the Nobel prize for literature. Forever known for her Sardinian origins through her writings, although not particularly liked by her compatriots because of her archaic view of the region, Deledda died in Rome, a city where she lived most of her life, and was buried in the Cimitero al Verano, which boasts a long list of famous tombs. Years later, however, some political and literary admirers figured that it would be better if she were reburied in her native place so that Sardinia could bask in her glory. Thus, *ipso facto*, in 1952 they moved her body to a church in Sardinia appropriately named after her last novel, *The Church of Solitude* (1936), which centers on a young woman felled by a fatal disease. However, Deledda's corpse could not be placed inside the basalt sarcophagus made for the occasion because it soon became clear that it was too small – the builder had taken too literally the notion that

17 As it is annotated in the chronicles of July 15, 1600, Tagliacozzi's body "was restored to its first place of burial with all the rightful solemnities and with the complete re-establishment of his fame, name and official dignity, and he was completely exonerated of the charges." In Gnudi and Webster 1950: 235–36.

18 The analyses in the painting are done by the J. Craig Venter Institute of California under the auspices of the Richard Lounsbery Foundation. See Knapton 2016.

19 The goal was to have some positive results by the 500th anniversary of da Vinci's death in 2019. See Lorenzi 2017.

women's tombs should be diminutive. To quickly obviate the failing, a tunnel was excavated to connect the outside to the catafalque; the casket was then placed underneath it so that visitors could be duped into thinking that the writer's body was literally where it was supposed to be, rather than below. And there it stayed until the church underwent restauration fifty years later and Deledda's corpse could finally be placed in a properly larger sepulcher.[20]

What these stories illustrate, among others that would have proven just as pertinent, is that dying, preparing to pass away, wishing to die, disposing of someone's bones or ashes, and enacting funerary practices are a complicated, often ingenious, and at times agonizing and flawed business. Culturally inflected customs, religious beliefs, moral values, political advisability, and social conventions accompany death and determine each step of a corpse's disposal. If Della Valle's choices described earlier foreground his fear of decomposition and denial of the nothingness that dying forebodes, the narratives of Falloppio, Petrarch, Galileo, Tagliacozzi, Da Vinci and Deledda show a postmortem desire on the part of the living to commemorate death by actualizing not the wishes of the grave's occupant, but more personal fantasies, hopes and fears of the living. Corpses can take on a life of their own.

Take corpse medicine, an eagerly sought therapeutic treatment marrying the living and the dead: *mumia*, defined by the French surgeon Ambroise Paré (1510–1590) as "man's flesh from . . . Arabia," was a bestseller in all European markets to fight disease or humoral imbalance, even though, given the lack of supply, pharmacists readily concocted it from the flesh of executed criminals or local vagabonds, and even then, there were fakes sold by charlatans. Already in the 16th century physicians were hotly debating whether the healing virtues of the desiccated body were tangible, yet this did not stop patients from asking for it at whatever price, in the belief that the mummy, that is bitumen, was an antidote to poison, and thus a recalibrator of the humoral body.[21] Just like today, when medical technologies can maintain a person's alive through the use of body parts from cadavers. Or just alive through machines. Let's recall the uproar following the death or non-death in Madrid in 1975 of Generalissimo Franco, who as Michel Foucault put it, "didn't even realize that he was dead and was being kept alive after his death."[22]

It is difficult to imagine how all-pervasive the experience of death was in the premodern and early modern world, even though our Covid-19 times have reacquainted

20 See *Vistanet.it* 2019. Deledda was later transferred temporarily to a cemetery when the church needed renovations. See Serroni, 2007. https://www.vistanet.it/cagliari/2019/06/20/accadde-oggi-20-giugno-59-la-salma-di-grazia-deledda-da-roma-arriva-in-sardegna/.

21 As Karl Dannenfeldt 1985: 165, explains it, "[O]riginally the mumia of the tombs was considered to be the resinous, aromatic exudate which came from the bodies of ancient Egyptians, and which received a special virtue because it contained the fluids of the body." See also Guerrini 2012.

22 Foucault 2003: 249.

many of us moderns with the overpresence of mortality. Life expectancy, for one, was seriously diminished then by poor hygiene, gastrointestinal disease, and lack of adequate medical know-how; approximately half of newborn babies were buried by their grieving parents within three years of their birth; recurring epidemics decimated entire regions or put in motion the famine that would dispatch many folks in later years; and even successful deliveries hardly kept mothers safe from deadly puerperal fevers. The majority of Christians in premodern times believed that the corruption of a cadaver was the inevitable result of the original sin and could be retarded by treating it with the preserving agent of medicinal herbs. Embalming was done routinely for the body of supposed saints; even today the Italian "in odore di santità" means both "saintly" and "smelling good."

To return to the case of Della Valle, we know that although he was interested mostly in his wife's heart, he also wanted to keep her body whole, following the 1299 Bull of Pope Boniface VIII, *Detestande feritatis*, which urged the faithful to conserve a cadaver intact for burial, without eviscerating it, separating parts, or boiling them in order to extract the bones.[23] This technique was especially used in northern Europe, when entrails were interred close to the site where death had taken place (for example, during crusades), while dry bones, previously boiled in water, were sent to the final destination for proper burial.[24] It was also used, of course, to prepare corpses for the dissecting table.

Royalty often had their body parts separated by embalmers preparing the corpse for lying in state; while the cadaver would be interred in the royal crypt, the heart was usually buried elsewhere, and the viscera could even be kept in jars.[25] This practice intriguingly comes to light now as body parts of the Medici family are discovered in various sites in Tuscany and are reunited by paleopathologists bent on reconstructing not only the lineage's health status, but also the occasional scandals that the sudden death of some of them instigated in the past. Such is the case of Granduke Francesco de' Medici (1541–1587) and of his Venetian wife, Bianca Capello (1548–1587), whose double passing in the space of a day in their villa of Poggio a Caiano was attributed to poison and gave room to countless gossips as to who could have murdered whom. Was Bianca trying to poison her visiting brother-in-law, Cardinal Ferdinando de Medici, who heartily disliked her, and the poison was by mistake tragically drunk by Francesco, which then made her decide to do the same, or was Ferdinando instead trying to poison Bianca, but Francesco became the unintended victim in her place? Francesco was given appropriate funerals and magnificently buried in the Basilica of San Lorenzo in Florence, while Bianca's despised body was rumored to have been interred in a common burial site in the same

23 For a thorough investigation of Pope Boniface's bull, see Brown 1981, and Alston 1944.
24 See Park 1995, Schmitz-Esser 2020.
25 On the sacrality of the body of the prince, see Ricci 1998, and the classic study by Kantorowicz 1957.

church. Recently terracotta jars containing supposedly the viscera of Bianca and Francesco have been found in the crypt of that church and are being studied to confirm or dispel once and for all whether arsenic or some other trigger, rather than the plausible culprit of malaria, caused their sudden deaths. The saga continues.[26]

It is also true that in family vaults bodies could be creatively arranged, if necessary, and indeed many people in the Middle Ages believed that Charlemagne, king of the Franks, was buried not lying flat with arms and legs straight as it is commonly done, but sitting on a throne, as emperors do, an idea that may well have influenced Duke Vincenzo Gonzaga of Mantua's final request in his will of 1612 not to be laid in a wooden coffin ("*nullo autem modo in arca lignea*") but to be composed on a marble throne with his sword alongside ("*sed sedendo cum suo ense apposito super chatedra marmorea ad hoc parata*").[27] The duke's body has not yet been found to confirm whether the Gonzaga family dutifully respected this idiosyncratic, if narcissistic and grandiose, directive for burial.

Likewise, the famous castrato Filippo Balatri (1682–1756) left precise information on how to prepare his body for burial at the time of death, in the hope that his directives were binding and his gravesite could be left undisturbed. In his "Testamento (burlesco) o sia ultima volontà di Filippo Balatri nativo Alfeo," the old singer left detailed instructions on who was to wash his body before enterrement: not any woman, he pleaded, because "beside the indecency I see, I do not want that they enjoy themselves in examining how sopranos are made."[28] His request begs the question of whether women or men would be more interested in seeing – and judging – male sexual anomalies, natural or surgically acquired. Given the fact that men taunted castrati in every possible way, while women were known instead to occasionally pursue erotic escapades with them since they could enjoy sex while freed from reproduction, I would suggest that Balatri was barking at the wrong tree.[29]

Balatri's tomb is still apparently untouched, but there is a fearsome search these days for burial grounds of male sopranos in order to study markers of castration, now that the practice of castrating boys for the sake of their high voice range has blessedly gone out of fashion. Today not the castrato's sexual organ, but his high stature, lower cortical and mineral bone density, thinning of the bones of the skull, large chest size, and most importantly his level of fusion of the cranial structure are eagerly measured in order to determine the role of hormones in shaping bodies, or as it happens, in determining why the castrato, and he alone within the male population at large, enjoyed such a long life. It is well known that most

26 See Mari et al. 2006.

27 On Charlemagne's body, see Chabannes 1999: 153–154 (3.31); on Vincenzo Gonzaga's body, see Finucci 2015a: 2 and 157.

28 Balatri 1755 (translation mine).

29 Finucci 2003.

castrati died in their eighties. Such is the recent case of Gaspare Pacchierotti (1740–1821), who was castrated when he was twelve years old, debuted on stage at nineteen in a woman's role, and became one of the most acclaimed male soprano stars in Italy and England. Pacchierotti died at eighty-one of dropsy and since his tomb had not been disturbed from the time of burial, unlike that of the famous castrato Farinelli in Bologna, a team of paleopathologists at the University of Padua had the perfect chance to examine the physical transformations of a male body that cannot be manufactured anymore through prepubertal excision or torsion of testicles.[30] Luckily, they found confirmation of most, if not all, what they were researching, once more concentrating on what the skull can reveal.[31] Given the fear that the castrato Balatri had to have his corpse exposed, we can safely say that neither Farinelli nor Pacchierotti would have ever consented to the reopening of their casket, whatever personal, musical, or scientific purpose were to be advanced.

In the Medieval period the cult of saints, male and female, meant that some of their body parts were violently or routinely broken and taken to be displayed in churches or simply were bartered away, but the reason for finding famous men's parts today in university settings or museums is culturally unsettling.[32] To return to Galileo, his middle finger, thumb and digits, together with a tooth, were removed from his corpse by devotees and kept in a 17th century wooden case that stayed with the same family for century. That case went missing in 1905. Found subsequently at an auction by the Florentine art critic Alberto Bruschi, those body parts are now displayed in the refurbished Museo Galileo in Florence, while one of the scientist's vertebrae is at the University of Padua, where he originally taught.[33]

Along similar lines, the heart of the sculptor Antonio Canova (1757–1822) was separated from his corpse soon after death in order to be sent to Venice, where much of his Neoclassical art was created, and is now buried in the Church of Santa Maria de' Frari, close to Titian's tomb. The right hand was kept in a lined green velvet case for almost two centuries at the Gallerie dell'Accademia in Venice, which requested it from the artist's brother because the art historian Leopoldo Cicognara wanted it for the glory of the museum. A few years ago, it was finally moved to the Gipsoteca Canoviana in Possagno, the birthplace of this most illustrious artist.[34] Was the heart the emblem of the artist's "spirit" that the Serenissima wanted to keep floating in the city and the hand a memento of this hands-on genius or were they another way to check what whole-hearted dedication to teaching and what

---

30 Farinelli's body in Bologna was exhumed in 2006, but since he was not buried alone, not much can medically be ascertained as unique.

31 Zanatta, Zambieri, Scattolin and Rippa Bonati 2016.

32 Bentley 1985.

33 Aloisi 2010; Natali and Gamba 1993–94.

34 *StileArte.it* 2018.

shape of hand can make an artist great? Is "who we are" recognizable, even after we cease to be who we truly are?

In a room at the University of Padua next door to the Anatomical Theater, the room in fact where until recently most students were awarded a medical degree, visitors can still see fully displayed eight skulls. They are those, we are told, of eight university professors who had generously given the students their own head to study postmortem. Until, that is, a couple of years ago a team of researchers reconstructed through Xrays and CT scans the chain of events that made a professor and rector of the university there, Francesco Cortese, collect skulls in order to further his own interest in phrenology, the now discarded science that makes one understand the intellectual and moral capabilities of an individual through precise measurements of that body part.[35] All those skulls – the head was presumably boiled to get rid of soft tissues and then set on a wooden base for display – have attached to it the name of a professor of medicine, including that of Santorio Santorio (1561–1636), the father of metabolic balance, who had dedicated his life to the understanding of the relationship between food intake and body measurements. Yet, no preserved will has any reference to a professor's generous donation of himself. Rather, we now know that Cortese collected most skulls for his own scientific purposes after he performed their autopsy; some skulls were given to him by other doctors aware of his desire to eventually assemble a museum of anthropology.[36] The scans confirm that all the skulls belong to male individuals who died at the age marked in their postmortem, with surprisingly one exception, that of Santorio, whose donated cranium started the collection. The one on display instead belongs to a man who died between the age of 41 and 60, while we know that Santorio was 75 at the time of his death.[37] Fact and fiction, prohibition and transgression, are often mixed up in historical accounts, but the cultural value they carry (for example, in the gender-biased agenda that craniology and phrenology had regarding the size of male and female skulls; or in the man-centric poetics of place, as "geniuses" are assembled for the glory of the university in which they taught) makes us appreciate differently periods that have been much less squeamish than ours regarding the use and misuse of cadavers and body parts.[38]

Let's take as a final example of biocultural practices the superstitious use of skulls recorded in the Cimitero delle Fontanelle in Naples, where thousands of victims of pandemic diseases, such as plague, smallpox, cholera, and flu, were summarily buried since the 17th century. Through the years the local population had embraced a strategy of adoption of a small skull (*capuzzelle*) for the purpose of

35 The phrenological theory was fully illustrated by its founder, Gall 1825.

36 Ricciardi 1884.

37 Zanatta, Scattolin, Thiene and Zampieri 2016.

38 For a recent study, see Beretta, Conforti, and Mazzarello (eds) 2016. For how the spaces of the dead can be socially and politically reinvented see Legacey 2019.

praying for the deceased's soul, who had presumably died too quickly to atone for sins committed in life and was often not given last rites. Soon enough, however, some craniums started to be venerated and given specific properties, such as that of Brother Pasquale, thought to help in winning money at lotteries, or that of Donna Concetta, believed to exude if the wishes of the petitioner were to be guaranteed.[39] In the end, the church had to intervene to stop this new cult and closed the site in 1966, thus putting an end to the holistic pair of care and cure.[40]

With this excursus on the different ways of thinking through and reacting to death and dying, corpses and burial sites I aimed at offering a way to understand the intrinsic value of probing for cultural purposes case histories that are present in an array of literary and historical pages. By observing the educational, legal, and spiritual valuations of postmortems and autopsies, I claim, by studying bequests accommodating family loyalties in drawing up wills, or cadaver stories that center on the manipulation of body parts for the purpose of shocking and inciting ridicule as in gallows humor, we can gain a better understanding of the psychology of death or of patronage choices. By reflecting on why, for example, separating the holistic corpse of the saint from the disowned carcass of the heretic was so important to medieval people, or why investment in good deeds during life was perceived as preparation for the salvation of the soul, we may better appreciate how the obsession with the "displeasure of death," in the words of the physician Fabio Glissenti (1542?–1615), was accommodated by the living then and may inflect patients' thinking today.[41] By examining how people chose to unite cultural practices and art in the representation of a horrific death, as in the pictorial illustrations of infants murdered by their own mothers or of saints skinned and quartered alive, we can appreciate in new ways why premodernists composed consolatory poetry and took great measures to contain their dread over the disposal of human remains. Finally, by examining how society fostered models of heroic suicide while condemning self-destructive voluntary death, we can again come to a better sense of the past's valuation of life, death, funerary practices of inhumation, commodification of body parts, and rites of healing.

Loss and mourning examined through the creative imagination of the living can bridge a gulf between material culture and medicine, and be used to instill, say, empathy today in the personnel of hospitals and nursing homes toward a patient's death experience or communicate the multifaced ways in which personal deliberations on end-of-life choices are regularly inflected by religious, ethical, cultural,

39 Naples has always been a place where too many dubious relics found their home, as testified by the list made as early as 1569 by the Venetian diplomat Luigi Contarino. See Contarino 1569 (now in Rossi, ed., 1998: 158).

40 The cemetery has only recently been reopened to tourists. It was made famous by Roberto Rossellini in his 1953 film *Voyage to Italy*. More generally on ossuaries, see Koudounaris 2011.

41 Glissenti 1596. More generally, Ariès 1981, and Alston 1944.

philosophical, even political drives. Medical training programs, assisted living personnel, and clinical aides at hospice facilities aimed at providing a whole-patient approach should value a different, novel or even holistic perspective of what the body, even a dead body, may mean imaginatively to those coping with grief and mourning. Medical practitioners trained on patient-centered care should be instructed, for instance, to have more compassionate awareness toward the occasional lack of understanding by lay people of the difference, say, between legal and medical death, if they are familiar with the array of narratives about the fear of premature burial, as reflected in some Renaissance novellas by Matteo Bandello, or in the tragedies of Shakespeare and the work of Edgar Allan Poe. "Mourning rituals," Robert Pogue Harrison reminds us, "would be feckless if they did not provide the means, or language, to cope with one's mortality even if they help one cope with the death of others too."[42] Thus we should not disregard the fascination with "the art of living and dying" as we remember and dismember, in fantasy to be sure, the death of historical figures and picture what their body, even their carcass, means today to us as we consider our own specter of death.

42 Harrison 2003: 70. See also Laqueur 2015.

Rinaldo F. Canalis

The Fatal Disease of the Last Reigning Inca: A Historical and Clinical Study

Introduction

Among several factors believed to have contributed to the remarkably rapid fall of Tawantinsuyo, the empire of the Incas, singular importance has been given to the depopulation caused by the introduction of European diseases to immunity-lacking indigenous groups. In this context a majority of historians have adhered to the hypothesis that Huayna Capac, the last reigning Inca, died of smallpox during an epidemic that killed thousands. This view, that considerably dampens the brutality of the conquest as a factor in the empire's collapse, was first effectively challenged by McCaa, Nimlos and Hampe Martinez nearly two decades ago.[1] In a comprehensive review of the Conquista's chronicles, the characteristics of smallpox in XVI century Europe and a linguistic analysis of early Quechua dictionaries, these authors concluded that the evidence that Huayna Capac fell victim to a smallpox epidemic is weak. Consequently, they advocated a skeptical approach in the assessment of this event and were hopeful that their argument could be strengthened, should work to be undertaken in Lima lead to the discovery of the Inca's mummy. But that effort failed and the likelihood that remains identifiable as his will be found is now considered remote.[2]

McCaa et al. did not offer alternatives to smallpox as the Inca's killer, focusing their study on the exclusion of an epidemic cause. Other investigators, generally assuming that he was the victim of an acute infectious proces, have explored various diagnostic possibilities but, with few recent exceptions,[3] their studies were undertaken before the epidemiology and pathogenesis of many diseases now known to have existed in Tawantinsuyo were identified. Moreover, they were published no less than two decades ahead of new discoveries and modern analytical works of the Conquest that further challenge the more commonly accepted views of this event.[4] Since this information has not been fully used in the assessment of this critical moment in the history of ancient Peru, a study was structured to harvest, from the historical record, clues useful to develop a clinical picture that could allow identification of the more likely agent of the Inca's demise and whether, native or imported, it could have been a factor in the rapid fall of his empire. The results of this endeavor are the subject of this essay.

1 McCaa and Martinez 2004: 1–24.

2 Daza and Barrera 2001: 18–35.

3 Kirakofe and Maar 2002: 145–164.

4 Principally, the complete text of de Betanzos' chronicle and the historical re-evaluations derived from it.

https://doi.org/10.1515/9783110788501-012

Material and Methods

To begin with, it is essential to acknowledge that the scant clinical information included in the chronicles requires careful and tedious analysis. It was frequently collected decades after an event had taken place, rarely from eyewitnesses and, for the most part, related by soldiers and clerics unable to understand the local languages. Not surprisingly then, accounts addressing specialized medical events are often confusing and inaccurate. *Viruela, sarampion, lepra, bubas, sarna* 'smallpox, measles, leprosy, blisters, scabies' and other terms referring to diseases with cutaneous manifestations, were not used specifically but generically and only reflect a chronicler's subjective impression of what his informants described.

In an attempt to overcome these and other interpretative difficulties inherent to the chronicles and obtain acceptably reliable information, this study was principally based on accounts deemed helpful to develop a clinical picture of Huayna Capac's last disease and obtained by Quechua speakers from witnesses likely alive during the last days of Tawantinsuyo.[5] Of the seventeen primary sources identified by McCaa et al.[6] as giving a cause for the Inca's death, only those of Juan Diez de Betanzos (1519–1576), Garcilaso de la Vega Inca (1536–1616) and Felipe Guaman Poma de Ayala (c. 1535–1618) meet these conditions, though not equally, with de Betanzos' being by far the more valuable. Works by most other chroniclers (up to the mid 1600s) were used to the extent that they were helpful to advance a possible diagnosis of the Inca's ailment. Descriptions of mummified remains believed his were reviewed with the same intent. The early accounts upon which the smallpox epidemic theory first found a foothold are addressed first as an essential point of reference for the analysis of the selected material.

Origins of the Epidemic Theory

Cristobal Vaca de Castro's (1492–1566) *Informaciones acerca de la descendencia y gobierno de los Incas (1540–1542)*[7] (*Information(s) regarding the royal descendants of the Incas and their government*) is the earliest document where *viruela* 'smallpox'

5 The importance of an in-depth knowledge of the many nuances of Quechua to obtain accurate information has been discussed in cogent detail by Cerron-Plomino 2010: 369–381.
6 McCaa, Nimlos and Hampe-Martinez 2008. Note that Atahualpa, when referring to his father's passing states, that he died *de aquella enfermedad* "of that disease," perhaps suggesting a malady that he was familiar with, p.19. See also de Xeres 1891: 115.
7 The original manuscript of Vaca de Castro's *Informaciones* has been lost. There exists only a late summary signed by an otherwise unknown Fray Antonio in 1608. It was edited and published in 1892 by Marcos Jimenez de la Espada (1831–1898) as *Una Antigualla Peruana* 1892. See also Pease 1988: 35.

is recorded as the cause of Huayna Capac's death. Vaca de Castro was sent to Peru in 1542 as '*Juez Pesquisidor*' or Inquiring judge, later to become governor. During his less than four years tenure, he gathered from a group of old Quipucamayos information on the political structure of the empire and the lives of its rulers.[8] He was aided in this endeavor by Juan Diez de Betanzos and Francisco de Villacastin (?–1549), though the official translator was Pedro Escalante (unknown life dates), an Indian fluent in Spanish. The *Informaciones*' account of the Inca's death was published by Pablo Patron (1855–1910) in 1894 as follows:

> En la informacion sobre los Incas hecha en la epoca de Vaca de Castro
> se lee que Guaina Capac Inca sabiendo de como habian
> entrado los cristianos en la tierra y le dieron noticia de ellos, luego
> dijo que habia de haber grande trabajo en la tierra y grandes
> novedades, y al mismo tiempo estaba muriendo de la pestilencia
> de viruelas que fue al año siguiente.[9]
>
> [In the information over the Incas done in the time of Vaca de Castro
> we read that the Inca Huayna Capac, aware that the Christians had
> entered the land and having later had news about them said that there
> was going to be a great deal of labor (strife?) in the land and big news
> (great changes), while at the same time he was dying of the smallpox
> pestilence, which happened the following year.]

The smallpox hypothesis was further advanced a decade later (1553) by Pedro Cieza de Leon (1520–1554),[10] one of the earliest and, although not a Quechua speaker,[11] more frequently cited and trusted chroniclers of the Conquista, and the first to state that the Inca's death was attended by a great mortality:

---

8 Dominguez-Favra 2008: 155–192. *Quipucamyos* or *readers of Quipus*: colored, knotted strings mostly used for accounting and much less effectively to record notable events. Termed historical *quipus* by Porras Barrenechea 1986: 10–11; their reliability has been discussed and questioned by Pease 1988: pp. 35–36.

9 Patron 1894: 179–183. Note that the Inca was said to be "*dying of smallpox*" and passed away "*the following year*," perhaps implying a somewhat lengthier and less acute malady than generally assumed.

10 Cieza de León (1554) 1992: 199–200.

11 Cerron-Palomino 2010: 369–81. Rodolfo Cerron-Palomino In his assessment of Cieza's chronicles, Cerron-Palomino, a distinguished authority in Quechua and its variants, notes that "Cieza de Leon, quien tambien consigna de vez en cuando frases y expresiones en Quechua para demostrarnos lo poco que habia aprendido en sus viajes y bajo la tutela de su maestro Fray Diaz de Santo Tomas," [Cieza de Leon who, from time to time, also includes Quechua sentences and expressions (in his texts) demonstrates how little he had learned (of the language) from his travels under the mentorship of Fray Domingo Diaz de Santo Tomas]. Cieza de León (1554):1922: 199–200. Cerron-Palomino 2015: 33. Diaz de Santo Tomas (1499–1570) was the complier of the first Quechua grammar (1560).

> quentan que vino una gran pestilencia de viruelas tan contagiosa que murieron mas de dozientas mil animas en todas las comarcas, porque fue general y dandole a el el mal no fue parte todo lo dicho para librarlo de la muerte, porque el gran Dios no era de ello servido[12]
>
> [they say that there arrived a great pestilence of smallpox which was so contagious that more than two hundred thousand souls died in all regions, because it was everywhere and having, he fallen ill with it, all that was done (the Inca had ordered prayers and sacrifices) was in vain to save him from death because the great God would not have been served by it.]

de Betanzos' Account

Juan Diez de Betanzos arrived in Peru in 1536, four years after the Conquista's initial phase, when he was about 14 years old. He learned to speak Quechua fluently and worked as a freelance *lengua* (tongue) or translator until he was appointed official interpreter by Vaca de Castro in 1540/41.[13] In 1544 he married Angelina Yupanke (Curixirimoy Ocllo), daughter of Yanke Yupanke Inca, grandniece of Huayna Capac, first to become one of Atahualpa's wives and later Francisco Pizarro's (1441–1571) concubine. In 1551, Viceroy Antonio de Mendoza (1495–1552), aware of de Betanzos' linguistic skills and special relationship with the Incan nobility, ordered him to write a chronicle of their history. de Betanzos completed *Suma y Narracion de los Incas* in 1552, a year before Cieza's *Cronica*, but the text remained unpublished until 1880 when Marcos Jimenez de la Espada (1831–1898) had an incomplete version printed in Madrid. A complete manuscript was discovered in 1987 by Maria del Carmen Martin Rubio in the library of the Bartolome March Foundation in Palma de Mayorca.[14] The huge historical significance of this finding is self-evident, but its late discovery is particularly important for studies of the late empire and the conquest. In the incomplete version the rulers' biographies stop with Pachacutec's, the ninth Inca, therefore all previous conclusions regarding Huayna Capac's life and death were reached without the essential information contained in the full text.

de Betanzos' *Suma* is among the closest in time to Huayna Capac's death, believed to have occurred sometime between 1525 and 1530 (see below).[15] It has two

12 Cieza de León (1554) 1992:199 tells us that the epidemic extended to all regions of the empire, an unlikely event when considering that Tawantinsuyo was over 4000 km long with many towns and settlements separated by extensive uninhabited regions. Even later in the well documented smallpox epidemic of 1558 the outbreak was limited to central Peru. See Lovell 1992.

13 Dominguez Favra 2008: 15. Cerron 2010 conducted a detailed linguistic analysis of the *Suma* and concluded that, at difference from early tentative translations de Betanzos' must be taken as the one that best reflects the speech of the descendants of the imperial caste.

14 de Betanzos (1800) 2004: 21–39.

15 de Betanzos (1987) 2015: 9–11.

parts, the first dedicated to the foundation, expansion and rulers of the empire and the second to its fall, the war between its conquerors and the establishment of the viceroyalty. Nine of the forty-eight chapters of the first part are dedicated to Huayna Capac, whose passing is the only Inca's death described in detail.

de Betanzos tells us that, at age sixty, the Inca left Cuzco with fifty thousand men to subdue several warrying tribes in the northern limits of the empire and that, following a successful campaign, he settled in Quito. In the sixth year of his stay there he fell ill of a disease that made him loose his mind:

> En fin de los quales seis años que en el Quito estuvo le dio una enfermedad, la qual enfermedad le quito el juicio y el entendimiento y diole una lepra y una sarna que lo puso muy debilitado. Y biendole los señores tan al cabo, entraron a el paresciendoles que estaba un poco en su juizio y pidieronle que nombrase señor pues estava tan alcabo de sus dias, a los quales dijo que nombrava por señor a su hijo Ninancuyochi, el qual avia un mes que avia nacido y estava en los cañares.Y biendo los señores que aquel tan niño nombrava, vieron vien que no estva en su juizio natural y dejaronle y salieronse y embiaron luego por el niño Ninancuyochi que avia nombrado por señor. Y otro dia tornaron a entrar a el y preguntaronle de nuebo que quien dexaba y nombrava por señor; y respondioles que nombrava por señor a Ataguallpa, su hijo, no acordandose que el dia antes avia nombrado al niño ya nombrado.[16]
>
> [At the end of the mentioned six years of permanence in Quito he got a disease that took away his judgement and comprehension and gave him a leprosy and a scabies that markedly weakened him. Having the lords noted that he was so finished but somewhat in his reason they asked him, since he was at the end of his days, to name a heir to become their lord and he told them that he was naming his son Ninancuyochi, who was a month old, as lord. The nobles left him and sent for the newborn Ninancoyuchi, that he had named as his successor. And another day they again came to visit and asked him once more who did he want to name as lord; he answered that he was naming Atahualpa, his son, not remembering that the day before he had named the mentioned child.]

At this point in the narrative, de Betanzos writes that members of Huayna Capac's *panaqa* (lineage/dynasty) went to inform Atahualpa that his father had named him as his successor, but he declined to accept. They returned to the Inca's quarters and, without telling of his son's refusal, they again asked him who was to be his successor and this time he chose his son Huascar. The Inca died four days later while the envoys that had been sent for Ninancuyochi, *que estaba en los cañares* 'who was in the Cañaris' (a region, near modern Cuenca, Ecuador, found him dead *de la misma lepra* 'of the same leprosy' that had killed his father.[17]

16 de Betanzos (1987) 2015: 305–325.
17 de Betanzos (1987) 2015: 319–320.

The first notable fact in de Betanzos' account is that after having participated in the retrieval of information for Vaca de Castro's report ten years earlier (see above) and eight after marrying Angelina, he no longer defines the Inca's disease as smallpox but as one characterized by a fluctuating mental condition and a cutaneous eruption he terms "a leprosy and a scabies." His account is focused on the Inca's malady and does not mention a great mortality among his subjects. These changes, likely the product of further inquiry from different sources, better line up with a protracted, non-febrile ailment, than with a virulent epidemic. Throughout the account of the event all members of the Inca's entourage, including those that repeatedly visited him to enquire about his successor, are presented as healthy individuals and Atahualpa, who ritually kept nail and flesh fragments of his father's remains, his captains Quizquiz, Chalcuchima, Unan Chullo and Rumiñaqui, whose historical deeds took place much later, are never mentioned as becoming ill.[18] Finally, it is difficult to conceive that, amidst the chaos resulting from a highly lethal infectious outbreak, the mummification, elaborate mourning ceremonies and procession assembled to transport the Inca's remains to Cuzco proceeded in the orderly way described in the chronicles.

It is reasonable to assume that much of the information included in his chronicle was given to de Betanzos by his wife and her relatives, a factor that strengthens its level of accuracy. It may also be assumed that, given their recent occurrence and dire consequences for her people, Angelina, though at the time still a child, would have remembered the events surrounding the Inca's death. de Betanzos states that, shortly after leaving Cuzco to join Atahualpa in Quito, she was ten years old, therefore old enough to recall the events of the time and able, with her family members' help, to relate the dramatic moments they had lived. It defies reason to think that, if an epidemic had spread throughout the empire when its ruler and granduncle died, she (and them) would have omitted relating its devastating effects.[19]

Garcilaso

Garcilaso de la Vega Inca (Gomez Suarez de Figueroa) was born in Cuzco on April 12, 1539 and raised there by his mother, Palla Chimpu Ocllo, granddaughter of Huallpa Tupac Inca Yupanqui and Huayna Capac's grandniece. Garcilaso was bilingual since

18 The location of his captains at the time of Huayna Capac's death is not known, but Atahualpa had a large army stationed in the northern provinces and the war against his brother Huascar was yet to begin. It is highly probable that his closest captains were at hand.

19 Huayna Capac left Cuzco to pacify the northern provinces after Angelina's one year old hair shearing ceremony and died in Quito, at least six years after his arrival there. This, his last stay, was preceded by a period of calm followed by a victorious campaign (both of unspecified duration) against rebellious tribes of the province of Yaguacoche. de Betanzos (1987) 2015: 316.

early childhood, but his first language was Quechua.[20] He began *Comentarios Reales* (Lisbon 1609) forty years after leaving Peru, writing what he had been related by his mother, members of her family and surviving elders of the Inca nobility. Given the many years thence, his recollection must not have been always factual but, even if his accounts sometimes show bias, he generally tried to be accurate, often contrasting his version with that of other chroniclers, sometimes inserting entire paragraphs of theirs within his text.

Garcilaso is often specific about how he obtained information and by whom. The events that preceded Huayna Capac's death were related to him by two of the Inca's former captains: Juan Pechuta, known only by his baptismal name, and Chauca Rimachi. Those related to the Inca's testament, his death and the events that followed were described by his mother, one of her brothers, Fernando Huallpa Tupac Inca Yupanqui, and by an old Inca named Cusi Huallpa.[21] Garcilaso consulted several chronicles, including Cieza's and Jose de Acosta's, when writing his great granduncle's death's account. It is different from all others; it does not mention an epidemic and is a prime example of his elegant prose:

> Estando Huayna Capac en el reino de Quitu, un dia de los ultimos de su vida, se entro en un lago a bañar, por su reacreacion y deleite; de donde salio con frio, que los indios llaman cucchu, que es temblar, y como sobreviniese calentura, la cual llaman rupa, que es quemarse, y otro dia y los siguientes se sintiese peor y peor, sintio que su mal era de muerte, porque de años tenia pronosticos della, sacados de las hechizerias y agueros y de las interpretaciones que largamente tuvieron esos gentiles.[22]
>
> [One day, one of the last of his life, when he was in the kingdom of Quito, Huayna Capac went into a lake to take a bath for his enjoyment and pleasure; when he came out, he felt cold, which the Indians call *cucchu* that means to shiver and following it a fever, which is called *rupa*, or to burn, and since the next day and the ones that followed he felt worse and worse he sensed that his disease was mortal, because for years he had received warnings of it from sorceries and the interpretations of omens that these pagan peoples had for long believed in.]

Garcilaso, ignoring the chronicles he consulted, does not refer to smallpox, measles, rashes or other skin lesions and follows this paragraph with a romanticized account of Huayna Capac's last days, relating how he ordered for the members of his lineage, captains and general entourage to come to him, hear his last testament and share his wisdom. Although he does not explicitly state so, Garcilaso appears

20 Garcilaso de la Vega (1609) 1945: 1.9 (Preface).
21 Garcilaso de la Vega (1609) 1945: 2.248
22 Garcilaso de la Vega (1609) 1945: 2.249.

to imply that the Inca, while at death's door, had enough vigor and survived long enough to carry on these, his last acts of statesmanship.

Felipe Guaman Poma (de Ayala) was born in about 1535, during the earliest phase of the Conquista, in San Cristobal de Suntuntu in what is now Ayacucho. He was the son of Martin Guaman Mallque de Ayala and Curi Ocllo, daughter of Inca Tupac Yupanqui. Guaman Poma probably completed *El primer nueva crònica y buen gobierno* (*Nueva Cronica y Buen Gobierno*) (The First New Chronicle and Good Government) in 1615, after Spanish domination had been generally accepted by the native population and more than half a century after de Betanzos' account. But he was a native Quechua speaker and as a descendant of a noble Inca family he had access to direct information from members of the conquered elite. His chronicle counts with 1180 pages and 398 drawings. It includes comparatively short biographies of the fourteen Incas and many other historical figures. Huayna Capac's death is briefly described:

> murio en la ciudad de Tumi Pampa[23] victima de la pestilencia del sarampion, viruelas. Por temor a la muerte se escapo de la compañia de los hombres y se refugio dentro de una gruta de piedra, y alli dentro murio sin que nadie lo supiera, pues dijo a sus servidores que no divulgaran la noticia de su muerte. Y dijeron que estaba vivo y lo trajeron al Cuzco en andas, tal como debieron traer su cuerpo vivo.[24]
>
> [died in the city of TumiPampa victim of the measles, smallpox pestilence. For fear of death he shun the company of men, took refuge in a cave and died there without anybody knowing he had, because he told his servants not to spread the news of his death. And they said that he was alive and took him to Cuzco in a litter, as they would have done with his living body.]

At first read this account does not appear particularly informative. What is apparent is that Guaman Poma did not know what the Inca's final illness was, ambiguously calling it, as other chroniclers at sea of its cause, smallpox /measles.[25] There is no Quechua word for either of these diseases, so his informants probably used *muru oncoy*, "spots' malady" (see below *mal de manchas*) a term applicable to many undefined cutaneous eruptions. He also called it *pestilencia*, a term usually equated with an epidemic, but given the general tenor of his chronicle it is surprising that he did not elaborate further on its effects. Guaman Poma's purpose in writing his chronicle was, in no small measure, to inform the King of his people's history and of the many ills brought upon them by the Spanish invasion. Should have he known of the

23 Most chroniclers name Quito as the city where Huayna Capac died. Tumi Pampa, today's Cuenca, is located about 200 miles from Quito.

24 Guaman Poma de Ayala (1615–1616) 1993: 71.

25 The smallpox-measles "diagnosis" reflecting the uncertain nature of the disease, was used as late as the early twentieth century by Jose Polo (see note 57).

devastation resulting from an imported European disease, it would have been a departure from his intentions not to describe it as he did, in some detail, for an epidemic that happened during the *purunruna* or 'third age of the Indians':

> Mira quanta suma de indios podia haber en el reino, dicen que una vez con una pestilencia se murieron muy mucha gente, y que seis meses comieron los buitres
> a esta gente, y no la podian acabar todos los buitres de este reino.[26]
>
> [Look (imagine) how many Indians there could have been in this kingdom, (that)
> they say that once a plague killed many, many people, and that for six
> months buzzards fed of this people and they could not have finished eating them
> all the buzzards of this kingdom.]

Guaman Poma's account is also of clinical interest because it includes what may be a paranoid reaction of the Inca to his illness and notes that one of his wives, Raua Ocllo Coya, died of the same ambiguously defined disease that killed him: "murió con las virguelas y sarampion, se murió esta señora de casi noventa años en Tumi casi junto con su marido" [this almost ninety years old lady died in Tumi (Tumibamba, now Cuenca) with smallpox and measles almost at the same time than her husband.][27] Guaman Poma does not mention the death of the Inca's son, but his chronicle includes an illustration of the mummified remains of Huayna Capac, his wife's and those of a child in a litter.

The Inca's Mummy

Among events related to Huayna Capac's death, Garcilaso reported that before leaving for Spain in 1560 he was invited by Juan Polo de Ondergardo y Zárate (1500–1575), *Corregidor* of Cuzco, to see the mummies of three men and two women.[28] Garcilaso makes it a point to note that he had learned the identities of the remains, not from Ondegardo, but from the local Indians '*dezian los indios que*' (the Indians said that) and goes on to tell that the three males, whose only visible body parts were their heads, faces and hands, were Viracocha, Tupac Yupanqui and Huayna Capac. He noted that "los cuerpos etaban tan enteros que no les faltaba cabello, ceja ni pestaña" (the bodies were so complete that they were not missing hair,

---

26 Guaman Poma de Ayala (1615–1616) 1993: 71.

27 Guaman Poma de Ayala is the only chronicler that tells us that the Inca's principal wife Rawa Ocllo Coya died of the same disease as her husband's (Guaman Poma de Ayala [1615–1616] 1993: 71. de Betanzos states that she and her daughter Chuquihuipa and her brothers Xauxigualpa and Acaurimachi participated in the Inca's mourning ceremonies and later travelled to Cuzco with his mummy. See de Betanzos (1987) 2015: 234.

28 Garcilaso de la Vega (1609) 1945: 1.286–287.

eyebrows nor eyelashes.)[29] He touched one of the fingers of the mummy he was told was Huayna Capac's, noting that "parecia una estatua de palo" [seemed a wooden statue], but otherwise neglected to further examine the remains. Later he ruefully wrote:

> Yo confieso mi descuido que no los mire tanto, y fue porque no pensaba escribir de ellos; que si lo pensara, mirara mas por entero como estaban y supiera con que y como los embalsamaran.
>
> [I confess my carelessness that I did not look at them so much, and it was because I did not plan to write of them; if I had thought about it I would have looked more carefully at how they were and would now know how and with what they embalmed them.]

Garcilaso's telling of this episode has been discussed as a flawed account,[30] a testimonial[31] and perhaps a missed opportunity to find clues regarding the disease that killed the Inca.[32] But regardless of these views, the account of his visit is controversial and according to Rostworwoski less than truthful; in fact, her analysis of the records demonstrates that Viracocha's and Tupac Yupanki's mummies had been burned and 'it was certain' that the ones Garcilaso was shown were Pachacutec's and Amaru Inca's.[33]

In apparently tacit acceptance of Garcilaso's assertion that it was Huayna Capac's, Rostworworski does not offer comments regarding the identity of the third male mummy. The earliest accounts regarding the location of the Inca's remains are Cristobal de Mena's (1492-?) *La Conquista del Peru*[34] and Pedro Sancho de la Hoz's (1514–1547) *Relación a su Majestad de lo que Sucedió en la Conquista*[35] [Report to his Majesty of what happened in the Conquest], both written in 1534. De Mena's description of the mummy is brief and second hand for he left for Spain before Pizarro's army entered Cuzco on November 15, 1533.[36] He knew of its existence through the eyewitness account of Martin Bueno, Pedro Martin and one among the three Zárates

29 Garcilaso de la Vega (1609) 1945: 1.289.
30 Rostworowski 2001: 67–69.
31 Hernandez 2016: 1–3.
32 McCaa, Nimlos, and Hampe-Martinez 2004: 1–2.
33 Rostworowski 2001b: 66.
34 Cristobal de Mena's was the first published account of the early Conquest. It saw the light in Seville in 1534.
35 Pedro Sancho de la Hoz (?–1547) was Francisco Pizarro's secretary from 1533 to 1534. Pizarro ordered him to write an account of this early phase of the Conquest. *Relacion para su Majestad de lo sucedido en la Conquista* was sent to the king in July of 1534.
36 Cristobal de Mena, *La Conquista del Peru, llamada la Nueva Castilla*, was among the first group of Conquistadores to return to Panama (1533) and onto Spain after the division of Atahuallpa's ramsom.

in Pizarro's band of adventurers, who volunteered to collect the first part of Atahualpa's ransom. Sancho de la Hoz did see the mummy and described it as follows:

> Este Guaynacapac que fue nombrado y temido, y lo es hasta hoy dia asi muerto como esta, fue muy amado de sus vasallos . . . su cuerpo esta en la ciudad del Cuzco, muy entero envuelto en ricos paños y solamente le falta la punta de la nariz. Hay otras imagenes hechas de yeso o de barro las que solamente tienen los cabellos y uñas que se cortaba y los vestidos que se ponia en vida . . . lo sacaban con frequencia a la plaza con musica y danzas.[37]
>
> [This Guaynacapac had been very famous and feared, and he still is today dead as he is, he was much loved by his subjects . . . his body is in the city . . . of Cuzco very whole wrapped in rich clothes and it only lacks the tip of his nose. There are other images made of clay or mud that only contain the hairs and nails that he had cut and the vestments he used while alive. they frequently took him out in the plaza amid music and dance.]

That the mummy described by Pedro Sancho was Huayna Capac's appears unassailable. Sancho was among the first soldiers to enter Cuzco and his was the first description of that magnificent city until then undisturbed by the invaders. The mummy's precise location had been given to the conquerors by Atahualpa himself, whose earnest plea for his father's body to be left untouched was partly honored by Bueno, Martin, and Zarate. Sancho was writing at most ten years after the Inca's death (see below) and, although the nose, eyes and ears of mummies retract through desiccation, it is highly unlikely that in that time the nose would have flattened so that it's tip didn't appear to be there. There are no accounts beyond Sancho's of the Inca's nasal defect. Whether it was simply ignored or later descriptions were of mummies assumed to be his is discussed below.

Much of the confusion regarding the identification of the Inca's remains rests on Ondegardo's assertion that Huayna Capac was one of the bodies he sent to Viceroy Hurtado de Mendoza.[38] It was written ten years after Garcilaso's visit and its accuracy has been questioned by authorities like Riva Aguero, Porras Barrenechea and Guillen Guillen.[39] To complicate matters further, Fray Luis de Morales, who had been in Cuzco since before 1539 claims, in his rarely cited *Relación* (account), to have been given by Paullu Inca (1518–1549) "el cuerpo de Huayna Capac y los cuerpos de muchos otros señores antiguos que luego enterre en el Cuzco" [the body of Huayna Capac and the bodies of many other ancient lords that I then had buried

37 de la Hoz (1534) 1938: 306.
38 Ondegardo wrote: "Guaynacapa que fue uno de los cuerpos que yo halle embalsamado y envie al Marquez" [Guayna Capac that was one of the bodies I found embalmed and sent to the Marquise] in Ondegardo 1940: 125–196.
39 De la Riva Aguero 2003: 28.386–387; Guillen 1983: 29–30.

in Cuzco].[40] Morales left Peru in about 1541, but was back in Cuzco as *Provisor Eclesiastico* ("Ecclesiastic Overseer") in 1545, the last year of Vaca de Castro's tenure (see above).[41] Assuming that the dates are correct, Morales would have buried Huayna Capac's mummy after Vaca de Castro left Cuzco and before Paullu's death in 1549. The time coincides with the Church's increasing fervor to obliterate idolatry, a policy strongly resisted by the conquered who persistently disinterred mummies, frequently changing their hiding sites to prevent them to fall again in the conquerors' hands.[42] This may have been the fate of Huayna Capac's mummy, since Sarmiento de Gamboa states that it was later found near the center of Cuzco guarded by two attendants.[43] To conclude: except for Sancho de la Hoz's account no description of the Inca's mummy is fully credible or clinically useful.

Huayna Capac's Killer

The material so far presented corroborates McCaa et al.'s contention that the evidence for a smallpox epidemic as the cause of Huayna Capac's death is weak. However, several authorities have, until recently, favored this view and their opinion cannot be ignored without offering more likely diagnostic alternatives. Although still incomplete, a list of the maladies suffered by the inhabitants of Tawantinsuyo and their predecessors has been established through the study of osseous and mummified remains, artistic representations of pathologies and philological evaluations of early Quechua dictionaries. Among the infectious processes found to exist before the conquest, pulmonary and osseous tuberculosis, American trypanosomiasis (Chagas' disease), uta (leishmaniasis), bartonellosis, trepanomatosis (yaws, pinta) and syphilis have been documented radiographically and/or histologically.[44] In addition, strong evidence for epidemic typhus to have been frequent in Tawantinsuyo has been found in chroniclers' accounts, linguistic analyses and biological studies of the illness' vector (see below; no evidence of smallpox has so far been uncovered in mummies dating from the Conquista and Pesce, in a landmark study, concluded that leprosy was first a European import and later a consequence of the

40 Porras 1986: 727. Paullu was one of Huayna Capac's sons. He sided with the Spaniards during Manco Inca's upraising (1536–44) and as a reward received a portion of Huascar's properties.

41 Porras 1986: 728–729.

42 Bauer 2018: 406–407. Ziemendorff 2018 has recently reviewed the fate of the Inca's mummies in detail and presented chronicles-based evidence that supports but does not prove the theory that Huayna Capac's was among the mummies sent to Viceroy Hurtado de Mendoza by Ondegardo. This position is held also by McCaa, Nimlos and HampeMartinez 2004: 1–29.

43 Sarmiento de Gamboa 2019: 150.

44 Guillen 1983.

slave trade, as was malaria.[45] Given that it is highly probable that Huayna Capac succumbed to an infection, only three of the ones above listed can be considered as possible culprits: epidemic typhus, bartonellosis and syphilis.

Epidemic Typhus

Epidemic or exanthematic typhus has been considered among the possible causes of Huayna Capac's death, but without a cogent discussion and not after recent research has enhanced our understanding of its pathophysiology.[46] The infective agent is *Rickettsia Prowesekii* and its vector *pediculus humanus corporis,* the human body louse. Contagion occurs when the louse releases *Rickettsiae*-laden feces as it feeds and these, aided by the host's scratching, are rubbed into the bite facilitating the pathogen's entry into the blood stream. The disease has a short prodromal phase with flu-like symptoms followed, two to three days later, by stupor, confusion and delirium and, four to seven thereafter, by a truncal rash that gradually extends to the extremities and occasionally involves the palate and conjunctivae but almost never the face.

The histopathology of typhus is characterized by intravascular proliferation of the causative organism, endothelial cellular enlargement and multi-organ vasculitis, which can lead to thrombosis and gangrene of areas under pressure and in the distal portions of the extremities, the ear lobes and the nasal tip. Death generally follows multisystem organ failure.[47] The disease's overall mortality is between 15 and 20%, but higher in patients older than fifty years[48] and, under certain conditions, may reach 50% to 60% as occurred in the Eastern Front during the Great War, when it accounted for three million deaths.[49] It was also cause for high mortality in colonial Mexico, where it was called *Tabardillo* or *mal de manchas* (spotted fever) during the epidemics of 1526, 1536 and 1576.[50]

According to Guaman Poma the Incas had special annual rites, usually held in early spring, to forestall the periodic occurrence of a highly lethal disease called *accelasta.*[51] This was in all probability epidemic typhus whose ravages, as noted above, he dramatically described. The pervasive presence of lice among the inhabitants

45 Pesce 1955:48–64.
46 Avendaño 1908: 189–191.
47 Bechah 2008: 417–426.
48 Pamo Reyna 1993: 33–47; Duran 1944: 791–795.
49 Holmes 2014.
50 Avendaño 1908: 189–191.
51 Human Poma also relates of severe epidemics during the reign of Pachacutec. September was the month during which special rites to prevent diseases were carried on.

of Tawantinsuyo was remarked upon by Cieza de Leon and Garcilaso,[52] who tells us that the infestation in some towns was so extensive that Tupac Inca Yuapanqui (Huayna Capac's father) "mandó imponer el tributo que los desastrados pagavan de sus piojos, por obligarles a que se limpiasen y no se dexacen comer por ellos" [ordered a tax to be imposed upon the unkempt because of their lice, to force them to clean themselves and not let them (the lice) eat at them]. Recently, the presence of the disease's vector has been documented by Raoult et al. in the hair of Peruvian mummies.[53] These authors have shown that the DNA of these parasites matches that of the most common louse known to exist worldwide and postulated that, since there are no definite European reports of the disease until the sixteenth century, *Rickettzia Prowesekii* could be native to the Americas.[54]

The chronicles contain two accounts of a pre-Conquest epidemic occurring in Cuzco at about the time of Pizarro's second trip (1526–1528). The episode was first described by Sarmiento de Gamboa (1532–1592) in 1572 and in nearly identical terms in 1585 by Miguel Cabello Balboa (1535–1608), who tells us that while Huayna Capac was resting in the island of Puna:

> recibio noticias del Cuzco donde le avisaban que reinaba una peste
> general y cruel de la que habian muerto Auqui-Topa-Inga su hermano
> *y Apoc Lliaquita, su tio a los quales habia dejado como gobernantes'*[55]
>
> [received news from Cuzco notifying him that there was a generalized
> and cruel plague from which his brother Auqui-Topa-Inka and his uncle
> Apoc Lliaquita, that he had left there as rulers, had died.]

This event has been subject to several modern interpretations, nearly all taking as a point of reference the first chapter of José Toribio Polo's 1913 compendium of the epidemics that afflicted Peru from the conquest to the independence (1821). On the basis of Sarmiento de Gamboa's account and two chronicles-based reviews from the eighteen hundred, Polo theorized that 'probably' the first Peruvian epidemic was a "1525 or a year or two earlier outbreaks of smallpox /measles of which Huayna Capac died."[56] Sarmiento had written that a 'great pestilence' began in Cuzco but

52 Cieza de León (1554) 1922: 199–200. For further information regarding typhus in Peru, see also Avendaño,1908; Duran Martinez 1944; Bechah 2008.

53 Raoult, Reed and Dittmar 2008: 534–543.

54 It was first described by Cardano 1535, and Fracastoro 1546. Note that the first recorded Mexican epidemic anteceded Cardano's descripition by a decade. See Avendaño 1908: 189–191.

55 Cabello Balboa 1576–1586: 113. Several, perhaps many, of the Inca's relatives, notably de Betanzos' wife Angelina, did survive, a highly improbable outcome if the event was caused by smallpox. Since Huayna Capac had presided over her second-year hair shearing ceremony before leaving Cuzco and was resting after six years of war against the northern tribes she was then seven or eight years old.

56 Polo 1913: 4–7.

did not associate it to the Inca's death, nor to a deadly plague extending throughout the empire. Nonetheless, Polo's speculative account has routinely been accepted as 'the 1524/25 smallpox epidemic' that killed the Inca, a highly improbable sequence of events for the virus would have had to travel from Central America, through sparsely populated and isolated regions, to Cuzco bypassing Puná, Quito and Tumibamba to later return north along the *Qhapac-Ñan* (the Inca road) to find Huayna Capac in Quito (2900 km) or while travelling from Tumibamba to Quito (2600 km).[57]

The "cruel plague" described by Sarmiento and Cabello Balboa very likely was an episode of epidemic typhus limited to the empire's capital and its environs. The same arguments that negate a smallpox epidemic as the Inca's killer apply to typhus and are further supported by the clinical features of the disease and the circumstantial information found in the chronicles under study. Firstly, propagation of typhus depends upon temporal conditions favorable to the dissemination of lice that, unable to fly, are inefficient, temperature sensitive vectors, quick to abandon the host when febrile or dead.[58] Effective transport of the disease would have required a large contingent of infected people travelling from Cuzco to Quito in close interpersonal contact such as it has occurred in our own times, in armies mobilized for war. This was certainly not the case at the time of Huayna Capac's death when Tawantinsuyo was at its most peaceful and its soldiers at rest.[59] Secondly, it is extremely unlikely for the Inca, who removed from the common people lived in thoroughly cleaned palatial quarters, was daily bathed and his clothing changed by attendants entirely dedicated to his care, to have been exposed to a disease brought about by overcrowding and filth. Thirdly, although Sancho's description of his missing nasal tip could be attributed to distal gangrene, this complication of typhus is seldom limited to a single body structure. It is an extensive process where necrosis of fingers and toes and pressure areas are prevalent, and apparently absent from the mummy described as intact by Sancho and perhaps seen by Garcilaso. Finally, and more importantly, the mortality of untreated typhus ranges between twenty and fifty percent but, even assuming that a majority of the Inca's large entourage survived because they were relatively young, healthy and lice-free, it is impossible

---

57 The Inca Road was exclusively used for sociopolitical and military reasons. Given the imperial policies of territorial redistribution of the population, most activities that involved large movements of people in peaceful times were limited to inter-village displacements along the road. Travel between cities was restricted to dignitaries and *Chasquis*, the empire's especially trained relay messengers who could cover, depending on terrain conditions, up to 300 km a day: no less than nine days between Cuzco and Quito.

58 http://phthiraptera.info/sites/phthiraptera.info/files/61235.pdf

59 Cieza de León's *El Señorio de los Incas*, quoted by Rostwoworski 2001b: 155; "no se hallo en tierra tan grande quien ozase alzar la cabeza para mover guerra ni dejar de obedecer" [there was nobody in in such a large territory that dared to raise his head to incite war or not submit]. This is Cieza being quoted by Rostwoworski.

to explain how typhus would have killed him, his newborn son and perhaps one of his wives, while sparing many historical figures then residing in the northern region of the empire.

Bartonellosis

Bartonellosis or disease of Carrion is an endemic infection prevalent along the western slope of the Andes, between 550 and 3200 meters above sea level in Peru, Ecuador and Colombia. It results from the bite of the blood feeding female of the *Lutzomya verrucarum* (sand fly), infected with *Bartonella Bacilliformis* an aerobic, Gram negative protobacterium. The disease generally presents in two distinct forms, occurring independently or sequentially. The first, acute phase of the infection is known as Oroya fever. It follows an incubation period that varies between seven days and nine months (average sixty-one days) and is characterized by fever and severe hemolytic anemia. Lopez de Guimaraes et al. have reported a 26.5% incidence of central nervous system compromise in a review of seventeen patients.[60] These authors and Lastres, in an earlier study, have recorded seizures, coma, somnolence, delirium, disorientation and psychomotor excitement during the acute phase.[61] The hemolytic phase is usually followed, after a latent period of several weeks, by a generalized eruption of angiomatous and nodular skin lesions of variable size, known as *Verruga Peruana* or Peruvian wart. Some of these may grow to over three centimeters in size and more frequently appear in the extremities, followed by the trunk the head and the face. In some cases, the oral, ocular and nasal mucosae may be involved. The histopathology of the verrucous nodules shows severe granulomatous changes with extensive infiltrates of various reactive cells and proliferation of capillary-like vessels in the papillary dermis. These lesions may mimic malignancies and ulcerate, so that their avulsion will leave a defect.[62] Superinfection with aggressive pathogens (especially of the salmonella group) is frequent in the hemolytic phase and potentially lethal. The mortality of Oroya fever in the pre-antibiotic era has been estimated at between forty and ninety percent.[63]

The existence of the disease in pre-Columbian Peru is well documented.[64] The term *sikii* for *verruga* can be found in the earliest Quechua dictionaries[65] and there are several examples of pre-Inca ceramics and stone statuettes depicting skin

60 Lopez-de-Guimaraes, Vera-Guzmán, Menacho-López, Avila-Polo, Loarte-López 2017: 8–15.
61 Lastres 1956: 73–104.
62 Arias-Stella 1987: 279–284.
63 Arias-Stella 1987: 121–134; Maguiña and Gotuzzo 2001: 1–3.
64 Arrese Estrada, Maguiña, and Perrard 1992: 350–353; Maguiña and Gotuzzo 2001: 1–22.
65 Garcia Caceres 1991: 58–66. In contrast to smallpox, for which a Quechua word did not exist until after the epidemic of 1558.

Figure 1A: Chancay Culture (ca. 1000–1400 CE), Ceramic depicting the eruptive phase of bartonellosis (Frontal view).
Courtesy of Museo Nacional de Arqueologia, Antropologia e Historia, Lima.
By permission carta No. 000013–2017/MNAAHP/MC.

lesions compatible with it.[66] Two Chancay's culture figurines, currently housed at the National Museum of Archeology, Anthropology and History in Lima are among the notable pre-Colombian representations of the disease (Figures 1A and 1B).[67]

An illness characterized by hemorrhagic skin lesions called *bubas* 'blisters' was recordedby several chroniclers as having afflicted a group of conquistadors during Pizarro's second trip in the island of Coaque (Ecuador), a sea level setting,

66 Most artistic expressions of disease in ancient Peru are found in ceramics from the Moche culture (first century to about 700 C.E). However, those showing skin lesions need to be evaluated carefully as they are not likely representations of Bartonellosis as this culture flourished in the northern coast of Peru, away from endemic areas.
67 Chancay was a predominantly coastal civilization dating from about 1200 to 1400 C.E. It evolved at about 50 meters above sea level but extended to the Huaral province of the department of Lima, thereby reaching an altitude of about 1800 meters, where the disease's vector can thrive.

Figure 1B: Chancay Culture (ca. 1000–1400 CE), Ceramic depicting the eruptive phase of bartonellosis (postero-lateral view).
Courtesy of Museo Nacional de Arqueologia, Antropologia e Historia, Lima.
By permission carta No. 000013–2017/MNAAHP/MC.

and as such not considered a common habitat for the *Lutzomia*.[68] The disease killed several Spaniards while local inhabitants were spared. This episode lends additional evidence to the pre-Hispanic existence of Carrion's disease, though Pinta, caused by *Treponema carateum* could also be considered as the culprit. Definite proof of the Pre-Columbian presence of bartonellosis in Peru has been demonstrated by Dalton et al.[69] and Allison et al.[70] who have shown Bartonella bacilliformis in the re-hydrated skin of Peruvian mummies.

Bartonellosis could have been the cause of Huayna Capac's demise, either as one of a cluster of isolated cases or, less likely, during a regional outbreak. He would have been infected after leaving the island of Puna, in the Cañari region where his

68 This may not have been an exceptional occurrence. A recent review of Bartonellosis in Alexander 1995: 354–359, documents a significant incidence of Bartonellosis in low lying areas of Equador, particularly in the Manabi province. For a detailed discussion of this incident see Pamo Reyna 1993: 33–47.

69 Dalton, Allison, and Pezia 1976: 43–48: "The documentation of communicable diseases in Peruvian mummies."

70 Dalton, Allison, and Pezia 1976; 295–300: "A case of Carrion's disease associated with human sacrifice from the Huari culture of southern Peru."

Figure 2: Chancay Culture (ca. 1000–1400 CE), Ceramic representation of a high-ranking personage being transported in a litter.
Courtesy of the Museo Nacional de Antropologia, Arqueologia e Historia, Lima.
By permission carta No. 000013–2017/MNAAHP/MC.

son Ninancoyuchi, was later found dead of the same "*sarna y lepra*" that killed his father (see above). This is an endemic area of Carrion's disease that harbors several small and medium sized bodies of water. Interestingly, Garcilaso tells us that the Inca became ill after bathing in a lake (see above), a favorable environment for the vector to thrive. Regional outbreaks of the disease are common and it is more frequent in children younger than five years as was Ninancoyuchi (figure 2).

Recently, Kirakofe[71] and Kirakofe and Maar[72] have presented evidence favoring bartonellosis as the Inca's cause of death, arguing that he was a victim of the "1524/25 epidemic" discussed above. Basing their hypothesis on epidemiological and historical grounds they proposed that this event was a virulent outbreak of Carrion's disease that, transported by large contingents of the Inca's armies, extended from Cuzco to the northern reaches of the empire. For the reasons above discussed regarding epidemic typhus, this is an unlikely sequence of events. Furthermore, the need to mobilize large armies to gain control of the northern empire had ceased and as noted by de Betanzos and others, the Inca and presumably his army had been at rest for six years before he became ill (see note 60). Kirakofe and Maar also proposed that recent reports of outbreaks of Bartonellosis in the vicinity of Cuzco, and the local presence of potentially alternative bacterial vectors and of carriers among its inhabitants, support the theory that the empire's capital was the epicenter of an epidemic of Carrion's disease. However, it is apparent that these modern human carriers, not being always native to the site, actively participate in inter-

71 Kirakofe and Maar 2002.
72 Kirakofe and Maar 2008: 145–186.

village migrations and that the appearance of vectors beyond known endemic areas is probably a recent adaptation of various sand fly species to non-habitual environments, a phenomenon unlikely to have begun in pre-conquest times.[73]

As noted, bartonellosis has a pre-eruptive phase with a significant incidence of central nervous system symptoms including disorientation, delirium and psychological disturbances, all related by de Betanzos and Guaman Poma de Ayala to have afflicted the Inca. However, there are several factors that, although not definitely excluding it, make Bartonellosis an unlikely cause of the Inca's death. He was a native of Tumibamba and, having lived in the area for many years, probably had a degree of immunity to the disease that would allow him to survive the hemolytic stage and pass to the eruptive phase where fatalities are very rare. Although it may be argued that a rash could have occurred during the Oroya fever stage, recent reviews encompassing several hundred cases of the disease report only pallor and jaundice as common skin findings. Petechiae may be present, but they have an incidence of less than 1.75% and it is highly unlikely that these minute, predominantly truncal, flat reddish lesions would have been referred to as *viruela, sarna* or *lepra*. Finally, even if ulceration and necrosis do occur in the exophytic lesions of bartonellosis, they don't lend an easy explanation for Huayna Capac's mummy's nasal tip absence. Avulsion of a necrotic mucosal nodule would have left a large defect as would postmortem trauma, an unlikely cause, given the veneration and care royal remains received from ever present, especially chosen members of their lineage.

Beyond those above discussed, two additional facts further militate against the theory that the Inca's death was a consequence of a 1524/5 epidemic of Carrion's disease. Firstly, the chronology of the Inca's death has been studied in detail by Bravo Guerrera,[74] who has convincingly argued that in all probability happened in late 1529 or early 1530, an opinion advanced by Ballesteros Gaibrois[75] and later shared by Rowe.[76] de Betanzos and several other chroniclers tell us that it occurred soon after the first Indian contact with Pizarro's band, an event now believed to have taken place in late 1528, so that an early 1529 date for the Inca's death may be more accurate. Secondly, bartonellosis in endemic areas generally presents as localized outbreaks of variable lethality. The historical record contains but two episodes that exemplify the disease's behavior in populations lacking immunity to the causative agent, a far more deadly and prolonged event than those occurring in endemic regions. The earlier of these killed no less than seven thousand workers during the

73 Only recently there have been several reports of bartonellosis in non-endemic areas, including the Peruvian high forest. See Tejada, Vizcarra and Perez 2003: 211–216; Maguiña and Maguiña 1984: 47–51. See also Maguiña Vargas, Peña, and Ponce 2008: 1–6.

74 Bravo Guerreira 1977.

75 Ballesteros 1963: 62.

76 Rowe 1978.

construction of the Peruvian central railroad at La Oroya in 1875.[77] La Oroya is a bartonellosis endemic area, but the vast majority of workers that perished during this event came from non-endemic areas of Peru and as far as Chile, Bolivia, and China. The second episode is also extraordinary because it took place in a non-endemic area of Colombia.[78] As a gesture of good will, following the resolution of the 1938–41 Peruvian-Colombian border conflict, two thousand Peruvian recruits remained for five months in the department of Nariño to help in the reconstruction effort of the area. Later, as an unprecedented epidemic of Carrion's disease spread through the region, it transpired that some of the Peruvian soldiers were from endemic areas and, in all probability, asymptomatic bartonella carriers. The epidemic resulted in over forty-five hundred documented (six thousand estimated) deaths. It lasted for over ten years ending, as it was near certainly the case in the Oroya epidemic, when bartonella human reservoirs became extinguished.

Syphilis

Syphilis, as is well known, is a venereal disease caused by *Treponema pallidum*, subspecies *pallidum*, whose existence in pre-Columbian America, for long a source of debate, is now generally accepted.[79] Syphilis is known in Quechua as *vanti* or *huanti* and the archeological record contains several pre-Inca ceramic representations of cutaneous and neurological lesions as well as osseous paleopathological specimens compatible with it (figure 3). The disease in XVI century Europe, where the population had not been previously exposed to the treponeme, was far more severe and more rapidly fatal than in modern times.

Whether this was true for the Amerindians in general or for the Quechua and Aymara in particular is not known, though, the ceramic examples believed to represent it suggest an aggressive ailment, as do several osseous specimens.

Syphilis has four stages: primary, secondary, latent and tertiary. Untreated it becomes a chronic, multi-organ fatal disease. Syphilis as the probable cause of Huayna Capac's demise has been previously considered and indeed several of the clinical features described in the chronicles suggest it. Most, if not all studies, that have addressed his mysterious death, assume that it was the result of an acute infection. However, in the earliest available account, Vaca de Castro reported (see above) that, when the Inca was sent news of the Spaniards' arrival to the northern limits of Tawantinsuyo, he was ill with smallpox and died of it the following year. Such timeframe would imply a lengthier process than the most common form of

77 Maguiña and Maguiña 1984.

78 Arroyo 1938.

79 Harper, Zuckerman, Kingston, and Armelagos 2011.

Figure 3: Moche culture (100–700 CE), Ceramic figurine showing multiple perineal lesions characteristic of syphilitic condylomata lata not to be confused with venereal warts.
Museo Larco, Lima. Catalog # ML 004247.
By kind permission of Museo Larco, Lima.

smallpox which may be lethal in 12 to 16 days, while the less frequent, but deadlier flat and hemorrhagic variants are even more rapidly fatal as is also the case in epidemic typhus.[80]

Further information suggesting a subacute process afflicting the Inca may be inferred, as already noted from de Betanzos' labelling it scabies/leprosy, both chronic ailments, and from his account of the days preceding his passing (see above). Members of his entourage aware of the severity of his condition approached him, sometime after he fell ill, to ask for the name of a successor. They did so several times, after judging that he was more 'in his reason,' that is more lucid than earlier, in the final stages of the disease. Although the Inca died four days after their last visit the sequence suggests a malady more protracted than a virulent fatal infection.

In addition to these facts, suggestive of a lengthier illness, Huayna Capac's psychological and neurological signs and symptoms: memory loss, disorientation, bizarre responses, visual hallucinations (seeing dwarves heralding his death) and paranoia (escaping from the court and hiding in a cave or building)[81] are all compatible with neurosyphilis, and more often described in the tertiary stage of the

80 Fenner, Henderson, Arita, Jezek, Ladnyi, Danilovihch 1988: I-170.
81 Guaman Poma de Ayala (1615–1616) 1993: 71–72.

disease, a generalized process during which the nervous system, the heart, the great vessels, the lymphatic system and the skin are commonly involved. A particular manifestation of this late stage is the formation of gummas, circumscribed nodules of inflammatory tissue with a necrotic center. They are prevalent in skin and bone, but may be present in any organ as isolated or, more frequently, clustered lesions. Involvement of facial structures resulting in punched out lesions such as the Inca's nasal defect is not rare.

Although central nervous system involvement is more frequently seen in the third stage of the disease, there is a significant incidence of syphilitic meningitis and meningovascular syphilis occurring in as little as twelve and six months respectively, after the primary infection. These complications of secondary syphilis may present in association with the often-dramatic maculo-papular skin rash of this stage of the disease, while neurovascular compromise induces, potentially lethal, increased intracranial pressure that may result in the signs and symptoms exhibited by the Inca. Of no lesser importance, when considering syphilis as Huayna Capac's cause of death, is that, of the three diseases under discussion, no other may be congenital and that his son died a neonate of the same illness. Congenital syphilis nearly always presents with neurological symptoms and a severe vesciculo-bullous skin rash. It remains one of the diseases leading mortality factors, causing stillbirth and infant death, even when treated with modern pharmacological agents.[82]

Huayna Capac is said to have sired more than three hundred children and, since his son was recently born when he died, was still sexually active. According to Cieza he was notoriously promiscuous: "vivia vicioso de mujeres," "women were his life's vice,"[83] and perhaps less credibly Oliva, writing three decades later tells us that in Quito "estuvose largo tiempo entretenido Guayna Capac en sus gustos de aquel reino hasta que le dio una grave dolencia 'que los indios llaman huanti y en nuestro romance bubas que le quito la vida."[84] (Huayna Capac stayed for a long time in that kingdom enjoying his pleasures until he got a grave malady that the Indians call *huanti* and in our language *bubas* that took his life.) Rostworowski quoting Santa Cruz Pachacuti, notes that Cusi Rimay, Ninoncoyuchi's mother died at childbirth,[85] while Raura Ocllo Coya, Huayna Capac's principal wife that according to Guaman Poma died of the same disease as her husband, travelled with his mummy to Cuzco (see above).

82 Arnold and Ford-Jones 2000.
83 Cieza de León (1544) 1922: 383.
84 Oliva, quoted by Cobo 2018: 189.
85 Rostworowski 2001b: 149.

Summary and Conclusions

A clinical approach was used to search for clues helpful in identifying Huayna Capac's fatal illness. Relevant information was principally obtained from the chronicles of de Betanzos, Garcilaso and Guaman Poma, selected because the authors were fluent Quechua speakers, had obtained their information from witnesses believed alive during the last days of the empire and contained data helpful to construct a clinical picture of the Inca's final ailment. Other chronicles were used to the extent that they contributed to a differential diagnosis of his disease or included descriptions of mummified remains believed to be Huayna Capac's. The more valuable clinical information was found in de Betanzos' chronicle.

The reviewed material corroborates McCaa et al. study (see introduction) advocating the exclusion of an epidemic, as the cause of Huayna Capac's demise. A virulent event occurring in Cuzco possibly in 1524/25, that has been interpreted as the departure point for an empire-wide epidemic was in all likelihood one of the episodes of epidemic typhus that periodically afflicted the populations of Tawantinsuyo. There are no evidences that it extended from Cuzco to Quito and recent research strongly suggests that the Inca died after members of Pizarro's band first reached the northern limits of the empire in late 1528 or early 1529. de Betanzos and Garcilaso do not describe a great mortality associated to the Inca's passing and neither includes a member of the Inca's lineage, other than his neonate son, as dying of the same disease.

Among the venues followed by historians searching for Huayna Capac's cause of death, descriptions of mummies assumed to be his have received considerable attention but have contributed more to confusion than to clarity. This notwithstanding, Pedro Sancho de la Hoz's account appears reliable and clinically useful. He saw the mummy before it was transferred, manipulated, buried or disinterred, gave it more than a cursory look and noted that it had a nasal tip defect. Garcilaso was shown five royal mummies twenty years before writing his chronicle. He stated that the remains had been identified by local Indians and that one was Huayna Capac's, but ruefully confessed that he had not looked at it carefully. His description of the mummy is not clinically helpful, nor is Guaman Poma's that is limited to a drawing of the Inca's flawless remains being transported to Cuzco.

It has been for long assumed that the disease that killed the Inca was an acute infection. The clinical picture that evolved throughout this study is compatible with an infectious process but not necessarily acute. It was characterized by a cutaneous eruption and a period of uncertain duration during which disorientation, dullness of mentation and hallucinations were prevalent. Among the infectious diseases known to have occurred in Peru before the Conquest only epidemic typhus, bartonellosis and syphilis may be considered as possible culprits. Of these, the dearth of evidences supporting an epidemic and the Inca's protracted symptoms exclude typhus.

In a non-epidemic scenario bartonellosis and syphilis appear as possible causes of Huayna Capac's death. The elements in favor of the former are: 1, the Inca and his son died in a region where it is endemic and, although clustering is more common, isolated cases do occur; 2, the disease has a higher incidence among children under five, such as Ninancoyuchi and is more lethal in untreated patients over fifty; 3, the central nervous system manifestations and psychological disturbances exhibited by the Inca are known to occur in about one fourth of patients. However, if bartonellosis was the Inca's cause of death he would have likely died in the hemolytic stage during which there are no skin changes compatible with a severe eruption. The Inca was born and lived for years in endemic areas (Tumibamba and Quito) and would likely have acquired a degree of immunity. If infected, he probably would have survived the deadly hemolytic phase, to present in the eruptive stage that is not associated with neurological symptoms and deaths are very infrequent. Bartonellosis does not explain Huayna Capac's mummy nasal defect, nor the absence of other facial lesions, nearly always multiple in the eruptive phase.

Syphilis has seldom been addressed in studies investigating the cause of Huayna Capac's death and when considered has been dismissed. However, these studies were undertaken when smallpox was generally accepted as its likely cause and before the complete version of de Betanzos' chronicle was found. The elements that favor syphilis as the Inca's killer are: 1, evidences of a protracted illness in its terminal stage; 2, severe, fluctuating central nervous system symptoms including dementia; 3, a nasal defect, possibly a consequence of a luetic gumma; 4, his neonate son, and possibly his mother, died of the same illness, indicating a likely congenital ailment associated with a severe skin rash. Against the diagnosis of syphilis are the unclear characteristics of the Inca's skin lesions and the timing of their appearance, though cutaneous gummas could certainly be called "una sarna y una lepra."

This study findings strongly suggest that bartonellosis and, more likely syphilis, should be considered as the probable causes of Huayna Capac's death. As a corollary, it can be ascertained with a high level of confidence that the Inca was not the victim of an imported epidemic malady and, more importantly, that the one that killed him did not contribute to the depopulation of the empire during the early Conquest. Although European diseases did later have a part in the Peruvian tragedy, the greed driven abuse of the indigenous peoples by the invaders, long ignored by the Spanish Crown, played the leading role.

Francesco Maria Galassi, Giovanni Spani and Elena Varotto

Paleopathology and Anthropology of the Renaissance: From the *Morbus Dominorum* to the Alleged 'Michelangelo's Shoes'

A substantial difference exists between contemporary medicine and paleopathology, particularly in the kind of information that can be retrieved from analyzed sources. Medical practice as we presently know it consists of a series of steps: anamnesis (the taking of a patient's medical history); physical examination (that leads to the first formulation of diagnostic hypotheses); obtaining ancillary tests (e.g., imaging, histology, genetics, etc.); establishing a diagnosis; initiation of therapy; follow up on the patient's conditions and response to treatment; reevaluation or post-mortem assessment in certain cases (e.g. anatomical pathology or forensic medicine).

Notably, the first two steps (anamnesis and physical examination) help to compile a vital body of semiological information regarding the patient's condition, one that allows for the discernment of the disease through characteristic combinations of signs and symptoms experienced by the sufferer. This essential context has already been irremediably lost by the time a scientific examination of ancient human remains (e.g., skeletons or mummies) begins. As a matter of fact, ancient bodies may present the traces and signs left upon them by pathological processes but they tell us nothing about how the disease was actually experienced by the patient at the moment of its clinical presentation. In order to overcome this obstacle, it is necessary to take advantage of both classical anthropological methodologies and a historical approach. That said, one would do well to avoid dramatic conflicts between science and the humanities, such as, mutatis mutandis, those imagined by Jonathan Swift (1667–1745), that recall the old motif of the "battle of books."

The Italian Renaissance, with its abundance of remarkable characters, offers valuable opportunities to investigate past pathological conditions through a combination of historical and paleopathological analyses.[1] Our first case is that of Duke Federico da Montefeltro (1422–1482, Figure 1), a philanthropist and military leader. Available historical sources (in particular Baldi) report that the duke was afflicted by gout but, by abstaining from the pleasures of life and by following the advice provided by his physicians, he managed to overcome it.[2]

"Gout" (Lat. "gutta") was an ambiguous word in the 15th century since it referred to a number of rheumatological conditions, and not necessarily to the hyperuricemia-

1 See Rühli 2016: 816–822, and Galassi 2018: 267–280.

2 See Baldi 1824: 270.

https://doi.org/10.1515/9783110788501-013

Figure 1: Justus van Ghent (fl. 1460–1480), *Federico da Montefeltro and his son.* Palazzo Ducale Urbino, Italy.
Image in the public domain from *Wikimedia Commons.*

related arthritis recognized by modern medicine.[3] Hence, any superficial interpretation must be supplemented with data from the autoptic examination of the ancient patient's mortal remains. A joint project with paleopathologists from the University of Pisa demonstrated that the Duke's first right metatarsal bone was characterized by a high degree of erosion, as typically seen in gouty arthritis.[4] These remains – a rather lucky discovery considering the high levels of humidity recorded in the church of San Bernardino in Urbino – were radiologically examined (X-rays and CT scans) and then juxtaposed with a unique piece of archival information. In a letter sent to his physician (Battiferro Battiferri da Mercatello), Federico da Montefeltro complained of an excruciating pain

3 Gout can be easily found in the general population nowadays but in the past it used to be thought of as a disease particularly affecting the affluent classes (from which the appellative "morbus dominorum et dominus morborum").

4 The project was led by Dr. Antonio Fornaciari (University of Pisa) and Dr. Francesco Maria Galassi (at the time at the University of Zurich).

in his right foot, precisely the one affected by the aforementioned erosive process. This is one of those remarkable instances in which anatomical evidence is clearly supplemented by primary and secondary historical sources. Indeed, the Duke's correspondence even provides a glimpse into patient-physician interaction and the selection of a treatment to address the symptomatologic presentation of the disease. He was given a medicine known as mithridate, not necessarily effective but probably in this case to elicit a placebo effect.[5]

The second case is that of Sigismondo Pandolfo Malatesta (1417–1468), lord of Rimini and Federico da Montefeltro's archenemy.

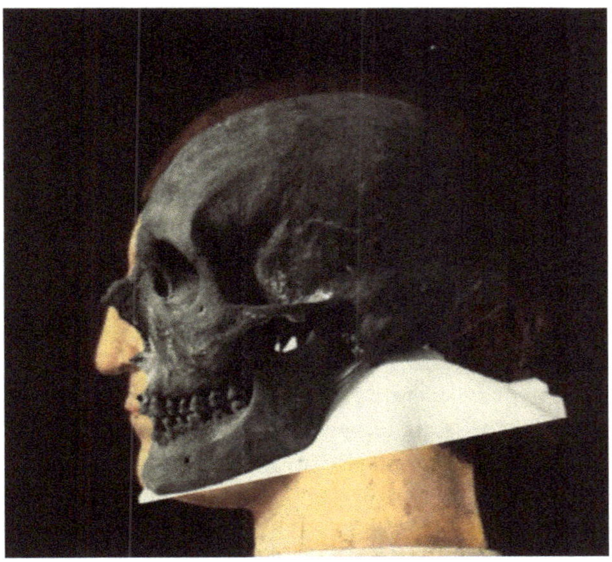

Figure 2: Superimposition of a picture of Sigismondo Pandolfo Malatesta's skull onto a lateral view of his face (the shape of his nose sharply differs from that seen in the artwork). The skull is visible on a photograph taken during the early 20th century exploration of his tomb.
Permission to reprint from Biblioteca Gambalunga, Rimini.
Louvre portrait by Piero della Francesca (ca. 1451). RF 1978 1.
Image in the public domain from *Wikimedia Commons*.

American poet Ezra Pound (1885–1972) called Sigismondo "the best loser in history," in reference perhaps to the ideal Prince that Machiavelli would see in Cesare

5 More details on this case study can be found in Fornaciari 2018: 15–20. For the historical aspects of gout and uric acid-related diseases, see Porter and Rousseau 1999, and Galassi and Borghi 2015: 373. Historico-medical considerations on a suggestive hereditary nature of gout within the Montefeltro family and the presence of gout in Federico's son, Guidobaldo (1472–1508), can be found in Bianucci 2016: e28–e30.

Borgia (1475–1507).[6] Several medical conditions have been suggested with regard to this important historical figure, of which two stand out in particular. The first is a severe form of depression experienced after being militarily defeated by Federico da Montefeltro in 1463, as we learn from an epistle addressed to Pier Francesco de' Medici (1430–1476) and dated December 5th of that same year.[7] In this letter, Sigismondo asks his fellow nobleman for a pair of greyhounds that might provide some solace in his sorrows ("bizarie et melanconie"). As most of his power and glory was by then gone, his depression seems to describe his own bitter reflections:

> retrovo la conditione mia essere l'opposito de quello che volgarmente se sole dire, che chi ha poca roba ha pochi pensieri; a mi è remasto poca roba e assai pensieri.
>
> [I find my condition to be the opposite of what is commonly said – that he who owns little has few concerns; I am left with little but I have many concerns].[8]

The second disease that likely destroyed his health was malaria, an infectious disease that he contracted during a campaign in Morea (Peloponnese) while serving in the Venetian army against the Ottomans (1464–1466). Once Malatesta returned to Italy, he was treated by Nicolò dal Dito (ca. 1416–1487),[9] a physician sent by the Pope specifically for this purpose. Unfortunately, dal Dito was never able to reduce Malatesta's high fever and the patient eventually died on October 9th, 1468. Further biological details regarding Malatesta can be drawn from the extant information gleaned from the four occasions on which his tomb was reopened.[10] Located within Rimini's so-called Tempio Malatestiano, a church built at Sigismondo's request by Leon Battista Alberti (1404–1478), who embellished it with all sorts of pagan symbolism. Indeed, Pope Pius II (1405–1464) scornfully derided it as a monument to Sigismondo's unrestrained ambition.[11] The 1920 investigation was by far the most important because it provided the bulk of what we know about Sigismondo's anthropological features. The study, led by archaeologist Corrado Ricci (1858–1934) and by physical anthropologist Giuseppe

---

6 See Paganelli 2013: 43–45.

7 See Archivio di Stato di Firenze. Archivio Mediceo avanti il Principato. Inventario, vol. 1. Filza II, Pandulfus Sigismundus de Malatestis. Arimini, 1463 dic. 5, n. 492.

8 Archivio di Stato di Firenze.

9 Dal Dito received a sum of 11 florins from the treasurer of the Apostolic Camera (i.e., Papal Treasury) to cover the expenses for a journey to Rieti. See Camera Apostolica *Introitus et exitus*, regesto 471, f. 137v. annuum 1468, in Archivio Segreto Vaticano.

10 August 21st, 1756; September 28th, 1920; February 24th, 1944 and May 11th, 1950.

11 One cannot be quite sure how much Pius II (born Silvio Enea Piccolomini), himself a humanist, really believed his own words, yet his anathema was sharply aggressive: " . . . ædificavit tamen nobile templum Arimini in honorem divi Francisci, verum ita gentilibus operibus implevit, ut non item christianorum quam infidelium dæmones adorantium templum esse videretur, atque in eo concubinæ suæ tumulum erexit et artificio et lapide pulcherrimum, adjecto titulo gentili more: Divæ Isottæ sacrum" (Pii II, *Commentarii rerum memorabilium que temporibus suis contingerunt*, liber II, p. 51), cited in Von Wessemberg 1845: 286.

Sergi (1841–1936), yielded the following results: Malatesta's stature was ca. 170 cm; his cranium was megalocephalic and it showed an aquiline morphology of the nasal bones; all teeth were still present with no signs of caries nor periodontal pathologies, though a precocious ossification of the cranial sutures was observed;[12] a peculiarly horn-shaped medially concave bone spur of the squamous part of the right temporal bone was also observed;[13] the right radius showed traces of an old diaphyseal fracture of the distal third, which meant that this skeletal segment was shorter than the left radius.[14] The ability to compare photographical documentation from this investigation with other forms of iconographic evidence teaches us still more. The superimposition of the photograph of the lateral view of Sigismondo's skull (his left hemicranium) upon the pictorial representations of him portrayed by Piero della Francesca (1416/1417–1492) (Figure 2) incontrovertibly demonstrates that the artist embellished his client's facial traits, as Malatesta's nose is considerably straighter (rather than aquiline) in the artwork.[15]

The last case[16] taken into consideration is that of Michelangelo Buonarroti (1475–1564), one of the icons of the Italian Renaissance. Many hypotheses about his precarious health have been proposed, ranging from a goiter to lead poisoning. Recently, it has also been suggested that his hands were severely affected by chronic arthritis in his final years.[17] Most of these interpretations are essentially retrospective

12 A rather trivial piece of evidence according to modern anthropology, to which perhaps an inappropriate interpretative weight was instead given back then.

13 This could be the result of a skull trauma followed by with muscular calcification, which resulted in such an exuberant horn. As a result of the 1944 bombing of Rimini, the Tempio Malatestiano and Sigismondo's sepulchre were heavily damaged. Sergio Zavoli (1923–2020) recalled this event and the image of this eye-catching horn was clearly stuck in his memory: "All'improvviso, con uno scoppio, andò in frantumi il cranio ardente di un soldato, un giorno si saprà che da uno squarcio del Tempio, rotolando nella voragine aperta da una bomba, era finito un teschio che conservava una protuberanza sulla fronte a destra: il cranio di Sigismondo Malatesta," in Zavoli 2011: 43.

14 This may have well resulted from military practice, potentially caused by a wound occurred to him in Fano in 1431 as he was assaulted during an uprising orchestrated by a certain don Matteo Buratelli da Cuccurano. For more information on Sigismondo's pathobiography and his mortal remains, see De Carolis 2001: 32–49; De Carolis 2002: 29–39; De Carolis and Galassi 2017: 79–82.

15 Piero della Francesca's artworks are the oil and tempera on canvas paintings exhibited at the Louvre Museum in Paris (ca. 1451) and the fresco in the Tempio Malatestiano (1451). Piero della Francesca seems to have been closer to a properly *ad verum* representation when he depicted Federico da Montefeltro. However, since the duke's skull had not been preserved when his burial was explored, this assumption remains only a mere speculation. See Fornaciari 2000: 211–218.

16 This final section of our paper on Michelangelo was also published in the journal *Anthropologie* as a reworked article in the interim between the 2018 Los Angeles meeting and the final publication of the present volume. See Galassi and Varotto 2021: 97–99.

17 See Lazzeri 2016a: 180–183; Lazzeri 2016b: 1–5; Montes-Santiago 2013: 223–240. Bianucci has recently added a new hypothesis for Michelangelo's visual loss while painting the ceiling of the Sistine Chapel by proposing the deconditioning syndrome. See Bianucci 2018: 13–14. However, it must be noted that Michelangelo did not paint the ceiling while lying on his back as some sources insist to suggest. See also Martini 2021: 1264–1265.

Figure 3: Measurement of one of Michelangelo's alleged shoes, housed in the Casa Buonarroti in Florence, using a sliding caliper. Inv. 1859, n.198; Inv. 1896, n.442.
Photo: FAPAB Research Center/Elena Varotto.

diagnoses based on indirect information, such as portraits of the artist or his own writings. Until now, no analysis had added further evidence based on biological evidence left by Michelangelo himself during his lifetime. The Casa Buonarroti Museum in Florence offers what might be the sole remaining clue. We present here an anthropological analysis of Michelangelo's alleged *babbucce* (shoes, Figure 3), consisting of two matching shoes (left and right) and one loose right slipper (both of leather), which were found in the house and subsequently musealized [Inv. 1859, n.198; Inv. 1896, n. 442].[18] They have been traditionally regarded as Michelangelo's, and indeed their style corresponds to that of the historical period in which he lived.[19] No C14 dating has been performed on them yet – nor will be – due to legitimate preservation concerns. For these reasons, no definitive conclusions can be drawn about their authenticity and it is possible that they might very well have belonged to other family members or perhaps to one of his descendants.[20]

18 Originally there were two pairs, one of shoes and one of slippers, but one of the slippers was stolen January 14th, 1873. See Procacci 1965: 210.

19 Personal communication by Dr. Elisa Tosi Brandi.

20 A possibility which is more strongly suggested for two walking canes exhibited in the same Museum. See Procacci 1965.

However, the possibility that they can help us to determine their wearer's stature is worth investigating. By applying the Uhrová anthropometric formulae to estimate stature from foot dimensions,[21] the results are:

a. matched left shoe = 158.2 cm ± 4.765 cm (standard error)
b. matched right shoe = 162.7 cm ± 4.652 cm
c. matched left + matched right shoes = 161.4 cm ± 4.668 cm

The average of the three measurements gives 160.7 cm. The unmatched right slipper gives instead the following stature: 159.2 cm ± 4.652 cm. The average of all the results is 160.3 cm.

The measurements of the two shoes and the single slipper are comparable. This similarity is reflected in the conjectured heights, which do not exclude their having belonged to the same person, at least from the perspective of their anthropometric characteristics. Moreover, as reported by Cardoso and Gomes, or by Giannecchini and Moggi-Cecchi specifically for Central Italy,[22] such an average is compatible with the mean for a chronological period that spans from the Middle Ages through the Renaissance. This observation is coherent with the claim of Michelangelo's biographer, Giorgio Vasari (1511–1574), that he was of average height.[23] While one cannot ultimately prove that these shoes were in fact Michelangelo's, multidisciplinary work of this nature underscores the importance of using direct biological information whenever possible. Indeed, we are continuing along this path in our ongoing analysis of fingerprints and facial morphology from a likely copy of his death mask.

In conclusion, the concurrent analyses of written records and extant biological evidence regarding prominent Renaissance figures can help science to shed light on past medical conditions and to answer anthropological questions that enrich our historical knowledge of this crucial period in recorded history.

Acknowledgements: The authors would like to thank all their colleagues (including the Paleopathology Unit of the University of Pisa and Stefano De Carolis, director of the School of Medical History of the Rimini Medical and Dental Association) who, in the past, contributed to the study and publication of the first two cases – Federico da Montefeltro and Sigismondo Pandolfo Malatesta – a concise summary of which has been presented here. We would also like to express our deepest gratitude to Roberta Ballestriero (Central Saint Martins College of Art and Design, University of the Arts London, UK), for her helpful stylistic remarks on Piero della Francesca's works, and to Maciej Henneberg (The University of Adelaide, Australia) for his help with the

21 Uhrová 2013: 448–451. Foot size is represented in our study by the length of the shoes.

22 Cardoso 2008: 711–725; Giannecchini and Moggi-Cecchi 2008: 284–292.

23 "Fu di statura mediocre, nelle spalle largo, ma ben proporzionato con tutto il resto del corpo," in Vasari 1832–1838: 1020. The word "mediocre" (as used by Vasari in 1568) should be read in its original Latin-derived meaning, thus not at all in a pejorative connotation.

facial superimposition briefly discussed in this context. With regard to the research on Michelangelo Buonarroti's stature, and the original material discussed here at length for the first time, special thanks go to the Casa Buonarroti Museum, its director, Alessandro Cecchi, and the staff for kindly allowing us to examine and measure "Michelangelo's slippers" in the spring of 2018 and for permitting us to publish a photograph of the shoes in this publication. The authors also express their gratitude to Elisa Tosi Brandi (University of Bologna) for her invaluable stylistic observations on the footwear. Last but not least, we would like to thank anthropologist Raffaella Bianucci (Universities of Turin and Warwick) for her interesting observations on Michelangelo's ailments during his work in the Sistine Chapel.

Reason, Affects and Madness

Remo Bodei†

The New World Opened by Madness

1. Let's submit to a simple mental experiment: think of the astronauts floating in the cosmic space. We were used to believing that the force of gravity possessed an absolute and necessary validity so as to keep the inhabitants of the antipodes with their feet on the ground, without suspecting that its absence, while not denying this principle, could give rise to the levitation of bodies. Similarly, when we reflect on madness, particularly on schizophrenia, we must free ourselves, conceptually, from that psychic "gravity force" that immediately holds us bound to our standardized image of rationality. We must, instead, face situations that, despite their absurd and twisted appearance, can make us glimpse other worlds that are not completely incompatible with ours. In this way, the force of gravity of reason is not denied by its absence, provided that this kind of reason that I call "hospitable" is able to consider and accommodate apparently absurd experiences and to analyze them without prejudices.

In order to understand these forms of abnormal logics of madness, we need to take into account a tradition that follows a line of thought that goes from William James, physician and philosopher, to Alfred Schütz, one of Husserl's most brilliant pupils. James theorized the idea of "sub-universes of reality," meaning that there is no a single universe, endowed with the same rules and the same criteria of relevance.[1] There are, instead, many universes of reality: for example, next to that of the "paramount reality" (the reality that appears superior to all others, that reality of everyday conscious and awake life), there is the sub-universe of the dream, that of madness, that of myth, that of science or that of religion. Each of them has specific rules and incomparable criteria of relevance. The time of dreams, for instance, does not coincide with the time measured by the clock, as well as the desire to treat the myth of Zeus with the same systems that are used in physics and mathematics is completely incongruous.

Schütz has translated and expanded this intuition of William James, speaking not of sub-universes of reality, but of "vital worlds."[2] They do not constitute

---

1 Cf. W. James (1890), *The Principles of Psychology*. Cambridge: Harvard University Press, 1981; W. James (1907), *Pragmatism: A New Name for Some Old Ways of Thinking*. New York-London-Toronto: Longmans, Green and Co., 1949. See also: W. James, The *Varieties of Religious Experience: A study in Human Nature*. New York-London: Longmans, Green and Co., 1902.
2 A. Schütz (1932), *Der sinnhafte Aufbau der sozialen Welt*. Frankfurt: Suhrkamp, 1993.

Note: Remo Bodei's essay was received, shortly before his passing, by the editors of this volume. Consequently, some notes to the text were added (cf. Afterword) and do not follow the bibliographical style and order of the rest of the book.

https://doi.org/10.1515/9783110788501-014

Husserl's "world of life," but "finite provinces of meaning." It is a vital world, for example, a child's game, in which he passes abruptly from one vital world to another imagining to be an ancient knight or comic book character. He then creates a world independent of the surrounding environment, provided with a kind of membrane that isolates it from the context. When this child is called for some reason to stop playing by the mother or teacher, the spell breaks, the membrane tears and the child is forced to return to the "paramount reality," to the world shared by others.

We too, in order to enter the sub-universe of madness (and more precisely of schizophrenia, as a member of the family of psychoses), must make a similar leap, similar to what we do every morning moving from dream to wakefulness. We must also stop thinking about experience as a homogeneous whole.[3] We do not realize, for example, that the same person, when he performs military service or takes religious orders, entering the barracks or in a convent, moves to a world in which the criteria of relevance change and space, time and rules have a qualitatively different meaning. In such contexts we are no longer the same person, we obey rigid norms, we lack the autonomous availability of our time, we wake up in the middle of the night for military exercises or prayers. Physical places seem equal, they have spatial contiguity with the environment that surrounds them, but they are also separated by thresholds that distinguish one world from another, or certain forms of identity from others.[4]

In an apparently paradoxical way, in its sub-universe of reality madness is creative: those who have seen the paramount world collapse around themselves are obliged to look for a substitute, surrogate, invented new world. When life becomes unlivable, psychosis takes the place of reality, reshaping and remodeling reality. It is induced to devise models corresponding to the new reality by means of perceptions and thoughts in conformity with it. Hallucinations and delusions thus appear as modalities of an inverted *adaequatio*: it is "external" reality that has at all costs to conform to "internal" one. Trust, then, in given reality diminishes and, quite often, hate and destructive fury against whatever can evoke it grows.

In other words, the psychotic individual breaks the agreement which demands of everyone that they conform to reality. In all these cases the concept of "reality" must be understood in a sense that is more prescriptive than descriptive. In fact, it indicates an obligation to be faithful to reality as the guarantor of the survival both of the species and of the individual. It indicates the discipline that has been and still is necessary in order to maintain a shared world and to bring each human being into tune with it, limiting the range of conceptual, perceptual and affective variation allowed.

---

3 See among others V. Melchiorre, (ed.), *I luoghi del comprendere*. Milan: Vita e Pensiero, 2000.

4 R. Bodei, *Destini personali. L'età della colonizzazione delle coscienze*. Milan: Feltrinelli, 2002.

2. This methodical construction of a new world, precisely because it is fragile, needs to constantly equip itself with new elements of defense (because it is better a wrong interpretation than any interpretation). Hence the forced creativity of psychosis. The psychotic subject believes indeed in the immediate truth of what is coercively revealed to him. The suspicion of its falseness or illusion does not arise in general, because he has remained intimately attached to the logic of desire, which erases any contradiction between the possible and the impossible. That is, he did not entirely complete the transition from the logic of desire to that of accepting constraints, to the logic of a relatively well-established rationality. Therefore, when the schizophrenic episode manifests itself, the false world of the previous, ungrateful and despised reality crumbles for him and in the real world forged by the logic of desire reappears, bright and triumphant, the truth that he had to reject. The "nucleus of truth" is now for him evident. It constitutes the rediscovery, the epiphany of a very ancient truth, which he has had forced to renounce in his process of growth and socialization. For him it is a return to his homeland, to the place where there were no obstacles to his desires. And yet, what appears to him as irrefutable evidence does not have anything original, but represents a reconstruction of the past, as mythical as the paradise lost.

I wish to put forward the hypothesis that psychoses arise when the suffering caused by what has been repressed provokes unbearable psychic tensions. Madness is therefore the result of an earthquake that wrecks the layers of personality that had been carefully laid one upon the other. A trauma, stress or a life event (or any quite ordinary, perhaps even joyous, matter that intimately involves the individual's existence: marriage, divorce, the birth or death of family members, moving house or a change of profession, unexpected financial gains or losses) may reopen wounds that had never completely healed, reactivate unsatisfied desires, renew old fears, feelings of guilt or misunderstandings, uncovering and aggravating latent cracks and old failings in the logical-affective delimitation of the internal and external worlds.

In the psychotic subject the old world not only vacillates, it is set aside and replaced by another one. However, its loss is counterbalanced and made good by the creation of a new and different reality, which does not present the same impediments to the satisfaction of desires. This is not a partial privation of reality: the whole universe as previously perceived, imagined, thought, as wrapped all around in passions and desires, seems suddenly to give way and must therefore be rebuilt as soon as possible. This is how the contents of delusions appear: like shreds or rags found – however and wherever – to plug the cracks in the relation between the self and the world. The fear of seeing one's own life sink increases with the recognition that the tears are concentrated where the dividing wall between the subject and the object is thinnest and most fragile. Madness is the last straw of an effort to somehow bring order out of the chaos through which an existence perceived as having no way out is plunging. In delusion, therefore it is not that logic proves defective or that the reality test has failed: it is the content gathered there that obeys a new, different, but coherent logic: "The specific form of the delusional idea . . . is nothing other than the attempt by thought, still intact, to establish a logical connection between the stones of a building in ruins."[5]

Grafting itself onto a past that has not been worked through, a real trauma serves to detonate deeper psychic charges that bring incomprehensible remains of what has already been crashing to the surface. These remains turn out to be just as incomprehensible when combined with new fragments of lived experience. The psychotic individual is in the middle of a tangle of logics that have structured his previous experience, but cannot now account for the new one. Caught in this vice-like grip, he must shape himself a personality and a reality that is synchronized intermittently with the shifting equilibrium reached in the struggle between these logics. His mind then becomes the matrix of a series of combinations that are inappropriate, absurd and bizarre, yet in conformity with the new world in which he wraps them.

3. But lack of contact with the world of others does not suffice to explain psychosis. You can flee the world, turn your back on it, refuse to share it with your fellows and behave like a hermit, without necessarily plunging into madness. It is not that the subject simply wishes to abandon a hostile reality, obstinately denying whatever contradicts his delusion. To seek shelter from suffering – in view of a catastrophe in progress or presaged as imminent – he rebuilds the world he has lived in until now, using whatever materials he has. He sets sail for what many have seen as a shipwreck of the mind, but which may be better described, according to Freud, as "another world in which its most unbearable features are eliminated and replaced by others that are in conformity with one's own wishes. But whoever, in desperate defiance, sets out upon this path to happiness will as a rule attain nothing. Reality is too strong for him. He becomes a madman, who for the most part finds no one to help him in carrying through his delusion."[6] He is condemned to symbiosis with his new world, from which he becomes inseparable to the point that psyche and world form a hendiadys in him. The delusional subject is thus not "an orchestra without a conductor" as Kraepelin believed, but rather a conductor who seeks to make his – for us cacophonic – orchestra function according to new, improvised programs.

5 E. Minkowski, "Étude psycologique et analyse phénoménologique d'un cas de mélancolie schizophrenique," *Journal de psychologie normale et pathologique*, 20(10) (1923), pp. 543–560, Italian translation "Studio psicologico e analisi fenomenologica di un caso di melancolia schizofrenica (1923)," in E. Minkowski, V.E. von Gebsattel, E.W. Strauss (eds.), *Antropologia e psicopatologia*, D. Cargnello (ed.). Milan: Bompiani, 1967, pp. 30–31.

6 S. Freud, *Civilization and its Discontents*, in *The Standard Edition of the Complete Psychological Works of Sigmund Freud*. London: Hogarth Press and the Institute of Psycho-Analysis, 1953–1974 [1978 reprint], vol. 21, p. 81.

In other words, the psychiatric patient breaks the agreement that commits everyone to a common reality. In all these cases the concept of "reality" refers to the discipline that is necessary in order to maintain a shared world and to bring each human subject into tune with it, limiting his range of conceptual, perceptive and affective oscillations. "Reality" represents the band of converging lines of perspective that frame the mental, affective and perceptive constructions always in progress in the building sites of different civilizations. It by no means constitutes a natural starting point, a given from which human subjectivity is expunged. We could say that all cultures, with different strategies and in different ways, reproduce the effort to keep individuals anchored to a common reality. Societies, languages, institutions create an orthodoxy of reality. Persons with mental disorder, instead of maintaining contact with the shared world, demand that everyone else share a world of his own making.

4. I will not dwell any longer on the solution I have offered to the question of the psychotic loss of a common world. I would like to end with a few remarks of an existential nature. In its banality and strangeness, madness reveals the latent fragility of everyone's experience, its reliance on assumptions that are uncorroborated, unanalyzed or simply forgotten.[7] One trusts in these invisible linchpins around which we have automatically made our thought and our life turn for so long: at least until they crack, dragging down the trust that we had in ourselves and in others as they give way. The desert, like polar solitude in which the psychotic individual encloses himself in the company only of his phantasies of persecution, jealousy and greatness; the visions and the voices; the anomalies in conceptualization and reasoning; the feelings of guilt, of shame or of emptiness; the suspicion or the garrulous rush; the ruin, the loss, the separation or the release from what one loves – all this cannot but drive him further from the path of common experience. Madness is disturbing and feared precisely because it threatens and puts shockingly into question the world of each and every one of us in all its supposed obviousness.

The question that I have implicitly posed is not so much why this loss of a common world takes place. With this inquiry I am continuing a program of research that began with a study of the passions and of those phenomena – such as political ideologies – in which rationality does not appear to enjoy the right of citizenship. Such a project is justified in my eyes because I am convinced that perhaps the most noted trend of modern philosophy, the introduction of so-called "rationalism" into common sense, in seeking to imitate the successes of the mathematical and physical sciences, has adopted a model that is strictly inappropriate to the human world. Unable to find anything corresponding with this model within its own boundaries, it has abandoned large and crucial areas of individual and social existence to the thorns and thickets of ignorance. It has thereby handed the task of establishing order here to political and religious power, to history, traditions, habits and fate. To

7 R. Bodei, *Le logiche del delirio. Ragione, affetti, follia*. Bari-Rome: Laterza, 2000.

paraphrase Lévi-Strauss (who speaks of *pensée sauvage*, meaning thought that is untamed, spontaneous, uncultivated, as the *viola tricolor* flower, the violet, called in French *pensée*),[8] I would call the *vie sauvage* that whole area of human experience – including our passions, phantasies, beliefs and anomalous logics – that is left to the mercy of the "irrational." It therefore seems indispensable today that we begin a long and exhaustive process aiming at the collective recognition of these terrains, their reintegration into the intellect and cultivated life, and their conversion into the seedbeds for the production and reproduction of meaning. The treatment of issues related to medical humanities is one of these seedbeds.

Afterword

A Philosopher and a Committed Intellectual

Manuela Gallerani

Since the text of Remo Bodei has remained unfinished, Manuela Gallerani has been so kind as to accept our invitation to comment on the content of the essay.

I. I wish to extend my heartfelt thanks to my colleague Massimo Ciavolella for having granted me the privilege and honour of editing this never-before-published essay by Remo Bodei. As one of his last works, it represents a priceless final bequest and gift. This premise is indispensable, for Ciavolella was like a brother to Bodei, and like him is endowed with uncommon empathy and generosity.

The unmistakable speculative thought of Remo Bodei, a fellow of the Accademia dei Lincei and fine scholar of the history of philosophy and aesthetics, is limpid and brilliant as the words he employs to express it.[9] Words so clear and incisive as to render the addition of notes or introductions superfluous. Nevertheless, it seemed meet to the distinguished editors of this volume to do so, as Remo Bodei is no longer physically among us following his passing on 7 November 2019. Moreover, the CMRS Ahmanson UCLA Conference entitled *Understanding Medical Humanities* gave rise to a fruitful dialogue of many voices: in the spirited debate that followed, Bodei himself expressed the wish to integrate his essay with notes that only a sudden and inexorable illness prevented him from writing.

The essay of Remo Bodei is written in an essential style, limpid and lacking redundancies of any kind. Bodei included notes no. 5, no. 6 and no. 8 in his own

8 See C. Lévi-Strauss, *La pensée sauvage*. Paris: Plon, 1962.

9 With reference, in particular, to the following studies by Remo Bodei: *Piramidi di tempo. Storie e teoria del déjà vu*. Bologna: il Mulino, 2006; *Paesaggi sublimi. Gli uomini davanti alla natura selvaggia*. Milan: Bompiani, 2008; *Le forme del bello*. Bologna: il Mulino, 1995 and 2017 (expanded edition).

hand in the text presented to the UCLA conference. I completed the other notes, incorporating the bibliographical references suggested or implied in the text, respecting *in toto* Bodei's thought and intervening unobtrusively, as is appropriate in a long-distance dialogue with an extraordinary, wonderful person.

Specifically, note 1 is reconstructed based on the following essay: R.Bodei (ed.), "Filosofia" in *La cultura del 900. Filosofia, Linguistica e Semiotica, Storiografia*. Milan: Gulliver, 1979, pp. 15–232; see p. 78 note 52 where the reference to William James' work *Pragmatism: A New Name for Some Old Ways of Thinking* (1949) appears. A further essay by William James, *The Varieties of Religious Experience: A study in Human Nature* (1902), is cited in Bodei's essay *Piramidi di tempo. Storie e teoria del déjà vu*. Bologna: il Mulino, 2006, pp. 61–62, note 11.

Alfred Schütz's 1932 text (see note 2) is a fundamental study that in some ways can be considered a "classic." The same goes for the note 8 on Claude Lévi-Strauss, *La pensée sauvage*.

Note 3 quotes a text also cited in R.Bodei, *Piramidi di tempo . . .*, p. 104, note 4.

Note 4 refers to an essay in which Remo Bodei addresses the theme of subjective experiences in the modern and contemporary world, developing his own critical reflection on the forms of individual and collective identity.

Note 5 refers to Eugène Minkowski's seminal essay entitled *Étude psycologique et analyse phénoménologique d'un cas de mélancolie schizophrenique*, often cited by Remo Bodei in bibliographies or in his book notes, as well as another well-known essay by the French psychiatrist, *Le temps vécu* (1933). Neuchâtel: Delachaux et Nestlé, 1968.

Concerning note 6 and the Freudian essay *Civilization and its Discontents,* it should be remembered that up to his final work entitled *Dominio e sottomissione. Schiavi, animali, macchine, Intelligenza Artificiale*, Bodei never fails to cite Sigmund Freud in his studies on passions and madness or in those pertaining to the power of political (or religious) ideologies.

Finally, note 7 refers to Bodei's rigorous analysis of the fascinating interweaving of reason, affects, and madness: a theme also shared by the above presented essay.

II. It is still difficult, for some of us, to reflect on two dates that remain etched in our memories for rather different reasons. The first is the fascinating, splendid experience of the International UCLA Conference held in Los Angeles on 2–3 November 2018, where a dense schedule of sessions allowed for high-level debate but also permitted speakers to converse amiably in the mood of convivial moments: for nothing could foreshadow such a tragic and imminent epilogue for one of us. The second date is the conclusion of Bodei's earthly sojourn, which has left his friends, colleagues, and students feeling a little more alone even as the world of culture has been impoverished by the loss of a brilliant intellectual. Bodei was a *contemporary* and *engaged* intellectual who believed in a militant philosophy.[10]

A loss all the deeper, and more melancholic, in the knowledge that we will no longer be able to meet him or call him on the phone to hear his voice again.[11] A calm and affable voice that revealed his irony and sober rigour, as well as the depth and authenticity of his critical and original thought. He was a rare person, able to engage listeners through storytelling and share the pleasure of philosophical debate,[12] used as a medium or instrument for both *understanding* the dark, hidden corners of existence and *narrating* small *things*:[13] a dialectic exchange of ideas, emotions, and thoughts that unfailingly generated mutual delight. For these and other, deeper, reasons, some of us continue to debate with Remo, remembering his luminous and illuminating *lesson*.

In conversing with him, one was immediately aware of participating in a relationship, an intense, culturally elevated, and absolutely *sui generis* dialogical exchange. Yet given his empathy and sincere immediacy, together with an innate curiosity to know others, the interlocutor – whatever his or her age, education level, or political beliefs – never risked feeling the slightest embarrassment. An omnivorous reader, he was never one to assume the air of an academic elite (which indeed he was); on the contrary, if he realized after a few exchanges that your

---

10 On the meaning and the role of *engaged intellectuals* it is possible to refer, among others, to the deep reflections and analyses of Antonio Gramsci, Raymond Aron, Edward Shils, Norberto Bobbio, Eugenio Garin and Michel Foucault carried out, for example, in: A.Gramsci, *Gli intellettuali e l'organizzazione della cultura*. Turin: Einaudi, 1949; R.Aron, *L'opium des intellectuels*. Paris: Calmann-Lévy, 1955; E.Shils, *The Intellectuals and the Powers, and Other Essays*. Chicago: University of Chicago Press, 1972; N. Bobbio, *Politica e cultura*. Turin: Einaudi, 1955; Idem, s.v. *Intellettuali*, in *Treccani Enciclopedia del Novecento* (1978) online, http://www.treccani.it/enciclopedia/ last accessed on 7/5/2021; N.Chomsky, *American Power and the New Mandarins*. New York: Pantheon, 1969; M.Foucault, *Il filosofo militante. Archivio Foucault*, vol. 2: *Interventi, colloqui, interviste. 1971–1977*, A.Dal Lago (ed.). Milan: Feltrinelli, 2017; E.Garin, *Intellettuali del XX secolo*. Rome: Editori Riuniti, 1974. On these issues see also: G.Vattimo, P.A.Rovatti (eds.), *Il pensiero debole*. Milan: Feltrinelli, 2010; P.Fabbri, *Biglietti d'invito per una semiotica marcata*, G. Marrone (ed.). Milan: Bompiani, 2021; P.Fabbri, *L'efficacia semiotica. Risposte e repliche*. Milan: Mimesis, 2017; R.Barthes, *Non si riesce mai a parlare di ciò che si ama*. Milan: Mimesis, 2017.

11 However, we can feel his *voice* in all his works, for instance, Bodei, *Immaginare altre vite. Realtà, progetti, desideri*. Milan: Feltrinelli, 2013; Idem, *Limite*. Bologna: il Mulino, 2016; Idem, *Comprendere, modificarsi. Modelli e prospettive di razionalita trasformatrice*, in A.Gargani, (ed.), *Crisi della ragione*. Turin: Einaudi, 1979.

12 These themes are explored in Remo Bodei's numerous reflective works, including: *Destini personali*. Milan: Feltrinelli, 2002; *Una scintilla di fuoco. Invito alla filosofia*. Bologna: Zanichelli, 2005; *La civetta e la talpa. Sistema ed epoca in Hegel*. Bologna: il Mulino, 2014 (1st edition published in 1975); *La filosofia del Novecento (e oltre)*. Milan: Feltrinelli, 2015.

13 R.Bodei, *La vita delle cose*. Rome-Bari: Laterza, 2009; Idem, *Generazioni. Età della vita, età delle cose*. Rome-Bari: Laterza, 2014.

reflections were personal and authentic, he would casually invite you to switch registers from the formal one to the one more informal and confidential. This posture (or attitude) was all the more welcome precisely because it was unexpected, though it was initially difficult to put into practice (at least for the younger colleagues among us) in light of the awe that his fame (though certainly not his manner) inspired. One could hardly forget that Bodei, in his time at the most prestigious universities in the West, frequented some of the most authoritative figures in contemporary European philosophy, as well as the most important intellectuals of the twentieth century and beyond. It was only natural that he mentions his meetings and discussions with such *maîtres à pense* as Ernst Bloch, Eugen Fink, and Karl Löwith, or his encounters with Michel Foucault and Jacques Derrida, to name but a few. This was his way of *being* and relating to others with affability and intelligence, with incomparable influential authority and acumen, united with a strong critical and self-critical sense. He himself was a *maître à penser*, to be aspired to first and foremost as a *person* for his remarkable rigour and ethical, cultural, and professional depth.

He was, more precisely, an all-encompassing professional of enquiry, passionate in the theoretical investigation of the philosophical problems of our time and of the past (as one of the greatest contemporary experts in classical German idealism and the philosophies of the Romantic age): which is why he chose to leave teaching and distance himself from the Italian academic establishment when he perceived a profound fracture among *paideia*, *polytheia*, and *episteme*. The pretext or antecedent coincided with the explicit request that he drastically "reduce" and "simplify" the syllabus of his courses to better "adapt" them to the needs of the reformed university in light of the reforms introduced by the two Ministers of Education, University, and Research Letizia Moratti (in 2003) and Mariastella Gelmini (from 2008 to 2020). Bodei opposed this *diktat* with a firm and decisive refusal: true to himself to the end, he wished to renounce neither the *freedom* of teaching nor the dignity and great sense of responsibility inherent in the teaching-learning process. Therefore, rather than accept such an arbitrary and autocratic imposition he preferred to leave the university system, which he had come to perceive as impoverished, constrictive, and not very attentive to the high *quality* of *knowledge*. This *new* way of conceiving of the public university profoundly clashed with his style and the ideals of education, training, and 'culture' (not merely philosophical) he believed in, adhered to, and committed himself to throughout his life, as a professor and citizen: his generous dedication as president of the Scientific Committee of the Consortium for Modena's festival *filosofia* (philosophy's meetings) which he took part in creating, is emblematic of this.

Bodei's divorce from a political and cultural trend that is supine to neo-liberal, corporate logic[14] was inevitable and benefitted the prestigious foreign universities where he taught in the latter phase of his long and brilliant career. Still, one cannot fail to recognise that any university reform must be rigorous if it is to exercise a real cultural impact. That is, to be effective it must first focus attention on *education* as a dynamic and ongoing process aimed at the high cultural, civil, and integral development of the individual and of *every* individual, in an inclusive spirit. Secondly, to be effective it must foster actions that promote active citizenship animated by a critical and emancipatory spirit. Such engagement is expressed in the development of individual and social empowerment, as an expression of both 'plural' and democratic thinking and 'complex knowledge'. However, all of this should be also interpreted and contextualised in a utopian perspective: that is, as a function of both social progress and the self-realisation and *happiness* (of people) that R. Bodei[15] not only theorised but also embodied as a person and as a committed and non-organic intellectual.

III. While formulating their hypotheses some scholars are particularly effective in describing and explaining phenomena, events and problems of factual reality. Not infrequently, they are able to do so through a clear analysis of the problem under scrutiny aided by an uncommon argumentative ability to make even the most complex passages of the discourse appear as self-evident affirmations. As a result of their analysis, the conceptual core of a problem – the *object* of investigation – is revealed and clarified by illuminating those aspects that not everyone is able to readily read or interpret. Within the context of their approach, they grasp the dynamic relationships existing between the many components at play in complex or hyper-complex social systems to interpret the processes that follow.

Thence, their ability to reflect not only on the present but also on the future, which can be imagined and built through actions, behaviours, relationships and daily positive postures guided by ethical thinking, as suggested by Remo Bodei. For the reasons above noted these scholars are called "intellectuals" and Bodei was an outstanding example of an intellectual engaged in the exercise of a militant philosophy.

---

14 R. Bodei, *Il noi diviso. Ethos e idee nell'Italia repubblicana*. Turin: Einaudi, 1998; Idem, *Dominio e sottomissione. Schiavi, animali, macchine, Intelligenza Artificiale*. Bologna: il Mulino, 2019.

15 Several seminal studies by the philosopher treat the latter themes: *Geometria delle passioni. Paura, speranza e felicità: filosofia e uso politico*. Milan: Feltrinelli, 1991; *Le logiche del delirio. Ragione, affetti, follia*. Bari-Rome: Laterza, 2000; *Destini personali. L'età della colonizzazione delle coscienze*. Milan: Feltrinelli, 2002; *Ordo amoris. Conflits terrestres et bonheurs célestes*. Paris: Belles Lettres, 2015; *Scomposizioni. Forme dell'individuo moderno*. Bologna: il Mulino, 2016. In addition, on the theme of *virtue*, cf. R. Bodei, G. Giorello, M. Marzano, S. Veca, *Le virtù cardinali. Prudenza, Temperanza, Fortezza, Giustizia*. Rome-Bari: Laterza, 2017.

The philosopher of law Norberto Bobbio (1909–2004), dwelling on the role of Twentieth-Century intellectuals, elaborates a rather effective and still current definition,[16] albeit with all due and appropriate historical-cultural contextualization. Above all, he states that anyone who intervenes in matters of concrete and political life can be considered an 'intellectual' only by having something important to say. He also specifies that the topics or issues to be discussed have progressively changed over time, as has the kind of civil and social *engagement* (commitment) of militant intellectuals who "get their hands dirty in reality" (per John Dewey's words). But, not to the extreme of having them fall into self-isolation: in disinterested or solitary meditations that prevent their participation in political debate and on issues and current events that are as urgent and concrete as they are convincing. On the other hand, the questions concern – yesterday as today – the *great alternatives* such as the dialogue between the capitalist countries and those of former communist regimes or rather between East and West and also between globalization and localism. These issues, which are not easy to solve, continue to interest philosophers, historians, social scientists and anybody involved in culture because due to their intrinsic complexity they, directly or indirectly, concern everybody, no-one excluded. A complexity that continues to raise new questions and problems even for those that, like intellectuals, are professionally doing the job of thinking on their own time. In fact, intellectuals are not a 'homogeneous group' nor a "professional category" (like medical doctors or journalists) but know well the inevitable intertwining between culture and political power, that is, the politics of culture.

Furthermore, Bobbio argues that the politicians of his time show that they are more and more searching for the support of experts, technicians and specialists to help them in their tasks, and it is precisely in this passage that a clear analogy can be traced to current and contemporary politicians. In other words, he warns us that the word "commitment" could be unsuitable to describe the relationship between the expert and the politician (or the political power). But, in fact, the opinions of the experts represent the usual practice in the relation between the political and the cultural world. Even if the themes and problems to be addressed politically and socially change over the course of time (and history), it may be observed that Bobbio's words show that, along this inevitable course, there is also an inevitable change in the people who debate them and face them in different contexts, with different methods of approach and dialogue.

Therefore, it emerges that the fulcrum of the discourse remains the indispensable commitment of the intellectuals (which cannot be renounced). A commitment that must always be present in the political-cultural debate of a country and must constantly be nourished for the effects, incisiveness and influence that this can have on the real socio-cultural and political contexts, and on the living conditions of people

16 N. Bobbio, *Politica e cultura*. Turin: Einaudi, 1955.

and the individual in a given society: as regards, for example, in the defence of fundamental individual freedoms (human rights), and in the defence of democracy. Once the reference framework has been outlined, some essential indications for current times (the twenties of the 21st Century) appear evident, since it is precisely in the most complex and critical situations that the role of each person's commitment, intellectual or otherwise, becomes crucial and indispensable. Taking up Bobbio's words that echo in Bodei's thought, the task of "men of culture" is today more than ever to sow doubts, not to offer certainties. In other words, complex or hyper-complex societies need "thinking people," who know how to ask questions to themselves and know how to cultivate *doubt* as an effective *method* to continue to know, such as to evolve and become a human community with a common destiny: a more aware interdependent, democratic and planetary civilization.

IV. At the end of this brief reflection, we take leave of a *maîtres à penser* (Bodei), a committed intellectual who through his studies has hypothesized a 'good life' conceived both as a proposal for his contemporaries and as a legacy for the new generations. Thus comes to mind a deep thought expressed by the British philosopher and logician Bertrand Russell, who summarizes in an admirable way the meaning of existence: "A life devoted to science is therefore a happy life, and its happiness is derived from the very best sources that are open to dwellers on this troubled and passionate planet."[17] Ultimately Russell suggests, sharing Bodei's thinking, that a good life is the one inspired by love and guided by knowledge. And as a man of science, he cannot better express this his intimate feeling by pointing out that an existence dedicated to study, knowledge and science – like Bodei's – is taking shape as a 'happy life' for at least two reasons: firstly, because study and knowledge provide gratification to those who dedicate themselves to it with interest, passion and dedication. And, secondly, because science contributes to creating new knowledge and making new discoveries which, in turn, generate cultural, civil and social progress for the community (not just for individuals). In summary, it can be concluded that knowledge and science tend to promote the well-being of people and promote a good life, understood precisely in that Bodeian meaning that shines through in all his latest works, including the one presented here.

Nonconformist and *unorthodox* – taking up the title of the original work by Friedrich Nietzsche *Unzeitgemässe Betrachtungen*/Unmodern Observations (1873–1876) – faithful to a heartfelt, deep ethical-moral imperative, his original epistemological approach can be traced in this brief essay, which offers a lucid reflection on the complex and unresolved interweaving of reason, affects, and madness. Psychic distress is understood as an expression of the disconnection between internal and external worlds (which gives rise to so-called deviant behaviour in relation to the *norm* and

---

17 B. Russell, *Mysticism and Logic and Other Essays*. Watford: Taylor Garnett Evans & Co., 1917, p. 30.

rationality); it is therefore, re-contextualised, historicised, and interpreted with subtlety and depth by Remo Bodei. In fact, Bodei's inquiry focuses on the relationship between rationality and truth and the link between reality and temporality, seen as interpretative and foundational categories of our society that may nonetheless be perceived and experienced in *another* and *different* way by schizophrenic or psychotic individuals. Bodei proposes that we accept these "apparently absurd experiences" and "abnormal logics of madness" or "alternative interpretations," driven by *other* logics (irrational logics, or deviant from the norm) with a greater willingness to listen, as they reveal *other* possible worlds: but only for those able to exercise and develop "hospitable reason" capable, first of all, of surpassing the rigid dictates of standardised rationality, which tends to align and anchor us to a common reality rather than respecting diversity. Nevertheless, every society creates and reproduces an orthodoxy of reality through the power of institutions (political, economic, or religious), culture, language, and traditions. An *hospitable reason* must recognise, secondly, that there are thresholds that distinguish one world (one experience) from another and every identity among different forms of identity.

In the final analysis, to better understand phenomena linked to madness it is necessary to learn to recognise, understand, and attribute new meaning to it, rather than labelling it – as western cultures often hasten to – as an irrational world. In adopting a new, authentic, and original *vie sauvage*, one recognises that madness reveals the "latent fragility of everyone's experience": as the very title of Bodei's essay, *The New World Opened by Madness*, evokes.

Sowon S. Park

Why Listen to the Mad? What *Schizophrenic Girl* Offers to Narrative Medicine

> You can't understand me. To understand my point of view, you have to believe it first.[1] *T. S. Eliot*

Listen to the Mad?

It is never easy to describe being ill.[2] But the task is almost impossible for schizophrenics because a key symptom of the condition is language impairment. One of the five diagnostics of schizophrenia listed in DSM-IV is "disorganized speech."[3] Consequently, first-person accounts of schizophrenia are rare, self-advocacy even rarer.[4]

In any case, schizophrenics' words are negated by the fact of their illness. Most people, including psychiatrists, consider it a mistake to attribute meaning to the utterances of the clinically mad. For theirs is speech that has no meaning. It has no message-bearing function, or no 'locutionary' force, as J L Austin would have put it.[5] This attitude is crisply captured in a review cited by Louise A Sass in his magisterial study *Madness and Modernism* (1992):

> Portrayal of this disorder (schizophrenia) as other than an unfortunate consequence of an aberrant processing deficit would be a serious mistake. There is no more wisdom to be gained in the communications of a psychotic patient than in the dementia of the Alzheimer's patient. Each form of disorder is a consequence of particular deficits. . . . In the schizophrenic thought disorder . . . there is little more than the tragic evidence of a mind wasted away in dementia.[6]

As this statement makes plain, schizophrenic speech does not qualify as speech and clinicians are trained not pay attention to gibberish.

---

1 Reported in Costello 1956: 76.

2 I would like to thank Julie Carlson and Daniel Martini Tybjerg for their astute comments on an earlier version of this chapter.

3 For a discussion of the relations between schizophrenia and language impairment, see Frith 2003: 56–57, and Kandel 2019: 86–87.

4 Schizophrenia has affected the lives of such notable people as Elyn Saks, a professor of law at University of Southern California, the novelist Jack Kerouac, the economist John Nash, and a member of the Beach Boys, Brian Wilson. With improved medication, more people with milder forms of schizo-affective disorder are able to lead fuller lives, including more prominent literary lives. For a contemporary memoir, see Wang 2019.

5 Locutionary means 'message-bearing' in Speech Act theory. See Austin 1962.

6 Sass 1992: 545.

https://doi.org/10.1515/9783110788501-015

This chapter runs somewhat contrary to this view. It makes a case for why we should actually listen to the mad and why this is important for the field of Medical Humanities. Not because there is validity in attributing meaning to schizophrenic speech in the usual sense of interpretation and communication. But because the dehumanizing mind-set that reduces patient speech to neural deficits has ramifications beyond the invalidation of locutionary force.

Message-bearing is not the only function that gets lost when schizophrenic speech is considered meaningless. A being who does not merit being heard is very close to a being who does not merit human attention. This attitude prizes apart a gap between doctor and patient that makes the possibility of a developing human relationship an impossibility. This then becomes a challenge to medical practice if the object of healthcare is to alleviate the suffering of the patient and to develop better awareness in the clinician.

In proposing that we listen to the mad, the aim is to add to the emerging movement called 'Narrative Medicine' that attempts to make patient narratives relevant to bio-medical discourses.[7] This chapter problematizes the occlusion of subjective testimony in quantitative and genetics-based accounts of schizophrenia and makes a case for patient self-presentation as a vital component in the broader production of medical knowledge.[8] The focal point of discussion is *Autobiography of a Schizophrenic Girl* (1950) by a "Renee" and her doctor. By analysing Renee's self-description and interpretation, this chapter discusses (1) *why* patient self-presentation is important for the psychotic, more so than for those with other disabilities (2) *how* patient narrative can be locked into the wider design of clinical and socio-political assessment and (3) *what* the implications of this case-study are for translational medicine and narrative medicine.[9] These three points offer "bottom-up" insights that acknowledge the problem and the value of patient self-report. In doing so, it attempts to realign the old dichotomy of patient and doctor, never sharper than in today's drug-based medical culture, and proposes a way to translate the voice of the medically disenfranchised into a comprehensible narrative. As such, it reflects many of the principles that have shaped the humanities historically.

---

7 See Charon 2006b and Charon 2017. Charon defines *Narrative Medicine* as a "medicine infused with respect for the narrative dimensions of illness and caregiving."

8 The ascendency of biomedical model of psychiatry is in contrast to the tradition of psychoanalytic treatment, which has, of course, been patient centred. Notable psychoanalytical work on schizophrenia is by Christopher Bollas. See, in particular, Bollas 2016, and also a special issue of *Schizophrenia Bulletin* 33 (1) (2006), *Phenomenology and Psychiatry for the 21st Century*, that takes a bidirectional approach to the patient-doctor relationship.

9 For a discussion on "translational" issues in biomedical research, see Wehling 2008: 6:31.

The Value of Patient Self-Presentation

Narratives of recovery from schizophrenia are rare but among the few that exist, *Autobiography of a Schizophrenic Girl* occupies a central place. It is, as the title announces, a first-hand account of schizophrenia written by a recovered patient, "Renee," whose real name was Louisa Düss, later Louisa Sechehaye-Düss (1912–2002).

Renee first started having hallucinations when she was five years old and for the next twenty-two years went through all the major symptoms of schizophrenia by current standards outlined in the DSM-IV of the American Psychiatric Association.[10] She has visual and auditory hallucinations, physically attacks people, harms herself, loses her capacity for speech, is unable to eat and alternates between catatonia and mania. At least fifteen psychiatrists diagnose her as beyond clinical treatment. She is given hypno-affective therapy, given tranquilizers, placed in solitary confinement, placed in restraints. She attempts suicide repeatedly. But Renee recovered. Not only did she recover enough to write this book, but she went on to train successfully as a psychoanalyst, even though she retained certain tendencies towards schizoid thinking for the rest of her life.

So, what is the particular relevance of this book for the field of Medical Humanities?

First is its descriptive value. In 100 pages of clear, unadorned prose, "Renee" transforms an ordinary landscape into a bizarre place where everything is seen through the distorting filter of psychosis. "(T)he enormous difficulty of such a task is obvious' as Frank Conroy has written. "She writes of scenes and sensations she could not clothe in language even at the time she experienced them."[11] This is a rare achievement. As mentioned earlier, loss of language is a typical impairment. Therefore, while we have detailed and numerous descriptions of other psychiatric conditions, such as manic-depression, this is not the case with schizophrenia.[12] Eminent neuroscientists, Chris Frith and Eve Johnstone, who have pioneered the cognitive basis of schizophrenia, have observed: "(T)he cognitive areas in which patients with schizophrenia show the greatest impairment – they are memory, attention, and executive

---

10 The *Diagnostic and Statistical Manual of Mental Disorders (DSM-IV)* currently recognises five subtypes of schizophrenia: paranoid, disorganized [ie, hebephrenic], catatonic, undifferentiated and residual.

11 Sechehaye 1951 (1971): 10.

12 Often written during the manic phase, there is a sizeable literature on and propelled by manic depression. See Redfield Jamison 1993. Manic depression in Lord Byron, Lord Tennyson, Emily Dickinson, William Blake, Samuel Taylor Coleridge, Ernest Hemingway, and Vincent van Gogh are given a sympathetic reading by Jamison.

function . . . . One test of executive function on which patients of schizophrenia often perform badly is 'verbal fluency'."[13]

Typically, Renee lost her grip on verbal fluency. But atypically, she was later able to recreate this language-deprived condition with clarity and resonance. Here is a remarkable section from the text when she has lost the capacity for thinking in language and is at her lowest ebb:

> In this eerie atmosphere of torpor and mutism the months passed, marked only by the wracking crises of guilt and antagonism. . . . The period of confounding guilt persisted along with the bitter moral pain and I shed tears for hours, crying . . . sorrowing in my own "language," in the meaningless, recurring syllables, "icthou, gao, itivare, giastow, ovede" and the like. In no way did I seek to create them; they came of themselves and by themselves meant nothing. Only the sound, the rhythm of the pronunciation had sense. Through them I lamented, pouring out the gruelling grief and the interminable sadness in my heart. I could not use ordinary words, for my pain and sorrow had no real basis. . . . Sunk beyond language, beyond thought, I reflected no more.[14]

Renee had the uncommon ability to retrieve the experience of when she was 'sunk beyond language, beyond thought' and has made it intelligible to the reader. The clarity with which she narrates this state is of inestimable value. The vivid portrayal of Renee's long descent into madness and recovery is compelling and is comparable to more widely known memoirs of schizophrenia, such as Joanne Greenberg's *I Never Promised You a Rose Garden* (1964) and Elyn R Saks' *The Centre Cannot Hold* (2007).

Communicating phenomenological experience is particularly valuable in the case of schizophrenia than for other disabilities. Because psychotic patients behave in unpredictable and inexplicable ways, which often alarm and frighten those who witness the behavior. It is very hard to attribute the inner motives of their behavior without the kind of report that patients like Renee can give. What Renee achieves is to communicate to the reader what generates her psychotic behavior. She lays bare the inner experience of behavior that comes across as irrational, frightening, and mad for those of us looking from the outside in. For example, the following is a description of when Renee burnt her hand on blazing coal. A baffling act of self-harm incomprehensible to most people, one that finally got her sectioned. She leads us through what led her to behave thus:

> I received orders from the System. I did not hear the orders as voices; yet they were as imperious as if uttered in a loud voice. While, for example, I was preparing to do some typing, suddenly, without any warning a force, which was not an impulse but rather resembled a command, ordered me to burn my right hand or the building in which I was. With all my strength I resisted the order. . . . The orders became more imperious, more demanding. I was

13 Frith 2003: 56–57.
14 Sechehaye 1951 (1971): 120–121.

> to burn my right hand, for the right hand was the hand of the law. . . . One day trembling, I placed the back of the right hand on the incandescent coals and held it there as long as possible. By thinking of my duty to the System, and that it might then stop issuing injunctions, I was able to stand the pain.[15]

Once we follow her behaviour from her inner point of view, the laying of her right hand on the burning coals, alarming though it is, becomes somewhat more comprehensible. While the notion of the 'System' may be hard to relate to, following the demands of a persistent inner voice is a thought-process that most of us can recognize. The clarity and immediacy with which she describes the commands in her head helps us understand not only how another mind thinks but how another body acts.

By guiding sane people to place themselves in the mental worlds of the afflicted, the testimony undoes the sharp divide between the behavior of the ill and the healthy. After all, the pain Renee suffers is not itself essentially different from pain others go through. When insane behaviour is seen in isolation, without the underlying process, it has the effect of setting apart the 1% of the population diagnosed with schizophrenia from the healthy neuro-typical population. Behaviour alone can only offer us the specifics of how we differ. Reading the patient's phenomenological description allows us to enter into a dimension of space where there is common ground.

The value of meeting on common ground is obvious. It provides the foundation on which understanding can be deepened between not only patient and physician but those indirectly affected such as friends and family and the wider public who are seeking to understand how this psychiatric disability throws light on the general human condition.[16] The suffering that Renee endures evokes feelings of shared emotion even if her acts are incomprehensible.

It thus provides invaluable material for medical teaching. Narrative has now become a subject of critical conversation in medicine. Pioneered by Rita Charon at Columbia University 'Narrative Medicine' has become a new framework for delivering patient-centred medicine. *Autobiography* is a dazzling example of how patient life-writing can mediate the relations between the doctor and patient and facilitate the acknowledgement of a schizophrenic's inner reality, ensuring these expressions are witnessed and heard.

---

15 Sechehaye 1951 (1971): 57, 60.
16 The website *Polyphony* (thepolyphony.org) produces up-to-date conversations around illness from a multidisciplinary perspective.

The Problem of Genre

Alongside Narrative Medicine, personal narrative of disability and illness as an identifiable genre of life-writing has grown in both size and significance in the last twenty years. Individual voices that share their struggles range widely over conditions that include but are not exclusive to anorexia, anxiety, insomnia, locked-in syndrome, obsessive compulsive disorder, Parkinson's, stuttering, stroke, cancer, amputation, chronic pain, depression, bulimia, substance addiction, alcoholism, cerebral palsy and body dysmorphia. A strong assumption that animates this field is that knowledge gleaned from patient self-report helps with clinical diagnosis, makes good teaching material, and generally expands the horizon of empathy in health practitioners.

But a weakness of this genre that demands our attention and discussion is the question of veracity. There is something clearly at stake when we take a narrative at its own value. When reading any text one must take into consideration how far the demands of form, genre and narrative expectation come into the writing. There are also the pressures of social conventions, not to mention financial and market calculations. The autobiographical form, since it is written sometimes years after the events, is necessarily inaccurate by virtue of how our memory works. On occasion, people will fabricate actual medical conditions, known as Factitious Disorders (FD) of which Munchhausen Syndrome is one. The hazy area of 'auto-fiction' also needs to be taken into account. There are varying degrees of reliability of voice. Thus, one needs to invest in a way of reading that is attentive both to the rhetorical and the factual.[17]

Naturally, the long list of questions that surround the genre of illness life writing should also be directed towards the *Autobiography*. Few narratives surpass it for its immediacy and potency in rendering schizoid cognition. But should its descriptions be uncontested? And what exactly is the validity of the way in which psychosis and hallucination is portrayed in the book other than it offers a fresh and fierce perspective of an individual? Since there are diverse accounts of the specific characteristics of any illness, what is the wider applicability of this single case-study? Is seeing through the eyes of the patient necessarily to see the condition more clearly? That is to say, are autobiographical texts more authentic than the works of "sane" authors also writing about schizophrenia?[18]

This takes us to the second reason why *Autobiography* is an exemplary case-study. And that is the way it is structured. Unlike *I Never Promised You a Rose*

17 In addition to those listed, there are privacy problems in representations of illness and disability in memoir and autobiography. See Couser 2004.

18 A powerful example of this genre is Powers 2017.

Garden and *The Centre Cannot Hold*, *Autobiography* has two authors.[19] The first author is 'Renee,' who writes the "Story." The second co-author is her doctor, Marguerite Sechehaye (nee Burdet, 1887–1964), who provides the second part called the "Interpretation."[20] What this unusual pairing effectively addresses is the problem of genre.

The dual-narrative structure offers a point from which to observe the same sequence of events from two different points of view, in two different kinds of language, and focalised by two different kinds of interpretive frame. The two narratives, when read side by side, allow the reader to compare them and notice the ways in which they each cast a light on the other. This is an example of Renee's self-presentation:

> Finally, Mama[21] returned and I saw her nearly every day. I was happy at her return for I felt abandoned and also because I was again experiencing the impulse to harm myself, to hit my head against the wall, to bite and mutilate myself.
>
> Mama brought me a gift – a little plush monkey of which I was at once afraid. When he had his arms up, I was anxious lest he hurt me; and then, he had a most shockingly unhappy expression. Oddly, at that very moment, I felt the impulse to strike myself. I realized full well that my own arms were delivering the blows, still I was sure the monkey was attacking me. . . . When I related my fears to Mama she did something extraordinary: She took the monkey's two arms, lowered them around his little knees . . . from that moment, the impulse to self-harm left me abruptly.[22]

And this is Marguerite's interpretation of Renee's experience in clinical language: "When the little monkey, her double, was ordered to keep his arm down, the impulses to self-harm disappeared. The patient projects her drives on the little monkey which is, at the same time, Renee herself."

The dual narrative yields a fascinating picture. The patient's voice is validated by the clinical case-history, providing a firmer grounding to subjective testimony. Likewise, the dry, exactness of the doctors' diagnosis is embodied by the personal storytelling. The juxtaposition of first-person report and forty-five page third-person clinical case-history allows each to support each other, producing a situated perspective as well as a translated continuum. To look to another example, this is when Renee describes psychosis:

> For me, madness was definitely not a condition of illness: I did not believe that I was ill. It was rather a country, opposed to Reality, where reigned an implacable light, blinding, leaving no place for shadow; an immense space without boundary, limitless, flat; a mineral lunar country, cold as the wastes of the North Pole. In this stretching emptiness, all is unchangeable,

19 Different interpretive questions arise according to whether the narrative is 'factual' first person (as in *Autobiography* Part I and *The Centre Cannot Hold*) or semi-fictional third person (I *Never Promised You a Rose Garden*), the discussion of which the scope of this paper does not permit.

20 For further details about Marguerite Sechehaye, see Balbuena 2014: 167–174.

21 Mama is what Renee calls her doctor, Marguerite.

22 Sechehaye 1951 (1971): 96–97.

> immobile, congealed, crystallized. Objects are stage trappings, placed here and there, geometric cubes without meaning.
>
> People turn weirdly about, they make gestures, movements without sense. . . . And I – I am lost in it, isolated, cold, stripped, purposeless under the light. A wall of brass separated me from everybody and everything.[23]

Of this experience, Marguerite analyses: "Renee's introspection reveals that the earliest disturbing subjective symptom bears uniquely on the perception of reality. Suddenly objects become enormous, cut off, detached, without relation to one another; space appears limitless and anxiety supervenes. This phenomenon of strangeness, of unreality is at first circumscribed, then progressively extends to the perception of objects, of people, of herself."[24] Renee's expressive account is reinforced by Marguerite's analytical diagnosis even as the clinical perspective is protected from dehumanizing reduction.

Finally, the dual structure helps us see that the voice of the patient and the doctor are not separable nor antithetical but complementary and intersubjective. As such, it provides a productive model for narrative medicine and translational medicine that overcomes many of the problems of genre and self-report in narrative medicine outlined above. It also levels the hierarchical relations between doctor and patient by acknowledging the validity of the inner experience of the mad while personalizing the clinical. This is a collaborative model that is under-utilized and under-theorized in narrative medicine currently. But it has a number of clear-cut benefits, as discussed, which may be used to facilitate better translational communications.

A Humanist Model of Medicine

The dual narrative structure also offers a provocative angle from which to view our current psychiatric practices. Though Marguerite's "Interpretation" section is written in scrupulously disinterested clinical prose, it becomes clear when the two narratives are read together that Marguerite is not only an administrator of cures based on objective data but also a person who is invested in a process that is mutual. In part I, we are shown the range and depth of pain and longing caused by schizophrenia. In part II, we are shown the depth of attentiveness and commitment that the clinician is called upon to demonstrate. The dual narrative that is *Autobiography* tells a story not just of an illness but of a human relationship.

A key symptom of schizophrenia is mental isolation – the shutting down, the lapsing into a condition of complete withdrawal from the world. It is typical for a schizophrenic patient to find it impossible to relate to or connect with another person

23 Sechehaye 1951 (1971): 44.
24 Sechehaye 1951 (1971):141.

on any level. Patients feel detached from themselves, and their bodies and a sense of "unreality" pervades their psyche. This is the condition Virginia Woolf portrayed vividly in the figure of Septimus Warren in *Mrs Dalloway*.[25] Just like Septimus, Renee experiences a "wall of brass" that separates her from everybody and everything. Renee calls this the world of "unreality." Iterations of Renee's "unreality" are abundant in first-person accounts of schizophrenia. For example, this typical description is by an eighteen-year-old boy who had been ill for about a year which is strikingly similar to Renee's description: "I am more and more losing contact with my environment and with myself. Instead of taking an interest in what goes on and caring about what happens with my illness, I am all the time losing my emotional contact with everything including myself."[26]

It was in this context that R D Laing declared "I have never known a schizophrenic who could say he was loved."[27] Laing also noted that recovery began when a patient is able to believe in the existence of the reality of another person. When a patient sees a glimmer of connection, it motivates them to engage with external reality from which they withdrawn. This is an indispensable insight that is clearly demonstrated in *Autobiography*. Though a strong sense of 'unreality' pervades Renee's mental state in the typical manner, her case is unusual in that she forms an unwavering belief in the reality of Marguerite. Marguerite as only one of her many doctors and not even the main one. The following section is Renee's reconstruction of the early phase of her therapy with Marguerite:

> Against this enlightenment (delusional psychosis) I waged a battle with the help of the analyst who later became my "Mama." Only near her I felt secure, especially from the time when she began to sit next to me on the couch and put her arm around my shoulders. Oh, what joy, what relief to feel the life, the warmth, the reality! From the moment I left her at the end of the session, I began to count the hours and the minutes: only twenty-four hours, only twenty-three and a half hours, only eighteen hours[28]

When under certain circumstances when Renee is not able to maintain contact with Marguerite, a sense of "unreality" dominates, and her condition deteriorates rapidly. It is only in the presence of Marguerite that Renee is able to feel real.

So, what allowed for Renee to form this attachment? What were the conditions that made this possible? Identifying what made it possible for Renee to forge such an unbreakable bond with Marguerite is key for understanding her recovery.

Marguerite was a unique figure to Renee because Marguerite was the only one who accepted her inner world. Marguerite demonstrated acceptance by showing an extraordinary willingness to provide Renee with what she needs, on her terms, no

25 For a discussion on literary representations of psychosis, see Ovaska 2016; Gaedke 2017.
26 Quoted in Lewis 1967: 16.
27 Laing 1959: 38.
28 Laing 1959: 45–46.

matter how unreasonable, strange or regressive her desires may be. Marguerite's primary goal was to reconnect Renee with the outside world from which she had escaped and to train her attention to maintain that connection against the background of extreme perceptual distortions and auditory hallucinations. After an important breakthrough, Marguerite wrote:

> Forced to recognize Mama as the source of her satisfaction her attention was directed outward and a link was forged with the environment. . . . This represented great progress. To continue on this road it was necessary to utilize this valuable insight, namely, that the first attention the patient gives to the outside world derives from the gratification of the primal need.[29]

For this end, through endless acts of perseverance, Marguerite repeatedly and consistently demonstrated to Renee that her existence is real by gratifying what Renee most needs, which was an acknowledgement of her inner experience. A powerful example is found when Renee writes of Marguerite's approach of using personal pronouns. "What did me the most amazing good was her use of the third person in speaking of herself, 'Mama and Renee' not 'I and you.' When by chance she used the first person, abruptly I no longer knew her, and I was angry that she had, by this error, broken my contact with her."[30] Similarly, towards the end, she writes:

> Of the greatest importance and a primary contribution to growing self-awareness and freedom from guilt in self-esteem, was Mama's way of speaking to me. I could never accept her addressing me in the second person, "*You* have a nice body, how clean *you* are." This would have aroused devastating anxiety and anger against Mama for laying such a sin on me. . . . "It is nice, this body that is going to be washed and scented," was to separate me from it. It became something independent of me. Mama washed it, found it nice; and little by little, like Mama I could wash it; I could do it since I was imitating Mama. Finally, when for some time I had done it with Mama, I was able to do it without her so long as I talked to it exactly as she did, to take the responsibility, to like it. Then I permitted myself to say, "*My* body, *I* am washing it, *I* am pretty."[31]

Marguerite took very seriously the extent to which Renee's psychic reality is grounded and enabled by the personal pronoun.[32] Because she understood that Renee manages the onslaught of overwhelming sensations by dissociating herself from them. Her main coping strategy was to dissociate and disown the flow of chaos that runs through her mind. Her use of language created a buffer zone between herself and her experiences. Marguerite accepted Renee's rejection of the psychological space of enunciation that the personal pronoun 'I' creates. She confirmed Renee's need to detach herself from her sensations.

29 Sechehaye 1951 (1971):160.

30 Sechehaye 1951 (1971): 52.

31 Sechehaye 1951 (1971): 127–128.

32 The power of personal deictics to organise cognitive-affective experience I have argued elsewhere. See Park 2022.

Here I am reminded of T.S. Eliot who once said to his friend Harry Costello, "To understand my point of view, you have to believe it first." Marguerite believed in Renee's reality first then came to understand her. By accepting Renee's reality first and maintaining a connection based on that blind acceptance, she laid the foundation upon which Renee can learn to be with others.

Readjusting the Frame of Interpretation

The intersubjective relationship between the schizophrenic girl and the doctor raises some active questions on the relations between patients and clinicians in today's health care system. Current models of care for schizophrenia derive from two major interpretive models – biomedical and psychoanalytical. This penultimate section will delineate and compare the two traditions to discuss ways of developing new forms of care in the final section.

A lot has happened in the treatment of schizophrenia in the last 100 years. While the basic construct of it as "splitting of the mind" has not changed fundamentally since 1908 when Paul E. Bleuler (1857–1939) coined the term, the thinking on its causes and treatment has changed quite dramatically.[33]

Since the 1980s, the dominant medical model for mental disorders has been drug therapy. But before the advent of genetics and neuropathology, the primary causes of schizophrenia were considered to be psychological or social. The central assumption of the psychodynamic school was that traumatic emotional events create psychic stresses that cause people to regress. Psychosis was seen as an accumulation of emotional stresses created by society.

The most radical of the psychodynamic approaches found its expression the ideas of the psychiatrist R.D. Laing who expounded the idea that the mentally ill are an oppressed social group.[34] Psychiatric illness is a condition produced by socio-political discourses, which police and patrol oppressive norms, according to this view. As Laing famously asserted "There is no such 'condition' as 'schizophrenia' but the label is a social fact and the social fact a *political event*."[35] Similarly, Thomas Szasz, also a psychiatrist, pressed on the view that schizophrenics are the sane in an insane world – they are merely those who "*have problems in living*."[36]

---

33 For a discussion on the reception of Bleuler's *Dementia Praecox* (1911), see Moskwitz and Heim 2011: 471–479.

34 This idea was popularised by Michel Foucault in *Madness and Civilisation* (1961. Engl. Trans. 1964).

35 See Laing, *The Politics of Experience*. In addition, see Andreason 1987: 1288–1292. She examines the link between creative writing and affective disorders. Also see Ludwig 1992): 330–356.

36 Szasz 1960. Szasz famously asserted that the "notion of mental illness has outlived whatever usefulness it might have had and that it now functions merely as a· convenient myth."

Likewise, Deleuze and Guattari argued in *Capitalism and Schizophrenia* that there is schizophrenic potential in everyone. They saw this potential as radical and revolutionary. In the dissolution of schizophrenic identity, they saw a subversive postmodern politics and schizophrenia was seen as exemplifying the culture of late capitalism. Louise A Sass in *Madness and Modernism* has also influentially signaled schizophrenia as a central experience of modernity.

For these theorists, schizophrenia extends beyond the illness to stand as a critique as well as a corrective to the material and political conditions of advanced capitalist societies. Given that typical symptoms of schizophrenia – such as dissociation and mental isolation – can also be expressed in quite ordinary ways, it is easy to see how the politico-cultural diagnoses of such critics answer very immediately to a set of lived experiences in modern atomized societies. And as a radical politics, these somewhat romanticized ideas on the illness have had a pervasive influence in humanist circles.

But, however durable the politically radical ideas on schizophrenia may have been in the humanities, they have had no bearing on clinical medicine. One reason is that humanist critical interventions are written at a very high level of generality. But there is a second, more immediate, reason. Experiments in care propelled by sociological antipsychiatry and psychoanalysis in the 1960s resulted in huge institutional disasters, which, in the words of the medical historian, Ann Harrington. amounted to nothing less than a "slow train wreck."[37] In her chronicle of modern psychiatry, *Mind Fixers: Psychiatry's Troubled Search for the Biology of Mental Illness* (2019) Harrington argues that the 'overreach of the 'Freudians' and social scientists drove psychiatry into a state of crisis producing, ironically, an exaggerated swing in the direction of quantitative, 'objective,' bio-medical, drug-based approach to mental illness.'

The new age of biological psychiatry from the late 1980s onwards considers schizophrenia as a genetics-based developmental disorder. And it is not concerned with whether schizophrenia occupies a central place in cultural theories of modernity or not. As Harrington notes, "(b)y the mid-1980s schizophrenia had officially stopped being a disorder of bad families and had become a disorder of damaged brains." The sociological understanding of mental illness has been superseded by neuroanatomical.

Among the strongest explanations of the etiology of schizophrenia is over-accelerated synaptic pruning in the pre-frontal cortex and dopamine-related issues. The Nobel laureate Eric Kandel writes:

> The overexpression of the C4-A variant leads to excess synaptic pruning . . . Carrying a gene variant that facilitates aggressive pruning is not enough in itself to cause schizophrenia; many

37 Harrington 2019: 181.

> other factors are also at work. But in a small subgroup of people, one specific gene – the C4-A gene – gives rise to anatomical changes that lead to schizophrenia.[38]

Today virtually all schizophrenic patients are prescribed anti-psychotic drugs such as clozapine or other dopamine blockers. And nearly all clinicians agree that genetic factors have an important role in the cause of schizophrenia.[39] As Kandel writes: "Schizophrenia, like autism spectrum disorders, involves an *anatomical* defect, in which certain neural circuits fail to develop correctly."[40]

However, equally, genetic factors cannot be the whole explanation. This is a position held widely, especially by neuroscientists and clinicians. As Chris Frith and others have argued, C4-A variant notwithstanding, the role of genetics in the etiology of schizophrenia is complex. And, as there is no diagnostic lab tests for schizophrenia, the definition, concept, the boundaries all remain somewhat arbitrary. The heritability of schizophrenia is estimated at 50%.[41] But these uncertainties have not prevented medicine being practiced on terms set by biological psychiatry. Harrington notes:

> Patients present with acute mental or emotional distress, and doctors look for a DSM diagnosis that will make sense of their suffering. They prescribe drugs because that is what insurance companies will pay for, and because they believe the drugs will take the edge off their patient's distress. By acting this way, general practitioners and psychiatrists perpetuate the fiction that the drugs they are prescribing are correcting a biochemical deficiency caused by disease, much as (say) a prescription of insulin corrects a biochemical deficiency caused by diabetes.[42]

"Perpetuating the fiction" is a verdict that may carry little weight in today's drug-based medical culture. But neither is it empty. One could make the parallel observation that today there is an over-extension of biomedical knowledge in the same way that anti-psychiatry overreached its scope in the 60s. The recognition that pills alone are insufficient for a long-term treatment of any mental illness is widespread whether that is anxiety, depression, or psychosis. So, drug treatment is often administered alongside therapy. The problem is that they are not always compatible. How to negotiate between them is a task that lies mostly ahead. What is needed, at this point in time, is an exploration of the kind of negotiations that feel valuable. In the last section, I discuss how *Autobiography of a Schizophrenic Girl* opens up a space for discussing what needs attending to in order to reach that point of negotiation.

---

38 Kandel 2019: 104.
39 Heckers 2000: 267–79.
40 Kandel 2019: 97.
41 See Plomin 2018: 6.
42 Harrington 2019: 273.

What the *Schizophrenic Girl* Offers to Narrative Medicine

Marguerite Sechehaye's method is humanistic in nature. Humanistic in the sense that the relationship between the doctor and the patient is not only the means of treatment but its very condition. Still, the devotion as shown by Marguerite to Renee cannot but be inimitable. Few have the luxury of being treated by a doctor who is so radically involved. To say that Marguerite contravened the rule of professional detachment is an understatement. She broke the medical code of practice many times to ensure that Renee continued to have faith in her existence and in their relationship.[43] Their relationship might be inspiring, but it isn't, and probably should not be, replicable.

But what one may take from her example is the insight into the role of language and narrative in the recovery process. What is distinct about this particular case is that unlike psychoanalysts to whom Marguerite was indebted, she pioneered a method of treatment for regression called "symbolic realization." This was based on Saussurean linguistics. Marguerite Sechehaye, as some may remember was one of the three editors of Sassure's groundbreaking *Course in General Linguistics*. Saussure actually never wrote the book that became the foundation for both structuralism and poststructuralism. *Course in General Linguistics* was produced from her class notes of his lectures at Geneva University.[44] Symbolic realization is an analytical method deeply rooted in the idea that symbols, language, and narrative play a critical role in organizing the cognitive and affective experiences of the individual, especially in the recovery process. This model of medicine highlights features that are noticeably absent in drugs-dominated treatment of mental disorders today. What is necessarily missing in today's dominant framework is language.

Language plays little to no role in the bio-medical conceptualization of schizophrenia because mental disorders are developed on animal models of brain function. Cellular and molecular biology and genetics are studied largely in mice and rats and within this frame, the role of language to create a unified sense of self is an irrelevance. Neural deficits in the brain are identified by knocking out a particular gene, and drugs are developed to treat the deficit. Accordingly, the interaction between the patient and other people cannot but be of minor consequence. In this frame, "icthou, gao, itivare, giastow, ovede" can be regarded as garbled speech generated by neural deficit.

43 This is an example: during the course of therapy, Marguerite legally adopted Renee.

44 *Course* was based on the lecture notes of Sechehaye, Charles Bally and her husband, Albert Sechehaye. She trained under Raymond de Saussure, Ferdinand's son.

Marguerite viewed schizophrenia in terms of psychic regression or 'ego sickness,' and the narrative realization of the "ego" the cure. The core of her thinking is captured in the following:

> Throughout this study, an effort has been made to demonstrate the similarities between certain aspects of schizophrenic and infantile thinking. . . . To reduce the schizophrenic mentality to that of a child would of course be an obvious mistake; too many dissimilarities intervene. . . . all schizophrenics avail themselves of the mechanisms of projection, participation, condensation and imitation to express the psychic life . . . these mechanisms can be drawn on and utilized in efforts to reconstruct the psychotic ego.[45]

Listening to Renee's garbled words attentively, as Marguerite did, produces a powerful shift in perspective. "(I)cthou, gao, itivare, giastow, ovede" are acknowledged in this humanist perspective as "the protests and wishes of the small child, and at the same time as participation in the reassuring and consoling maternal voice."[46] The difference the two approaches have on the patient is stark.

Autobiography is a story of psychiatric recovery from severe schizophrenia through psychoanalytical treatment – that is to say, through narrative engagement and sustained attention to the primal needs of the patient. Many will think that the idea that psychosis can be treated by symbolic realization has been rendered obsolete by the 'biological turn' in the late twentieth century.

What this historical text offers is an element that is missing in today's biomedical culture – the power of acknowledging patient's inner experience. The case of the schizophrenic girl shows up the way in which the biomedical model of the human effectively licenses disengagement by reducing patient speech to mechanical defects. It sanctifies the ways in which we avoid listening. But, ultimately, a collective, joined-up understanding of how best to treat patients will require a dimension of the patient's story in the physician's interpretation. In the example of the schizophrenic girl, there is a powerful recommendation to correct the marginalization of patient experience in drug-based medicine: listen to the mad.

45 Sechehaye 1951 (1971): 185–186.
46 Sechehaye 1951 (1971): 155.

Massimo Ciavolella

The Malady of Love in Early Modern Medical Thought

Between 1275 and 1280, a student of Arnaldus of Villanova (c. 123F5–c. 1313),[1] practicing medicine in Sardinia, encountered a mysterious disease seemingly caused by an excess of love. Puzzled by its significance, he wrote to Arnaldus on the matter thereby prompting the great Catalan physician and natural philosopher to write a short treatise on ". . . unde fiat quod tam vehemens irrationalisque concupiscentie motus in heroico amore causatur. . . ." [. . . how is it that such violent and irrational movement concupiscence is generated in *amor hereos* . . .].[2] Though the doctrine of erotic melancholy had already become a part of the culture of the Latin West, Arnaldus' *Tractatus de amore heroyco* shows the doubts and confusion still surrounding the nature of unrequited love. Physicians and natural philosophers agreed that it was an illness, for which medical writers had described a clinical picture, but had never discussed its etiology in any programmatic and scientific way. Even the identification of its chronic stages with other forms of mental aberrations, such as melancholy or madness, had been implied rather than stated. Furthermore, the many Arabic medical texts translated into Latin did not include any attempts to rectify these ambiguities, and the Western writers simply copied them without attempting to gloss the difficulties.[3]

There is little doubt that the credit for introducing the concept of erotic melancholy into the West goes to Constantinus Africanus' (d. ca. 1098) Latin translation and adaptation of the manual for travellers *Zād al-musāfir wa-qūt al-ḥāḍi* [Provisions for the Traveler and the Nourishment of the Sedentary] by Abū Jaʿfar Aḥmad ibn Ibrahim ibn Abi Khalid ibn al-Jazzār (Ibn al-Jazzār) (d. 979 or 1004–5), titled *Viaticum* or *Breviarium*, one of the most circulated books of the High Middle Ages.[4] It is believed that its translation also introduced the term *hereos* for the pathological condition.[5] Constantinus often took liberties with the content and arrangement of the parts of his sources, but in the case of the section dealing with *hereos* he was content to follow the original text.

Mary Wack and Michael R. McVaugh have repeatedly pointed out the importance of the commentaries on the *Viaticum* for the establishment of the etiology of

1 On Arnaldus, see McVaugh 1970.

2 Arnaldus de Villanova, *Tractatus de amore heroyco* (ed. McVaugh 1985: 43).

3 For a more thorough discussion on this topic, in addition to my earlier book, Ciavolella 1976. See also Wack 1990; Couliano 1984, and Beecher and Ciavolella in Ferrand 1990.

4 On the *Viaticum* see Wack 1990, and Brachter 2005.

5 Lowes 1914: 515; McVaugh 1985: 14.

https://doi.org/10.1515/9783110788501-016

erotic melancholy, with reference to a work written sometime before 1237 by Gérard de Béry.[6] Before glossing several phrases from the *Viaticum*, Gérard presents his own interpretation of *amor hereos*, drawn from the *Canon* of Avicenna. The central portion of his commentary attempts to explain the etiology of the disease:

> Que sit causa huius passionis qua uirtutes impediantur difficile est uidere. Causa ergo huius passionis est error uirtutis estimatiue que indicitur per intentiones sensatas ad apprehendenda accidencia insensata que forte non sunt in persona, Unde credit aliquam mulierem esse meliorem ac nobiliorem et magis appetendam omnibus aliis. Quod est ideo: quia aliquod sensatum aliquandum occurrit anime ualde gratum et acceptabile, unde estimat cetera sensata non esse consimilia, unde si qua sunt sensata et non conventia, occultantur a non sensatis intentionibus anime uehementer infixis. Estimatiua ergo, que est nobilior iudex inter apprehensiones ex parte anime sensibilis, imperat imaginationi ut defixum habeat intuitum in tali persona. Ymagionatiua vero imperat concupiscibili, unde concupiscibilis hoc solum concupiscit, quia sicut concupiscibilis ymaginatiue obedit, ita ymaginatiua estimatiue, ad cuius imperium cetera inclinantur ad personam quam estimatiua iudicat esse conuenientem, licet non sit. Ymaginatiua autem uirtus figitur circa illud propter malam complexionem frigidam et siccam que est in suo organo, quia ad mediam concauitatem ubi est stimatiua trahuntur spiritus et calor innatus ubi estimatiua fortiter operatur. Unde prior concauitas infrigidatur et dessicatur, unde remanet dispositio melancolica et sollicitudo. Ubi autem concupiscibilis sit sita non determino.[7]

> [What the cause of this malady is by which the powers of the soul are affected is difficult to understand. The cause of the malady is a defect in the estimative power, which is induced by sensed intentions to apprehend insensible accidents that may perhaps not be real, so that it may believe some women to be better and more desirable than all others. This happens, then, when something highly acceptable and pleasing strikes the soul, which then judges other sense objects to be dissimilar, so that if it should encounter sensations that are not desirable, they are obscured by the non-sensed intentions deeply fixed in the soul. The estimative [power] then, which is the noblest judge among the perceptions on the part of the sensible soul, commands the imagination to concentrate its gaze on that particular person. The imaginative [power] commands the concupiscible, thus the concupiscible desires this one alone, for just as the concupiscible [power] obeys the imaginative, so the imaginative obeys the estimative, at whose command the others are inclined towards the person whom the estimative judges to be fitting, even though it may not be so. Now the imaginative power is fixated on it on account of the unbalanced complexion in the organ, cold and dry, for the *spiritus* and innate heat are drawn to the middle ventricle [of the brain], where the estimative faculty of estimation is located. The first ventricle is chilled and dried, so that there remains a melancholy disposition and worry. Where the concupiscible power is located, however, I will not try to decide.]

Michael R. McVaugh, quoting this passage, finds its most provocative feature in "its attempt to fuse psychology, physiology, and ventricles physical causation in an account of *amor hereos*, one fully in the Galenic tradition of providing material explanations of mental states."[8] In fact, the fusion of psychology, physiology, and physical

6 Wack 1990: 51–73. For McVaugh, see below.

7 For the text, see Wack 1990: 198–205; see 198–199 for the passage cited here.

8 McVaugh ed. Arnaldus de Villanova, *Tractatus de amore heroyco*, 1985: 23.

causation already belonged to the Aristotelian system of natural philosophy, and it is through the revival of Aristotle that it entered the Latin West. What Gérard de Bérry understood well is the phantasmological nature of desire, that in the pathology of unrequited love the *phantasma* – the image of the object of desire perceived by the senses – plays a fundamental role in causing emotional and mental derangements.

Phantasia and the images, *phantasmata*, occupy an essential role in the process of thought. The body was believed to be simply a form of organization of the natural elements, and without the vitality that the soul guarantees it would immediately break apart. Yet, soul and body are by nature completely unaware of the other, just like they are blind of each other's sphere of influence. "The soul has no ontological aperture through which it can look down," writes Ioan Couliano.[9] In turn, the body, being a substance, cannot communicate with the soul. The instrument, the *proton organon* that allows the soul to communicate with the body is the *pneuma*, the spirit, located in the heart. Being made of the same substance of which stars are made, it is ethereal enough that it comes close to the immaterial nature of the soul. At the same time, being a body, it can come into contact with the sensible world. It is through the *proton organon* that the soul transmits to the body the vital activities. In turn "the body opens up to the soul a window to the world through the five sensory organs . . . . Called *phantasia*, or inner sense, the sidereal spirit transforms messages from the five senses in *phantasms* perceptible to the soul."[10] Without *phantasmata* the souls would remain completely blind, incapable of perceiving and understanding and therefore unable to set in motion, to activate the body. The Dominican Guillaume de Moerbecke (ca. 1215–ca. 1286) translates the passage in which Aristotle sets this fundamental concept in these words: "Nunquam sine phantasmate intelligit anima."[11] Saint Thomas Aquinas (1225–1274) appropriates this very concept, and in his *Summa theologiae* writes: "Intelligere sine conversione ad phantasmata est [animae] praeter naturam."[12] In other words, the image has the potential to become an object of intellection, but it requires the intervention of the active intellect to be actualized. That means that the image contains the form of the object perceived by the senses, that is then extracted and abstracted from its materiality by the active intellect through a process of "illumination" (Avicenna [below] speaks of the *denudatio* of the image according to the Latin medieval translation). When the image enters into contact with the active intellect, the form within the image finds its intrinsic, dinamogenic power. This dinamogenic power, which is communicated to the possible intellect through the image and under the guidance of the active intellect, is that *species intelligibilis* located within memory and whose task is to complete the process of cognition. Having thus

9 Couliano 1987: 4.
10 Couliano 1984: 4–6.
11 Translated in Foster and Humphries 1951: 442.
12 Thomas Aquinas, *Summa theologiae*, I.q.89.a.1.c.

become aware of the external object of perception, imagination presents it as a desirable or non-desirable end, as an object to pursue or to avoid, by means of the practical intellect.

Medicine taught that excessive desire pours its deleterious effects on the mind and body by altering the very regimen that regulates man's existence, as it contravenes the rules of balance and order that control such elements in the body as sleep, food and drink, exercise, and sexual relations. These rules were discussed in the Hippocratic treatise *Epidemics VI,* again by Aristotle (384–322 BC) in book III of the *Nichomachean Ethics*, and they were universally endorsed by the Arabic and Western physicians and natural philosophers.[13] Poets found in the medical discourse on love the language to express their passion and pains in a realistic manner. Christianity incorporated the concept of regimen into its ethical view concerning the choice of objects by holding as axiomatic that the degree of immorality in an act of pleasure is related to the intensity of the desire for the object of pleasure.[14] While pure love is regulated by reason and is a totally selfless love, all other forms of love are always egocentric, excessive, contrary to the rules of morality and to the regimen of health. For the theologians, pure love attempts to reproduce love as it must have existed before the Fall of Man; concupiscence is the desire accompanying love after the Fall – a perversion of the original love. Yet the matrix and the psychophysiology of these two manifestations of love are the same, because both are born out of the same loss: that of the object of desire. For the Christian, man's craving for love is a constant reminder of the loss of the earthly paradise. The true object of desire is not outside the self, no matter how beautiful or noble. It is only with the eyes of the mind that man can search out the half-forgotten traces of true Love that exist within. The guide is Christ, the Physician of the soul, who opened the way back to that paradise through the crucifixion.

I have, of course, been mixing two quite different things: the speculation of the theologians concerning the objects of desire pertaining to the hereafter, and the discourse of physicians or poets concerning the objects of desire in the here and now. Both physicians and poets speak of the same absence, and analyze this absence within the subject of desire, the distraught lover. The difference consists in the fact that the physician examines the negative effects of this absence in terms of the human organism, and tries to find a cure, a method of restoring organic harmony through the precepts of traditional medicine. The poet describes the experience itself that is based on the irremediable absence of the beloved because that void is the very *antefactum* of the poetic discourse. Within the conventions of poetry in which the Lady occupies a position loftier than the poets own, he can never seek or expect favor. He can only love the Lady as a *phantasm* of his imagination, speak to her

---

13 Corpus Hippocraticum, *Epidemics* (ed. Smith 1944); Aristotle, *Nicomachean Ethics III* (ed. H. Rackham, 1933).
14 For an in-depth overview see Brown 1988.

image inside himself, analyze the effects of that love on his entire being, and, within this solipsistic discourse, find his joy in loving from afar, in suffering, and in speaking of this constant suffering. The poet's sorrow and solace is his poetry, which somehow helps him exorcise the longing left by the unattainability of the object.

As St. Augustine (354–430) reproaches Petrarch (1304–1374) in the *Secretum* "Si enim non nisi quod oculis apparet amare potes corpus igitur amasti"[15] [If in fact you can love only what appears to sight, then you have loved the body.] Because human love, in all its manifestations, consists of speculation on bodily forms apprehended through the senses, it is concupiscence *tout court*; it is the same whether described by the natural philosophers, by the physicians, or by the poets. All are forms of the same love; they share a common causality and psycho-physiological development. Both the poets and the *physici* were interested in erotic obsession; ultimately, they share a common vocabulary. The poet adopted a "scientific" symptomatology to enhance the verisimilitude of his desire and longing. The physician, in his pragmatic way, considered this poetic record to be like any other phenomenon verifiable in nature. The frequency of pathological manifestations of erotic melancholy in literature became, for them, not only a further demonstration of the universality of the syndrome, but such literature constituted a source of historical documentation – a corpus of case histories understood in clinical rather than in literary terms.

This body of ideas was based on the Aristotelian-Galenic system of the soul and of the passions as it was espoused generally by the medical writers in the scholastic tradition. There were, to be sure, variations in its interpretation that will be evident even in the writers we have selected for documentation. But in its global design, this system remained remarkably uniform throughout that period of intellectual history. It was the basis for the analysis of the diseases arising from desire in the writings of those who deal with *amor hereos* and forms the intellectual substructure for all that writers from the 13th to the 17th century believed, and professed, concerning the etiology of erotic love.

Since human love – it was believed – is fundamentally *amor concupiscentiae*, carnal desire, it can cause states of disease because, being a *passio*, it could alter the balance of elements within the body that constitutes health. Love can be deleterious if it includes a component of desire, if it involves the opposite sex – if it is occasioned by "dispositiones corporis inclinantes ad talem concupiscentiam propter aliquam utilitatem sive necessitatem, sicut est inter virum et mulierem complexio venerea vel humiditas titillans in organis generationis"[16] [the dispositions of the body inclining toward such a desire (*concupiscentia*) because of some compulsion of necessity, such as a venereal complexion or moistness and tickling in the organs of generation.] Carnal desire arises where there is an excess of humors (especially blood) or of pneuma,

15 Petrarca, *De secreto conflictu curarum mearum III.*
16 Arnaldus de Villanova, *De parte operativa* (1532), fol. 146 verso, col. 2, ll. 46–49.

that is, when the body is in a hot and humid condition. It was believed semen is a product of burned blood, thus an excessive quantity of blood would result in a state of surfeit – the technical term is *complexio venerea* – responsible for the attraction between the sexes. Hence, those who possess a sanguine complexion are more prone to the stimuli of the flesh. Moreover, since a complexion can be acquired through a diet of rich, fatty foods, or through a sedentary life, it was believed that the wealthy classes and the nobility were more prone to the excesses of erotic desire.

Gérard de Bérry, in his commentary on the *Viaticum*, glosses the expression *amor hereos* as meaning "heroes dicuntur uiri nobiles qui propter diuicias et mollitiem uite tali pocius laborant passione"[17] [the noblemen who, because of their wealth and the softness of their life suffer this passion and that for this reason they are called heroic lovers.] This sense of *amor hereos* deriving from the Greek *eros*, or hero became a commonplace in subsequent treatises on erotic love and can also be verified in the alphabetical lexica of the thirteenth and fourteenth centuries. According to Danielle Jacquart and Claude Thomasset, in the lexica edited by Mario Roques, the following equivalences are given from Latin into French: MS Paris, Bibl. nat., lat. 13032, *heroicus* signifies baron, *heroys, baronesse, heronicus, id est heroicus heros, -ois*, baron; MS Paris, Bibl nat., lat. 7692, *heroys* signifies dame; MS Vatican lat. 2784, *hero* signifies dame.[18] Excessive love was the expression of the style of life of the ruling classes; that fact underscored the appropriateness of the association between *hero* and *hereos*. "The disease is called *hereos*, remarks the French physician Bernard of Gordon, "quia hereosim et nobiles propter affluentiam deliciarum, istam passionem consueverunt incurrere, quoniam sicut dicit Viaticus "Sicut felicitas est ultimum dilectionis, ita heroes ultimum dilectionis"[19] [because the hereosim and the nobles are more inclined to fall into this passion, given the abundance of delights, for as the *Viaticum* says, "Just as happiness is the highest level of pleasure, so hereos is the highest stage of pleasure"];

The main cause, however, is the form of a person or even an object – the Spanish Muslim scholar and poet Ibn Hazm (994–1064), in his *Ṭawq al-ḥamāmah* [The Dove's Neck-Ring], recounts the story of several friends who fell madly in love with various objects – a form perceived by the external senses, especially by sight, and judged to be overwhelmingly pleasing.[20] The Florentine physician Dino del Garbo (c. 1280–1327), commenting upon the poem *Donna me prega*, by Guido Cavalcanti (1250/59–1300), writes that "passio que est amor causatur ex apprehension alicuius

17 Wack 1990: 202.
18 Jacquart and Thommaset 1985: 152.
19 Bernardi Gordonii, *Opus, Lilium medicinae inscriptum . . ., Particula II, De passionibus capitis, cap. XX. De amore qui heroes dicitur*, 1574: 216, ll. 16–20.
20 English translation Arberry 1994.

forme visibilis, que quidem comprehenditur . . . sub ratione complacentie"[21] [the passion called love is caused by the apprehension of some visible form, which is apprehended . . . by reason of its excessive pleasantness]. And he adds:

> Nam amor est passio quedam appetitus, qui appetitus consequitur formam rei apprehense per sensum primo exteriorem et deinde per virtutes sensitivas interiore . . . unde in amore concurrit duplex potentia sensitiva, scilicet cogniscitiva et appetitiva, quia omnis appetitus qui est in nobis insequitur cognitionem.[22]

> [For love is a passion of the appetite and the appetite follows the form of the apprehended object first through the organ of external sensation, then through the internal powers [*virtutes*] of the senses . . . therefore in love the two-fold sensitive passion happens at the same time, that is to say, the cognitive and the appetitive, since every appetite which is in us follows cognition.]

The comment points clearly to the fusion of Aristotelian and Galenic thought that was the basis of scholastic medicine.

Arnaldus of Villanova specifies further that *hereos* is a "vehemens et assidua cogitatio supra rem desideratam cum confidentia obtinendi delectabile apprehensum ex ea"[23] [violent and obsessive cogitation upon the object of desire accompanied by the confidence of being able to obtain the pleasure perceived from it.] If the confidence is well placed, if the lover can satisfy his/her sexual appetites, the body is allowed to return to a state of normality. If that confidence is caused by a deranged reason, by *idée fixe*, and the lover insists upon encouraging that passion, the increased state of sexual tension can be ruinous. In medical terms, the solution is simple, or better simplistic: check the excess through a regimen capable of counteracting the cause of the excess. A surplus of semen – it was believed that both male and female has semen – requires evacuation, if possible, through sexual intercourse within the bond of marriage. The physician, however, is not concerned with the quality of the object of desire or with the social boundaries regulating the act itself. What counts is the quality of the act, which must be capable of reestablishing order out of disorder, and which must, therefore, be conducted in accordance with medical practice.

When someone is *philocaptus in amore* by a woman, explains Bernard de Gordon, (fl. 1270–1330):

> ita fortiter concipit formam et figuram et modum, quoniam credit et opinatur hanc esse meliorem, pulchriorem et magis venerabilem, magis speciosam et melius dotatam in naturalibus et morabilus, quam aliquam aliarum, et ideo ardenter concupiscit eam, et sine modo et mensura, opinans si posset finem attingere, quod haec esset sua felicitas, et beatitudo; et intantum

21 Edition and translation in Fenzi 1999: 68.

22 Fenzi 1999: 106.

23 Arnaldus de Villanova, *Tractatus de amore heroyco* (ed. McVaugh 1985: 46).

> corruptum est iudicium rationis, quod continiue cogitat de ea, et dimittit omnes suas operationes ita quod, si aliquis loquatur cum eo, vix intelligit aliqua alia"[24]
>
> [he dotes on her form, figure, and manner so intensely because he believes her to be the best, the most beautiful and most respectable, the most handsome and most gifted, both in moral and natural qualities, a woman without peer. Hence, he covets her ardently, without limit or measure, wondering if he can reach his goal, for this is his happiness and blessedness. Meanwhile, his reason becomes corrupt, for he thinks of her continuously leaving aside all other considerations.]

The nature of this derangement, as Bernard of Gordon explains, must be explained in terms of the physiology of the human brain, in terms of the nature and disposition of the internal senses: that body of internal functions that parallel the external senses, and oversee sensation, perception, and intellection [See figure 1]. The power of cognition apprehends the object of sensation through the external senses and prepares it by abstracting the form from its materiality through a progressive *denudatio*, or "unveiling," thereby rendering it manageable as an object of intellection. This *denudatio* is carried out by the internal senses. Very briefly: since the brain has cavities, and the brain resides in the highest part of the body, the internal senses must reside inside the cavities of the brain. The frontal part of the frontal ventricle of the brain is occupied by the *sensus communis*, the common sense, which is a vast depository of the forms of the object of perception received and transmitted by the external senses, especially by sight. Once the object of perception is no longer present the form is handed over to the power of fantasy (the Greek *phantasia*, which can be translated as retentive imagination), located in the dorsal lobe of the anterior ventricle of the brain. The forms, the impressions retained within fantasy are then separated or combined, if they belong to the same genre, by the *virtus cogitativa* or cogitative power, located in the anterior part of the middle ventricle of the brain. The dorsal part of the same ventricle is occupied by the *virtus aestimativa*, the estimative power: it is within this faculty that the object of sensation is freed from every trace of materiality, and the non-sensitive intentions within the single objects are finally extracted. These non-sensitive intentions perceived by faculty of estimation are preserved by the *virtus conservativa et memorialis*, by memory, which is in the posterior ventricle of the brain. In other words, memory parallels, but at a higher level, the power of fantasy, which preserve, as I mentioned before, the sense impressions received from the common sense.

This corruption of the judgment is caused by the derangement of the power of estimation (*virtus aestimativa*), one of the animal faculties (*virtutes animales*) of the soul. This concept has already been noted in the statement by Gérard de Béry quoted above. It takes us to the system of medieval and Renaissance psychology,

---

24 Bernardi Gordonii, *Opus, Lilium medicinae inscriptum . . ., Particula II, De passionibus capitis, cap. XX. De amore qui heroes dicitur*, 1574: 216–219, and 216, ll. 6–15 for this passage.

one that was characterized by a compliance with the Aristotelian doctrine of cognition and sensation, and that remained remarkably uniform throughout the Middle Ages and the Renaissance.

According to medieval and early modern medicine, the first stage of the torment of the obsessed lover is caused by the corruption of the faculty of the estimation, that power responsible for judging whether the non-sensitive intentions of the object of sensation are good or harmful. If the desire for the object is strong and persistent, this power can become confused, allowing the subject to believe that harmful things are good and unattainable things attainable. Arnaldus of Villanova sums up this state in the lover in the following way:

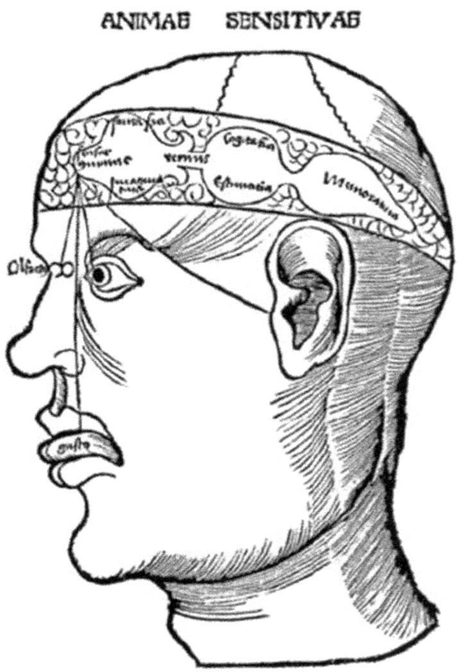

Figure 1: A map of the brain according to early modern medical doctrine.
From M. Qualle: *Habes Hic Amande Lector Textum Parvuli, Quod Aiunt, Philosophie Naturalis cum Commentariis* (Hagenaw: Impensis Ioannis Ryman, in officina Henrici Gran, 1513). Image taken from Daniel D. Cavalcanti, William Feindel, James T. Goodrich, Forcht Dagi, Charles J. Prestigiacomo, and Mark C. Preul, "Anatomy, technology, art, and culture: toward a realistic perspective of the brain," *Neurusurg Focus*, 27 (2009), p. 8.

> quod propter hoc rei desiderium vehemens eius formam impressam fantastice fortiter retinet et memoriam faciendo de re continue recordatur. Ex his vero duabus nascitur tertium consequenter, oritur etenim ex vehementi desiderio et recordatione assidue cogitationis impulsus;

> cogitat namque talis, qualiter et quibus ingeniis valeat rem ad libitum obtinere, ut nocivis delectabilis cultum possit assequi, quod concipit. . . .[25]
>
> [Because of the violent desire, he retains the form imprinted upon his mind by the fantasy, and because of memory, he is constantly reminded of the object. From these two actions a third follows: from the violent desire and from the constant recollection arises compulsive cogitation. The lover dwells on how and through which methods he will be able to obtain this object for his own pleasure so that he may come to the enjoyment of this destructive delight that he has formulated in his psyche . . . .]

The object of desire becomes an obsessive idea that polarizes all the cogitative activities, while the corruption of the power of estimation brings about the derangement of the remaining faculties of the soul, because all other powers, as we noted above, are subject to the *aestimativa*. In time, this obsession can darken and overpower reason itself, driving the lover to seek the gratification of his sexual impulses in opposition to all sense and good judgment. The faculties of the soul, however, cannot undergo change, and therefore they cannot be mistaken. The cause of error must lie with the instruments employed by the faculty to carry out its functions: the middle cavity of the brain and the pneuma or the spirits it contains.

According to medieval psychology, the well-being of man is controlled by the *pneuma*, or *spirits*, for they provide that essential link between the body and the soul. The spirits are the instruments of the faculties and correspond to the three parts of the souls: the natural, the vital, and animal (animus) parts. The *natural spirit* is generated in the liver from pure blood, and from the liver it circulates to all parts of the body through the veins. The *vital spirit* comprises two elements: inhaled air and the emanations of the blood. These two elements are blended in the heart as soon as the air has been transformed and purified in the lungs through a process quite like digestion. From the left ventricle of the heart, the vital spirit is passed to the arteries and to the "retiform plexus" at the base of the brain, where it undergoes a further transformation.[26] Finally, it enters the lateral ventricles of the brain where it joins with the air inhaled through the nostrils; the result is the *animal spirit* that occupies the major ventricle of the brain, the parencephalon. This spirit performs all the operations required by the rational soul and controls, through the nerves, the sensory activities of man, as well as voluntary motion.[27]

In brief, in every human being there are three vital centers: the liver, the heart, the brain. From these centers an interlacing network of vital currents branches out across the body by means of the veins, arteries, and nerves. These vital currents are critical to the efficiency and well-being of the organism. Being formed of air and

---

25 Arnaldus de Villanova, *Tractatus de amore heroyco*, (ed. McVaugh 1985: 46–47).
26 It was Andreas Vesalius who demonstrated that the *rete mirabile* does not exist. For the history of the retiform plexus, or *rete mirabile*, see de Gutiérrez-Mahoney and Schechter 1972.
27 Green 2002.

blood, they are susceptible to changes in the internal balance of the humors as well as to external conditions such as atmospheric pressure.

The psychological processes must be understood to operate by means of this pneumatic circulation. Avicenna [Abû-'Aly al-Husayn Ibn ʻAbdallah Ibn Sînâ: 980–1036], in the *De anima*, explains in these terms the process through which the form of the object passes from the external senses to the internal powers of the soul:

> unitur . . . forma similitudinaria una penes partem spiritus qui gestat virtutem videndi. Deinde . . . penetrat in spiritum qui est repositus in primo ventriculo cerebri, et imprimatur iterum forma visa in ipso spirito qui est gerens virtutem sensus communis . . . Deinde haec virtus quae est sensus communis reddit formam alii parti spiritus, quae est continua cum parte spiritus quae vehit ipsum, et imprimit in illam formam ipsam, et reponit eam ibi apud virtutem formalem, quae est imaginativa, sicut postea scies, quae recipit et conservat eam. Sensus etenim communis est recipiens formam, sed non retinens; imaginativa vero retinet quod recipit illa . . . . Deinde forma quae est in imaginatione penetrat posteriorem ventriculum . . . et coniungetur forma cum spiritu qui gerit virtutem aestimativam, quae vocatur in hominibusvirtus cogitationis, et forma quae erat in imaginativa imprimatur in spiritu virtutis aestimationis . . . .[28]

> [The similitude [of the object] is fused with the spirit that carries the power of vision and it penetrates the spirit that is located in the first ventricle of the brain. It is then imprinted upon this spirit, which is the one that carries the power of the common sense. . . . Then the common sense transmits the form to the neighboring spirit, imprinting it with the form, and thereby places the object in the imaginative power, that which creates forms. . . . Then the form that is in the imagination enters into the posterior ventricle of the brain and unites itself with the spirit that carries the power of estimation and the form that was in the imaginativa imprints itself upon the spirit of the power of estimation. . . .]

Now we are in the position to summarize the psycho-physiological process that leads to erotic melancholy. When a pleasing form reaches the internal powers of the soul, the sudden pleasure that accompanies it causes a rapid multiplication of the vital spirits which overheat and spread throughout the body, thereby overheating the entire organism:

> Cum enim anime gratum seu delectabile presentatur, ex gaudio delectabilis apprehensi spiritus in corde multiplicati subito calefiunt, et calefacti subito . . . delegantur ad membra corporis universa.[29]

> [When something pleasing or enjoyable is presented, the joy coming from the apprehended pleasure multiplies the spirits in the heart. Suddenly they heat up, and this heat . . . causes the spirits to be spread to all the members of the body.]

28 Avicenna, *De anima*, III.8 (ed. Van Riet 1972: 268–70).
29 Arnaldus de Villanova, *Tractatus de amore heroyco* (ed. McVaugh 1985: 49–50).

Because the vital spirits generate the animal spirits, these will also overheat. The receptacle of the faculty of estimation, the dorsal portion of the middle ventricle of the brain, being in contact with the burning spirits coming from the heart, also becomes inflamed, and from this state of inflammation the permanence of the phantasms of perception occurs.

The faculty of estimation controls the *imaginativa*, and the permanency of the *phantasmata* in the faculty of imagination depends upon its degree of dryness. The heating of the *aestimativa* through an overheating of the encephalic area of the *imaginativa* causes a state of excessive dryness:

> Cum itaque firma retentio formarum in multis quibuslibet nequaquam effici valeat sine sicco, necessario sequitur cerebellarem partem imaginative virtutis aliqualiter exsiccari. Hoc vero ex pretactis sic ostenditur: cum et fortis et frequens sit transitus calidorum spirituum ad cellam estimative fluentium ad iuditium celebrandum, pars anterior in qua virtus imaginativa residet propter humidi consumptionem a calore spirituum derelicta remanet necessario siccior seu minus humida quam fuerit per naturam.[30]

> [Since dryness is necessary for the fixed retention of forms, it follows by necessity that the encephalic part of the power of imagination must suffer considerable dehydration. This can be shown from what has been said above: the anterior section in which the imaginativa resides is abandoned by the warmth of the spirits as they flow toward the segment of the aestimativa to accompany the strong and persistent thought and reflection. Because this intense heat consumes the humor in the anterior section, it necessarily becomes less humid than it previously was.]

Once the imaginativa becomes dry, the *phantasma* remains firmly imprinted in the organ of memory as a seal in wax polarizing the attention of thought itself; the image of the desired object remains the only datum present to the consciousness of the lover. It is this obsessive presence of the phantasma that causes the pathological condition known as *amor hereos*.

The physician, of course, is mainly concerned with this pathological stage of *hereos* because, as Arnaldus of Villanova states it, "ex hoc igitur impetu cogitationis intense gravia quamplurimum accidentia patiuntur amantes heroyci"[31] [From the intense force of this cogitation, 'heroical' lovers suffer many accidents.] The first is insomnia, which wears the body out by producing an excessive evaporation of the vital humor. Because the health of the body depends upon the balanced combination of the four humors, the evaporation of the vital humor breaks this tenuous balance. The excessive dryness and the heating of the spirits prevent the instruments of the natural faculty, the natural spirit, and the liver, from performing their functions. As a result, the lover becomes anorexic. The body becomes gradually thinner and drier; the eyes become hollow and tearless, and with the increase of black bile the color of the skin turns green. The medieval medical treatises explain in detail

---

30 Arnaldus de Villanova, *Tractatus de amore heroyco* (ed. McVaugh 1985: 50).
31 Arnaldus de Villanova, *Tractatus de amore heryico* (ed. McVaugh 1985: 51).

the symptoms of this fearful disease: why lovers are pale, why they sigh, why the pulse slows down in moments of desperation and races with the sudden remembrance of a joyful moment (figure 2). They also explain, in detail, the therapy: light diet, sleep, frequent baths (to nourish and humidify the body), bloodletting (given the close relationship between blood and semen), intercourse, and the more "psychological" cures such as listening to music, distracting conversation, exercise, and travel.

If these therapies fail in returning the body to the required state of equilibrium, its continuous overheating and overdrying will produce an excessive quantity of melancholy humors (*melancholia adusta*), which will dry the body completely, darken the skin, and the lover, now melancholic, will ultimately become mad and die. If everything fails, desperate and tragic measures are often employed, as can be seen from the following two case histories, one concerning a young woman, the other a young man.

Figure 2: Jean-August-Dominique Ingres, *Antiochus and Stratonices.* Courtesy of Cleveland Museum of Art.

The German physician Michael Frederich von Lochner (1662–1720) recounts an example of its indiscriminate use of bloodletting in his medical dissertation defended at the end of the seventeenth century, and repeatedly quoted until the nineteenth century:

> A French girl, both noble and marriageable, was leading an idle life and following a warm diet. After she had a clandestine love affair with someone below her station, to whom her parents refused to give the nod, she began to be worn out by insomnia. For several days she began to call out loudly, to exhibit the part of her body that distinguishes the sexes, to sing lascivious songs, to look ferociously, and when someone tried to resist the woman in her

> sexual fury her excitement increased. In fact, if she were not bound with strong chains and held in bed by two or three men, she would conceive a fire in her joints and leap naked from the bed, and if she should encounter some man, she would rush fiercely at him and lustfully beg him to perform the rites of Venus with her. She was constantly awake, her eyes glowed, her intentions were bad, her speech was coaxing. Her face was wholly inflamed and swollen. A pungent, sticky mucous humor that almost rotted the bed linen flowed at irregular intervals through the portal of modesty. "And foul breath pouring from black jaws strikes the nostrils with its odor." The pulse was strong, the tongue dried out, and there was wasting of the entire body. When a doctor was called, he had the girl confined in a French medical torture house [carneficina] where after bloodletting, repeated 30 times in six days, he grew from the girl along with her blood at the same time her insane mind, mad love, dear life.[32]

Another possibility is for the doctor to experiment with therapies that act directly upon image of the loved one fixed upon the organ of memory. If the narcotic drugs and the other therapies cannot erase the object of desire, the lover must be surgically "shocked," and the organic substance of the encephalic area in which the power of imagination resides, and which imprisons the image of the object of desire, must be forcefully liquefied. And since the fixation of the image is caused by the excessive cooling of black bile created by the sudden overheating of the brain cavities due to erotic desire, the method adopted to "shock" and at the same time "liquefy" the receptacle of memory – a kind of hardened wax, according to a tradition that dates to ancient medicine – is to overhear artificially the cranium of the erotomaniac (See figure 3 for example cauterizing instruments). A classic case of misuse of this method is related by Pieter van Foreest (Petr Pieter van Foreest 21–97), one of the most prominent physicians of the Dutch Republic, in a chapter dedicated to erotic folly in his *Observationum medicinalium ac chirurgicarum opera omnia*:

> Another young man of Delphi also driven insane by love was lying bound to his bed where he lay neglected and wretched. After six weeks, when he was consumed by his insanity and by a wasting illness, I was summoned. We found him not only insane but so ill-treated from the various remedies of attending simple-minded women that I had never seen anything comparable or anything so horrible to describe. They had placed on his bare head a bronze device ordinarily used to warm the bed, so hot when lit that they burned the whole crown of his head.In the wound that followed the whole pericranium and skin were removed right down to the skull, so that the skull and cranium were stripped bare in an area the size of a crown . . . .[33]

The proof that not only the uneducated women mentioned by Forestus used this method to cure erotic melancholy, but that it was part of common medical practice against epilepsy, melancholy, and erotomania, can be found in the most important treatise written in the seventeenth century on erotomania, Jacques Ferrand (b. ca. 1575), *Traité de l'essence et guerison de l'amour, ou mélancolie erotique*:

---

32 Quoted by Diethelm 1971: 66.
33 Diethelm 1971: 63.

Me contentant de vous dire que si ceste enseigné maladie empiroit de telle façon qu'on doutast que la melancholie erotique devint loup-garrou, alors il le fault seigner des veins du bras jusques au syncope . . . et neantmoins on luy appliquera un cautere actuel, ou en son refus, ou defaut, un potential sur le davant de la teste, ainsi que nous ont enseigné de faire Paul Æginete, Oribase, Avicenne, et autres autheurs classiques. . . .[34]

[I will simply add here that if the condition grows worse in a way suggesting that the erotic melancholy could turn into lycanthropy, then the veins in the arms must be bled until the patient faints . . . and in spite of this, one must continue by cauterizing the front of the head with a searing iron [actual cautery], or if he refuses or cannot bear it, with a caustic compound [potential cautery], applied to the same place, as we have been taught by Paul of Aegina, Oribasius, Avicenna, and other classical authors. . . .]

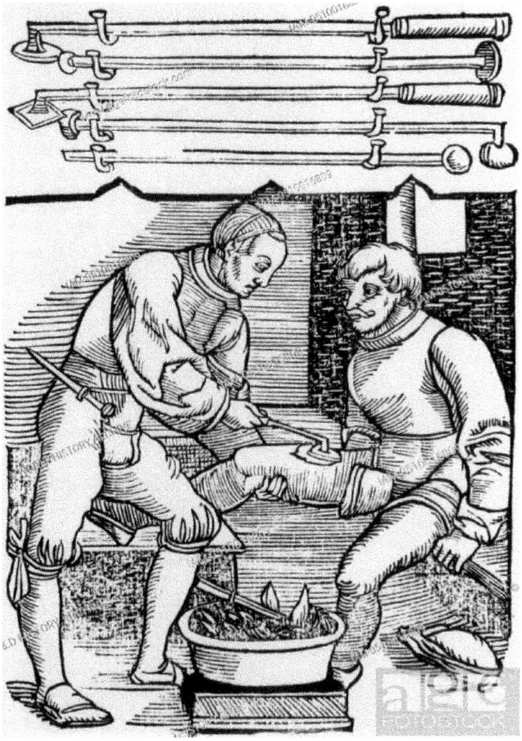

Figure 3: Hans Gersdorff, *Feldtbuch der Wundartzney* (Augburg: Getruckt durch Hainrich Stayner, 1542), p. XXVII.
Photo: Massimo Ciavolella.

34 Ferrand 2010: 363, and 1990: 357 (English edition).

The women described by Forestus quite obviously knew that erotomania can be cured by applying a burning object around the cranium. However, since they did not have the medical training necessary to carry our properly such an operation, and since they did not possess the proper surgical instruments, they used a household object which can easily be heated in a fireplace, an iron used to warm the bed sheets.

What these two case histories clearly show is that as lovesickness became part of the European Universities' medical curriculum and the subject of medical dissertations, it begun emphasizing the pathological side of excessive erotic desire, linking it to madness and erotomania.

The Humanities in Medical Education

Eduardo H. Rubinstein

Art Images And Medical Teaching

Introduction

Medical students are exposed all along their training, to images of the human body obtained using many different methods. Traditionally, guided exposure to those images has been at the core of the students' development of hand-eye coordination and also of the vision-memory processing that is essential for long term learning. But, the effectiveness of this approach has recently come into question; firstly, because it constantly demands the inclusion of challenging new teaching material on a pair with the rapid growth of applicable technology and secondly, because of the need to maintain a balance between the strain of these changes on the students' demanding schedule and their development of a compassionate approach to patient care. The first of these problems is routinely addressed by regularly changing the teaching curriculum, while the second is more difficult to confront because of the number and complexity of the components at play in a successful physician-patient relationship.

Over the past decade or so, novel approaches to these problems have been looked in the humanities in an effort to develop techniques that facilitate a less stressful and more efficacious way of learning how to view normal and pathological images and to enhance the students' sensitivity and their sense of empathy towards the ill. A particularly successful approach to achieve these goals has been the introduction of art courses in medical school curricula. In this essay the results of those strategies are briefly reviewed, and a new method is proposed, that based on the physiology of memory and vision, may enhance the ability of medical students to look at and remember key features of medical images.

The Teaching of Art Appreciation in Medical Schools

Several medical schools in the United States and Europe have recently added, as part of the years of clinical instruction, either elective or compulsory courses in art appreciation. For the most part, these are held at a museum where art works are presented to the students by docents and curators that, following a traditional approach, describe the overall organization of a painting or a sculpture emphasizing the aesthetic aspects of its components. This method is structured to be art oriented, without

https://doi.org/10.1515/9783110788501-017

any overt focus on the medical learning objectives.[1] The pictures, if well presented, should fulfill the students' experience objectives: a) to facilitate the perception of the emotional content of the presented art and become aware of their own response to it, and b) to enhance their image visualization skills. The first objective should enhance the students' ability to understand the patients' feelings as they communicate mostly through facial expressions and body language, and less so with clearly articulated statements about their predicament.

The goal to improve the students' observational skills through an art experience has been explored by several investigators. Naghshineh[2] et al. have combined the exposure to art works with a reference to specific medical images, such as those resulting from a dermatological examination. This simultaneous presentation of an art and a medical image, requires the art specialist to work with a physician, a setting that maybe more difficult to achieve in most medical schools. In this study the artwork varied between painting, sculpture, and photography while the styles presented oscillated between modern figurative art and abstract expressionism. Testing after the course revealed a comparatively better analysis of the pictures, normal or pathological, by students that had the art exposure.

In another study a group of medical educators reviewed the art courses offered at several American medical schools, seeking to identify a common theme that could be of value for other schools that may consider incorporating an art course in their curriculum.[3] Their main objective was the acquisition of visual skills and changes in the students' ability to rapidly identify and describe the important diagnostic elements of medical images presented after the art section of the course. Their performance was evaluated by a panel of physicians and curators that concluded that the students exposed to the art experience performed better than those that were not, when viewing identical medical images.

The obvious difference between the two methods discussed above is the time-in -between factor. The choice of one or the other maybe determined by the number and the availability of teachers and docents, but future studies should involve testing students after sequential time intervals, to explore whether the effect of the art appreciation experience is retained in the memory.

1 Mann 2018.

2 Formal Art Observation Training Improves Medical Students' Visual Diagnostic Skills. Naghshineh et al. 2008.

3 Mukunda et al. 2019. https://pubmed.ncbi.nlm.nih.gov/30810510/ Accessed: February 5, 2022.

Pairing of Art and Medical Images: A Hypothesis

We advance the hypothesis that art images presented along medical figures, enhance the ability to learn and analyze the latter. This proposal will be supported by reviewing the neural mechanisms of image formation and emphasizing those aspects of vision, that enhance the process of learning, especially in relation to attention. The latter is usually defined as the behavioral and cognitive process of selectively concentrating on a discrete aspect of information, whether *subjective* or *objective*, while ignoring other perceivable information. This practice is fundamental in *education*, *psychology*, *neuroscience*, *cognitive neuroscience*, and *neuropsychology*.

Figure 1 defines the steps that are involved in vision and image formation begin as a pure optical process in the eye, that transduces an accurate description of objects outside the body into the retina via a small opening-the pupil- and a lens. This route is similar to that taking place in a camera oscura, originally described by Alhazen in the tenth century C.E.

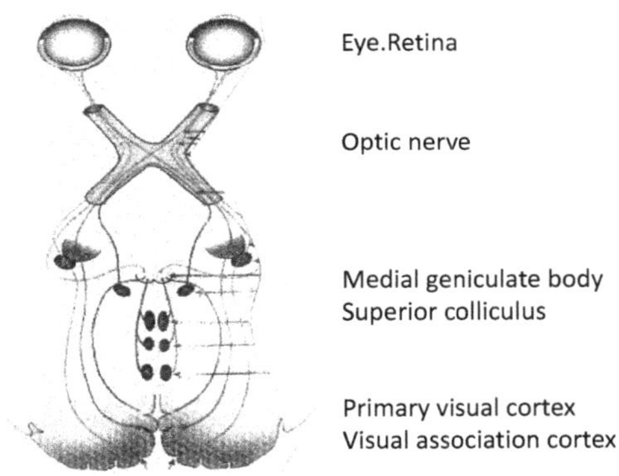

Figure 1: Schematic of the visual system (Courtesy of Creative Commons).

In the retina, neural cells convert the light signals into electrical pulses (cell and nerve action potentials). This activity courses along the optic nerve to two subthalamic structures: the lateral geniculate body and the superior colliculus nucleus. From there, a wide neural bundle, relays the electrical signals to all levels of the visual cortexes. This multilayered organization effectuates the creation of the images and the comparison with those stored in the memory, that are related to the recently formed one. These areas are interconnected and specialized and it has been proposed that the internal image results from the interaction of two “streams” or pathways: a ventral stream (the “what pathway”) that is involved with object and visual identification and recognition, and a dorsal stream (the “where pathway”) that is involved with processing the object’s spatial location relative to the viewer.[4]

This subdivision starts already at the level of the retina, where there are two types of ganglion cells that receive the light signals via photoreceptors: bipolar and horizontal cells. Large ganglion cells are fast responding to motion, spatial organization, depth perception and foreground/background separation. But they have no color sensitivity and are of low acuity.

At the cortical level they become part of the *where* system. In contrast, the small ganglion cells are slow and with low contrast sensitivity but high acuity. They are also very color selective and specialized in object and face recognition. These cells became part of the *what* system. It is interesting to remember that from the evolutionary point of view, this dual visual design is present in humans and primates but not in all mammals. This organization affects our perceptions and as a consequence there are contrasting characteristics such as: acuity versus color, acuity versus spatial resolution, luminance versus color evaluation. The interactions between the “what” and the “where” systems may be the basis for our perception of a foreground and a background when looking at an image. The image created by the processes described before is constantly changing because we explore the environment continuously during viewing, an action that requires movements of the eyes and the head, namely:
- rapid eye movements: saccades.
- brief interruptions of movement: fixations.
- slow scanning movement: smooth pursuit.
- gaze shift: convergence

Typically, when we start looking, long saccades and short fixations are followed by longer fixations that are controlled by a feed-forward mechanism-from the brain to the eye muscles-that modify the scanning pattern as a function of the internal “impression” created by the developing visual image.

To understand how we “organize” a visual journey over the surface of a painting or a radiograph, we shall consider only two of all possible viewing strategies:

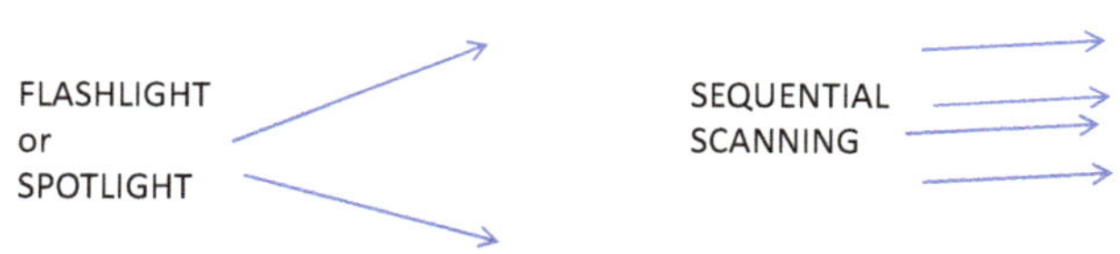

1 – the flashlight or spotlight. This a descriptive name for a global view, that is important for the initial recognition of style, theme, overall color balance, initial emotional reaction.

4 Mishkin and Ungerleider 1982: 57–77.

2 – the sequential scanning. This follows the previous view and focus on our evaluation of the points of interest, from high to low priority.
We have described the initial steps in the process of viewing, namely, image formation (the pathway from the eye to the brain) and external object exploration (scanning).

The acquired image is elaborated further at other levels of the cerebral cortex where it is "compared" with memories, "colored" with an emotional significance and briefly retained as working or short-term memory. This progression is modulated by the level of attention which is critical in gaze and focus adjustment on the important elements of the object that is been looked at. The eye movements are designed to expose the portion of the retina with the highest acuity – the fovea – to the "points of interest" on the visual target.

We have proposed that the simultaneous presentation of artistic and medical images, enhances the learning process. It may be argued that this combined presentation could in fact be a distractor and result in the opposite effect. However, such diversion of attention will be avoided if the selected art work elicits a positive emotional reaction that is maintained while considering the adjacent medical image. The proposed interaction has been studied as an example of tangential learning, a very common process during game playing.[5] The presentation of a visual image – a stimulus – before a response is required, is in effect, a mode of pairing found in associative learning and from this it could be construed, that pairing is a form of conditioning but, in fact, the presentation of contrasting images – artistic vs medical – should create a situation of newness, an important element in attention restoration theory.[6] The novelty of the pairing may be of benefit because stimulus-driven attention is minimally affected or not at all by mental fatigue.

The Process of Looking at Art Images

We have been discussing the presentation of two pictures side by side with the intent of increasing the level of attention by using the proposed facilitating influence of the art image. Interestingly, there is an example of a paired presentation in most figurative paintings, and this is the contrast between the foreground and the background, that may be considered as separate elements. Figure–ground organization is a type of perceptual grouping that is a vital necessity for recognizing objects through vision.

Subjective factors can also influence figure–ground perception. For instance, if a viewer has the intention to perceive one of the two regions as the figure, it will likely

5 Mozelius 2017.
6 Kaplan and Berman 2010: 43–57.

alter the ability to analyze the two regions objectively. In addition, if a viewer's gaze is fixated on a particular region, this area will be more likely to be considered as the figure.[7]

Figure–ground organization is used by artists and designers during the composition of a 2D piece. Figure–ground reversal may be also used as an intentional *visual design* technique in which an existing image's foreground and background colors are purposely swapped to create new images.

The separation between foreground and background depends on the relative luminance of each one and the sharpness of the edges. From a practical point of view, luminance (Lv) depends on the level of illumination (Ev) and the reflectance (R) of the surface and is expressed by the relationship Lv:Ev.R/π.

This is an important determinant for the selection of an art image, that should not be too different in luminance from the associated medical representation, otherwise the initial perception of the two figures (using the "searchlight," as discussed above) will be unbalanced.

And, at the same time, the characteristics of the colors on the pictures will also be changed by the level of luminance because the perception of a color starts in the retina when light strikes the red, green, or blue cones, that respond to the wavelength of the light and its intensity. The consequence is that an object characterized by a bright color at daylight, will be perceived dull to grayish at dusk.

The manner in which the foreground and background of a painting were managed, varied along the history of art. The very early and medieval works did not clearly differentiate the two components, but this changed dramatically in the pre-Renaissance period. This vision became more sophisticated with the Renaissance painters not only in Italy but in the rest of Europe.

Paired Presentations of Art and Medical Images

It is proposed that the characteristics of our visual system favor the hypothesis that pairing images is valuable for learning, provided that we select a quality of the art image that enhances the experience of looking at the medical picture. The images may be contrasting in color or complementary. They could share common angles or an obvious linearity. It is expected that when dealing with a medical book with many images, e.g., an anatomy textbook, only few art images may be included, enough to change the level of attention. In all cases, the selection of the art image should be determined by the nature of the medical figure, with the intent of obtaining a balance along the lines of the examples that follow.

7 Peterson and Salvagio 2010: 4320.

The pairing in Figure 2 is based, not on specific aesthetic or formal features, but because each represents a very characteristic image on their fields. In art: Leonardo's portrait of Ginevra de' Benci and in everyday medical practice: the chest X-ray.

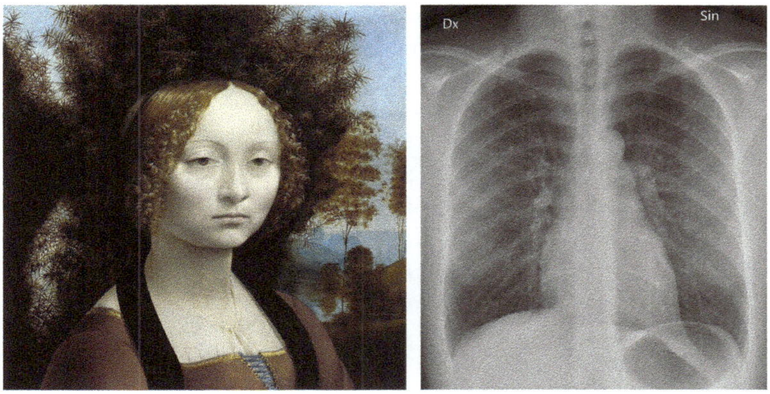

Figure 2: Left: Ginevre de' Benci, Leonardo da Vinci. 1474–1478 (National Gallery of Art, Washington, D.C.). From Wikipedia Commons.Google Art Project.Right: Chest X-ray. From Wikipedia Commons.

The attentional advantage of pairing becomes more apparent when it is used to present anatomical images (figure 3).

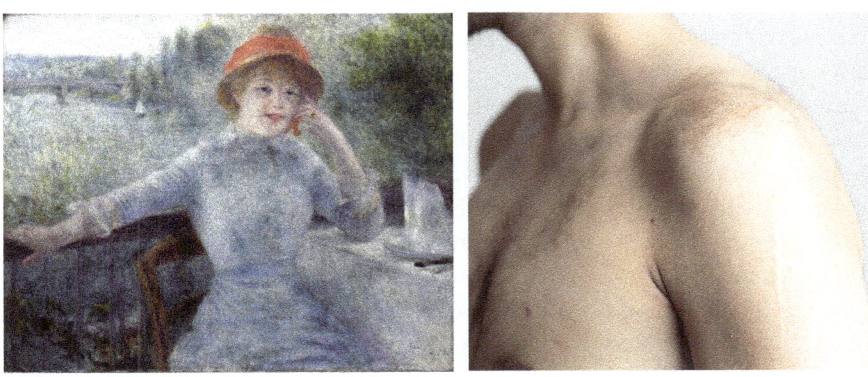

Figure 3: Left: Auguste Renoir, Young Spanish woman. 1989 (Courtesy of the National Gallery of Art, Washington, D.C.). Right: Brachial plexus. J.M. Bourgery and N.H. Jacob (1830–1850), Atlas of Human Anatomy and Surgery (Courtesy of Taschen GmbH (2005), p. 291).

The angles on the painted figure (left) and those of the exposed structures in the anatomical representation (right) are comparable and reinforcing.

The technical improvements that started with the Industrial Revolution, were paralleled by comparable advances in the development of diagnostic medical tools. We will apply the "paired view" to the analysis of images obtained using those devices.

One of those diagnostic techniques is radiographic angiography, where images are obtained after injecting a contrast medium into a blood vessel – usually a large artery – while suppressing the background structures.

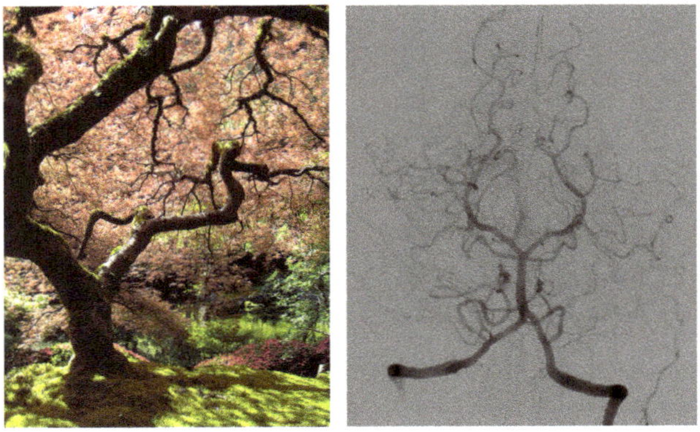

Figure 4: Left: Trees. Photograph (Courtesy of Creative Commons). Right: Posterior cerebral circulation showing two vertebral arteries joining one basilar artery (Wikipedia Commons).

Both images on Figure 4, emphasize the bifurcations of branches: on a tree (left) and in a major cerebralartery (right), exemplifying the reinforcing of images by the similarity of a repetitive pattern.

The electrocardiogram derives from the electrical activity that is generated during each heartbeat. This small voltage difference is picked up by electrodes on the skin of the chest and extremities and easily increased using electronic amplification. Tracings are obtained from different electrodes and the electrical signal is transduced into a graphic either on paper, via an electronic printer or directly on a video screen.

Figure 5, shows, at the top, the Roman aqueduct "Pont du Gard," near Nimes, France and a the bottom, a three lead electrocardiogram. Both share a "three level configuration" that reinforces the linearity of the representations.

Another important technique is magnetic resonance (MRI) that produces medical images of outstanding quality.

The realistic painting of a face (left) on Figure 6 is balanced by the detailed MRI image and both figures are enhanced by their common sharpness.

Positron emission tomography (PET) is an imaging technique in which a radioactive tracer is injected and its radiation is scanned using a circular approach comparable

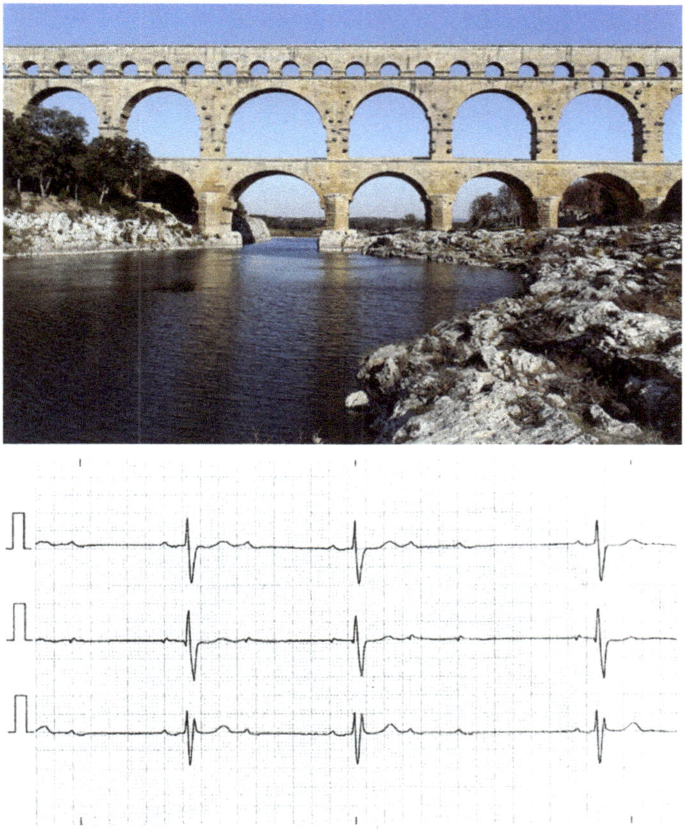

Figure 5: shows: on the top, the Roman aqueduct "Pont du Gard," near Nimes, France. Photo R.Ferrari,2007. On the bottom, an abnormal three lead electrocardiogram. From M.Rosengarten,M.D.

to that of the CAT and MRI. Varying the injected substance, oxygen, glucose, or others, it is possible to evaluate the flow of blood or the metabolic activity of selected organs.

The two pictures in Figure 7 represent a human face (left) against the depiction of the "inside of the head" (PET) on the right. Both images are polychromatic and of different brightness, the latter enhances their contrast.

An ultrasound beam is also used to create images by processing the echoes reflected from deep anatomical structures. A handheld transducer emits and receives the signals while scanning from the skin surface. Changing the placement and the angle of the transducer it is possible to "visualize" various aspects of a selected organ. This is of great value for the examination of the cardiac chambers.

In Figure 8 the painting complex outline (left) reinforces the equally complex view of the cardiac chambers (right in black and white), that differ in shape and

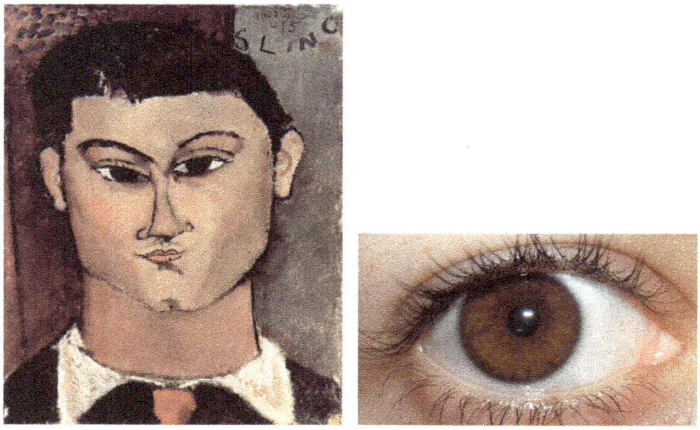

Figure 6: Left: Portrait of Moise Kisling.Amedeo Modigliani.1915. Mailand Collection. Direct Media. Right: Human eye. Light brown iris. Wikipedia Commons.

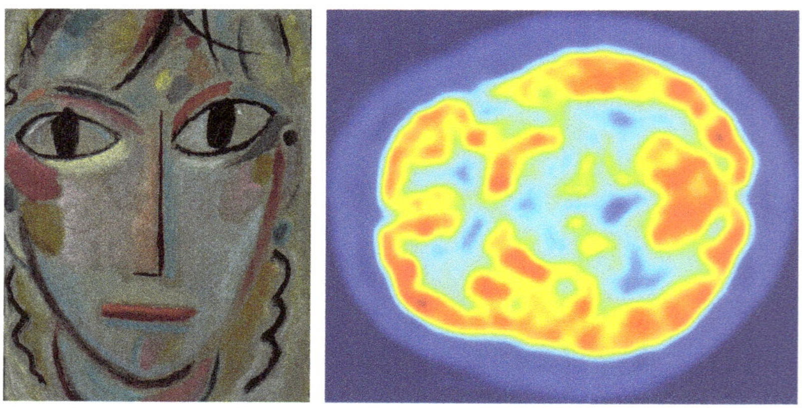

Figure 7: Left: Medusa. Alexev von Jawlensky. Musée des Beaux-Arts, Lyon,France. Right: PET scan of the human brain. Wikipedia Commons.

internal configuration. The artwork also gives an impression of movement, that is very appropriate when looking at the pictures of a potentially beating heart.

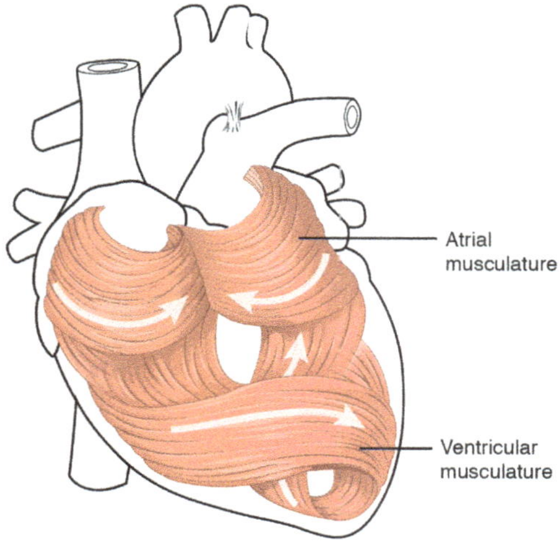

Figure 8: Left:Head of a sleeping woman. Pablo Picasso.1905.Museum of Modern Art, N.Y. Source: MOMA. Right: Diagram of the cardiac muscles.Wikipedia Commons.

The Support for Art and Medical Image Pairing

We have discussed several mechanisms by which an art picture may enhance the visual processing of an associated medical image, especially the modulation of the level of attention, a fundamental element in memory acquisition. The selected artwork should have the characteristics that conform with the process of looking attentively, such as the reciprocal scanning of the points of interest as presented in the examples shown.

Another aspect of memory acquisition is priming, that is, when the exposure to an image affects the perception of another presented simultaneously or shortly thereafter. This effect is greater when both objects are of the same nature. Priming works in both positive (enhancing) or negative (decreasing) modes with respect to memory formation. One characteristic positive influence is the emotional response triggered by the priming image. Thus, it is proposed that the art image will have this affective quality as it will attract, with its novelty, the students' attention especially in the context of the projection of slides during a medical lecture.

Presenting art works to medical students, even when not explained by a person or a text, may be the first step towards art appreciation and its associated positive humanistic value, e. g. the interpretation of the emotional content of facial expressions. And, for those with an above average knowledge of art and its history, the exposure will have a pleasant element of recognition that may add another motivational element to their learning experience.

Finally, the paired exposure may help the students to learn, while performing the various steps of a medical exam, such as auscultation, percussion and palpation, the ability to "see" the anatomical structures of the body. This process of creative visualization is common to the evaluation of medical images and the appreciation of art works and is yet another route for the proposed enhancing effect of pairing.

Francis C. Wells

Medical Humanities: A Tautology or a Necessity?

> "In this world nothing can be said to be certain, except death and taxes." We will all die. And we need to make sure that our taxes are spent wisely to ensure that when we do, wherever we do, we are able to die in comfort and with dignity, confident in the knowledge that those who love us are also being supported and cared for. Benjamin Franklin

This is a personal and heartfelt communication following 42 years of caring for my fellow human beings! I have sat at the side of dying fellow creatures, I have hugged survivors of dreadful disease and laughed in the face of despair with people of immense bravery at the most critical times of their lives.

Perhaps, Humanism may be defined as a philosophical and ethical stance that emphasizes the value of ourselves as human beings. In the context of medical humanity, it could be described as the way that we within the caring professions relate to, and fully appreciate the human needs of others in our professional relationships, including our professional colleagues.

I have framed the title of this essay as a question because it may reasonably be assumed that those entering the caring professions would of necessity be deeply caring by nature. It may also be assumed that just being of a caring nature is enough to carry us through our professional relationships with those people who need our help and support. Hence the idea of a 'tautology.'

It could be reasonably thought that the selection of medical students would be inherently biased toward those of a more altruistic nature. Altruism, the performance of cooperative unselfish acts beneficial to others, has been studied in several medical contexts, including the donation of organs and genetic material and patients' participation in potentially hazardous experiments and trials. Physicians' altruism towards their patients and others has been less well studied and is implicit, rather than explicit, in statements about medical professional values and attitudes. Altruism is, however, embodied in many cultural stereotypes of the "good doctor," such as John Berger's country practitioner in *A Fortunate Man*.

Altruistic behaviour by physicians might include, for example, continuing to work or providing informal medical advice outside contracted hours, giving free treatment to poor patients in fee for service healthcare systems, and a general willingness to go the extra mile in professional activities. There is much evidence that many doctors work beyond their contracted hours, but there is also a growing feeling that altruism in medicine, if not dying, is at least declining.

This might be expressed, for example, in the anesthetist's unwillingness to accept a final case on the list because the operation would run beyond the limit of the contracted session; in the general decline in home visiting rates by general practitioners;

https://doi.org/10.1515/9783110788501-018

or in the recent explicit choices now made by young doctors in balancing professional and domestic commitments. Generation X is making a cool appraisal of the costs and benefits of a medical career.

Explaining the emergence and maintenance of altruistic and cooperative social behaviour has been a longstanding problem in the biological and social sciences, and there is currently intense debate about the determinants of human nature. Darwin recognised altruism as a particular difficulty for his evolutionary concept, which was based on competition and the struggle for existence.

Thus, there is an inherent struggle within us all between giving of ourselves for others and fighting for our own place in the academic, social and financial hierarchy.

When I and my peer group went through the selection process for medical school aside from satisfactory grades in our examination results, an interview was 'de rigueur.' At that time, one could be admitted to medical school with less than straight 'A' grades on the basis of a good interview. Furthermore, there was a tendency towards taking potential students from medical families as there was a presumed understanding of the commitments involved in the profession. Did these presumptions lead to anymore empathic graduates? I suspect not but it was done with that in mind. This latter situation has largely disappeared today.

With regard to qualifications, the examination standards are generally higher with candidates required to attain straight A* grades in all subjects to allow them to even be considered. Several additional achievements are also looked for in different fields of endeavor often involving the sporting arena. With the extreme didactics of teaching programmes in most western countries today aimed at the achievement of academic success, there appears little time for reflection and lateral thought let alone encouragement for an appreciation of the Arts. A requirement for an appreciation of the classics is now even more remote. An understanding of Latin and Greek, once pre-requisites for medicine have long disappeared. The over-crowded curriculum of "hard" subjects in both the pre-clinical and the clinical courses leaves, it is argued, little or no time for interdigitation of humanist education.

However, there is a paradox in the system. In the United Kingdom, the negligence of successive governments in matching the number of doctors in training to the need has led to a significant shortage. This linked to the hostile litigious environment within which doctors now are expected to work, and the widening gap in remuneration between those in the medical profession and those in the "City" or in business has led to a relative diminution of applicants. In turn this has led to shortened training programmes for mature graduates who in their middle twenties and older can gain a place at fast-track courses. In addition to this, the training programmes for specialism have been dramatically reduced, again curtailing the experience of 'junior' doctors prior to their release on the ailing public as practitioners in their own right. Therefore, if the more mature student can be presumed to achieve the necessary skill sets in a reduced time frame and specialist training can be achieved in a shorter time, why cannot time be created in the basic curriculum for a more humanistic approach?

Humanism, or the humanities, cannot be assumed to be naturally inherent within the chosen professional, and even if the doctor is a deeply caring individual, is she/he equipped with the necessary emotional milieu to work from a perspective of understanding the human condition?

Naturally, the discussion has to go back even further in the maturation of the future doctor, and indeed any functioning member of a healthy society. The inculcation of an integrated science-arts programme is surely what is lacking. As school and University qualifications are aimed at producing young people for the perceived work force rather than for a position in society, education becomes ever more didactic, and examination orientated. Indeed, in the modern world, schools and Universities are judged by the grades achieved by the students rather than the quality of the human being in the eye of the storm. By the very nature of the time spent at school and scholastic activities, parental influence, even if positive in life skills is less and less. Even sadder is the fact that the classics, within the discipline of the humanities are seen as less and less relevant in the modern world.

So, can we assume that a lack of education in the Humanities is disadvantageous to today's doctors? Can caring be taught or rather instilled in other ways? Does, in fact, an exposure to a classical humanist style education convey any positive benefit in society at large, or more specifically in the medical profession? These are big and important questions to be considered before there is a whole-sale introduction of such subject matter into the curriculum, that as we have already observed is extremely crowded? In the modern world the practice of medicine is now a financially driven profession which revolves within a fiercely academic world; this within the setting of a highly litigious society. Medicine has all but had the humanistic element of life squeezed out of it. This leads to almost daily news items of disturbing events within the healthcare arena. Just in the last few days headline news stated that doctors no longer have the skills to speak to patients and their families of death and dying. Special courses are to be set up to "teach" these skills. The public react with horror expressing sincere concern that this sort of thing has not been taught in the process of a medical training. Yet there remains a humility in the general public towards their doctors who remain one of the most valued groups in society.

The National Health Service in Britain still is held in highest regard and any politician seen to meddle with it does so at his/her peril; and yet it is visibly failing in so many areas. Every day we hear of failing hospitals, negligence in many guises and inadequate numbers of beds, nurses, doctors and almost all of the support groups necessary to run the service. All professional groups are under enormous pressure to deliver ever better services for patients and their families with ever less resources and time. Doctors now have to be appraised annually, to revalidate every 5 years, job plan every year and be accountable through audit and mortality and morbidity reporting. In our specialty of cardiac surgery if our results fall outside an accepted norm set by risk stratification, we are likely to be suspended from practice pending investigation. In the midst of all this is there any surprise that humanism has left the building?

So, what of an education in the humanities for doctors at any or every stage of their training? I strongly argue that it is a necessity and not simply desirable.

First let us consider the origins of the concept that immersion in a humanities-based education has merit. The term 'Humanism' has its origins in the Latin *humanitas*, a term used by Cicero and others in what have become referred to as classical times, embracing Greco-Roman rhetorical and literal ideals. The concept arises from the perceived merits of a liberal education. *Studia humanitatis* constitutes the study of what today would be considered an Arts based curriculum, including literature, language, history, and moral philosophy. Perhaps it is the latter, as appertains to medical practice that may prove to be of the most relevance in the modern world. In the Renaissance, the term *Umanista* was used to describe a teacher or student of classical literature and the arts associated with it, including Rhetoric (the art of effective or persuasive speaking or writing; skills well suited to a career in clinical and academic medicine). It appeared in the English equivalent in the late sixteenth century. Humanism came to find expression in virtually all of the arts-based subjects, theology, the law, medicine, mathematics and the creative arts. It became fundamental to the continuity of European cultural and intellectual history.

What then was the role of Humanism in the Renaissance, the period in history where it so frequently appears in discursive texts? For it seems to have had different roles in different countries. For example, in France the study of classical texts was related to the better understanding and utilization of grammar. In Italy it developed along the lines of the use of rhetoric in daily professional life.[1] This extended into the widespread importance of letter writing in that time. Also, the utility of Roman law was expressed in the codices of the time, especially in Bologna, Pavia and Padua.[2] Widely regarded as the father of Humanism is Petrarch (Francesco Petrarca), whose writings and activities particularly in Avignon, the home of the so-called Babylonian exiled Papacy, largely defined the then accepted description of what it was to be a humanist: a scholar of the history of the classics in literature, rhetoric and language.

In 1576 Gabriel Harvey, the University praelector opened his spring lectures with one entitled "Ciceronianus." He spelt out his conversion from Ciceronianism, expression only in the words and phrases of their masters to the more profound expression of reasoning taken from his speeches, Ciceronianism:[3]

> Pay attention not only to the brilliant greenery of the words, but more to the ripe fruit of meaning and reasoning . . . . Remember that Homer described words as 'pteroneta,' that is, winged,

1 Mann 1986.
2 Mann 1986: 5–6.
3 Mack 2003: 82–98.

> because they easily fly away unless they are kept in balance by the weight of the subject matter. Unite dialectic and knowledge with rhetoric. Keep your tongue in step with your mind. Learn from Erasmus to combine an abundance of words with an abundance of matter; learn from Ramus to embrace a philosophy which has been allied to eloquence; learn from Homer's Phoenix to be doers of deeds as well as writers of words.[4]

A precept of this approach was to encourage the use of dialectics and rhetoric in the reading of classical texts. This process would enrich one's understanding of how to use words and arguments. The lack of this understanding of the use of dialectic in public speaking is seen frequently in modern scientific meetings and presentations where faulty conclusions can be heard to be drawn from inadequate and sometimes inappropriate use of available data and the lack of knowledge of, or understanding of, preceding and indeed current work.

A recent example is a presentation at the American Association of Thoracic Surgeons where the presenter described to the cardiac surgical audience how restoring the competence of the mitral valve of the heart caused a significant worsening of ventricular pump function in all patients, described as the ejection fraction. This derivative use of ejection fraction as a measure of cardiac function was flawed and revealed the simple re-iteration of previously stated knowledge in the absence of any reflective thought. The leaking mitral valve allows blood under pressure to escape the wrong way from the ventricle making the use of the ejection fraction of the ventricle completely inappropriate as a measure of ventricular function. Once the valve is rendered competent all of the blood has to flow into the much higher resistance of the peripheral circulation, making the ventricle appear a lot worse, although only reflecting the true function. This kind of superficial discussion of subjects is quite commonplace in the medical literature, and as has recently been shown departs into the realm of deceit and invention in rare cases (perhaps not so rare). This lack of discipline in the academic sphere perhaps could be addressed through an exposure to moral philosophy.

The education in and understanding of syllogistic logic as utilized by Aristotle develops expansive thought and as I shall explain in a moment directly altered my way of thinking about a surgical problem. This dialectic approach aims to persuade by argument, whereas rhetoric teaches a variety of means of persuasion, including self-presentation, audience manipulation, emotional appeal and the use of figures of speech. Provided that the basic subject matter is correct and properly researched the ability to persuade an audience of veracity remains important today.

In the scientific world of astronomy Sir Fred Hoyle, a notable British astronomer in the last century, argued strongly that life on earth could have arrived from

---

4 Gabriel Harvey, *Ciceronianus* (London: Bynneman, 1577) ed. Wilson and Forbes, 1945, 82. For the speech of Phoenix, see Homer, *Iliad*, I.201 (ed. Bloom 1996: 9.443).

other worlds by means of meteor strikes (panspermia; the theory that life exists everywhere in the universe). This was laughed out of court for many years but now with cogent argument and properly peer reviewed data the concept is gaining traction. It often seems that the messenger is more important than the message. Accepted authorities in the field are heard and given preferential 'air-time' because of who they are and not necessarily because of the veracity of their rhetoric.

I have taken a deep interest in early anatomical knowledge and its development. Central to this interest has been the work of Leonardo da Vinci. I am frequently asked if my reading of Leonardo has altered my approach to my surgery. The answer is, not directly in an iterative sense. There is no passage in Leonardo's notes that describes a modern cardiac surgical operation. However, the widespread reading that I undertook to understand his life and times and his own style of dialecticism led me to look at the anatomy of the Mitral valve in a very different way. Also, his unashamed use of lateral thinking and other disciplines in engineering and hydrodynamics urged me to think in different ways.

There is a very beautiful passage in Leonardo's notes where he uses a plant seed as an analogy for the heart and its place in the cardiovascular system. Galen had argued that the liver was the central component of the circulation with a continuous formation of blood arising there fuelling the continuous ebb and flow and consumption of the blood. As we now know, in what seems to be so obvious a statement, the heart is the seat and source of the circulation within the cardio-vascular system. To make his point, Leonardo argued that just as in a seed the roots grow down from the seed and the branches grow upwards, in the heart the veins flow into the heart and the arteries away. The simple drawing that accompanies this mind experiment is shown on the notebook page RL 19028 recto.[5]

My reading of his notes, and following as best one can his patterns of thought that reflect the logic of the time, led me to a different appreciation of the structure of the Mitral valve and the mechanism of disease affecting it and an approach to reconstruct this complex structure. The syllogisms and dialectic argument in his work and that of others of the period began to influence the way that I was thinking. The Mitral valve isn't just a mechanism to allow uni-directional flow but a structure that turns into a force field when the full force of systolic contraction is exerted upon it through the mass and flow of the blood within the ventricular chamber. These thoughts were bolstered by an appreciation of his application of architectural and engineering principles. The result for me was a different perspective on the biologically engineered valve through millions of years of evolution. The valve literally became alive in front of my eyes! A tangible result from the thoughts of the past.

5 Leonardo da Vinci, Notebook, Codex Arundel 1478–1518; Arundel MS 263; ed. O'Malley and J. B. de C. M. Saunders 2003, 282.

Can an exploration of the Humanities throughout an education, extending into the years as a medical student and beyond bring something important to the education of our future doctors? In this modern and fast developing world not only do I believe that it can, but I am certain that it is an absolute necessity.

In a recent book, *Re-engineering Humanity*, Brett Fischmann and Evan Selinger discuss the impact of the digital world on what it is to be human.[6] With contact with patients becoming ever more remote through computer-based data retention, recording and communication the patient becomes ever more remote. Walking onto the wards today instead of seeing nurses with the patients you are likely to see the nurses and the junior doctors crowded around a computer terminal. The patient seems ever more remote. In some hospitals it has been deemed bad practice for the doctor to sit alongside the patient on the bed and hold the hand of the patient. Yet these simple acts of human contact and humility can be all important for the state of mind of the patient.

The medical defense unions are constantly informing doctors that the commonest cause of litigation is poor, or the complete lack of, communication. Learning communications skills by rote is far less likely to succeed in my opinion than a steady insemination through an understanding of human nature that can be developed through an understanding of the humanities.

A further reason to embrace the humanities throughout and is the simple fact that 'Life' is enriched by a broad education firing a real and deep-seated interest in the Arts. Time spent immersed in all or any form of the arts is something that becomes ever more appreciated as time passes. The introduction of Humanities based subjects at an early stage of education provides an outlet for the pressures of a modern clinical life with all of its demands and rigours. Medical burn-out is a real phenomenon in the modern medical world and is causing the significant premature loss of doctors in all parts of the profession. It seems to be affecting the newer generations of doctors more, where expectations of a "normal" lifestyle is more accepted. The life skills maturity that a broad and deep education brings, must inevitably lead to a wider appreciation of ways to enjoy and to foster meaningful interests outside work, and to provide a release from the inevitable pressure of working in such an intense environment that is caring for one's fellow humankind. In the medical profession a longitudinal exposure to the humanities could, would, in my opinion broaden horizons, engender a greater understanding of life and human inter-relationships. At the very least it would allow the possibility of discovering other worlds that would allow professionals to derive other interests outside of medicine. Returning the human to the humanities, from which the term was derived, cannot be anything other than a good thing.

6 Frischmann and Selinger 2019.

The Late Renaissance and Early Enlightenment Arguments against the Humanities

Teaching the humanities has not always been accepted. The 17th century saw the beginnings of a change in the stance towards the Humanities. There were those who still maintained the importance of a deep education in the Humanities centred around a very sophisticated understanding of spoken and written Latin. Such a person was the distinguished academic Constantijne Huygens. On the other side were Michel de Montaigne, the French late Renaissance philosopher, Francis Bacon and René Descartes, all in favor of modernist scientific and analytical thought over classical revivalism through ancient texts. Although all classical scholars with an excellent command of Latin they also declared the utility of the vernacular in academic explanation.

Huygens, however, was living proof of the ability to exercise academic excellence and original thought with a deep Humanism. As a child he was encouraged to study science, music and painting. He was enrolled into Leiden, the most modern university of the age. He attended courses in Latin on canonical texts and courses in modern mathematics and military engineering. He was as modernist in his outlook as René Descartes, a revolutionary who had deliberately turned away from humanist learning and scholastic philosophy of the college system and built a new world of learning. Descartes did concede that "the reading of good books is like a conversation with the best minds of past centuries."[7] Huygens, in his autobiography,[8] was as delighted in his learning and use of Latin as he was in his discoveries with the use of the microscope and saw no academic distinction between the disciplines but only a complimentary role.

Montaigne who had benefitted from a classical humanist training revealed in his last essay "on experience" that the whole enterprise of trying to find guidance for modern behavior in classical texts required the reader to rip their supposed authorities out of time and context. Individual lives and situations, societies and religions differed so very much that one could not hope to have the past shed light on the present. He labelled the hours spent in Humanist education a waste of time.

Even more scathing was Bacon who treated Renaissance Humanism as scholasticism before it as a fatal disease of learning. He argued that Humanists had failed to see that the world had changed, that the modern world was more expansive and wide ranging in scope and therefore that the old ideas could not scope these developments. That 'they' had confused antiquity of the Greeks and Romans – the fact that their texts existed for a long time – with the authority that human beings gain as

---

7 René Descartes, *Discourse de la methode*. Leyde: de l'Imprimerie Ian Maire, 1637 (ed. Mclean, 2006, Part 1).
8 Huygens, 2018.

they age, an authority that can be invested only in people who continue to learn as they age, not in books which are impervious to experience.

Descartes admitted that the learning he had obtained through his Jesuit education was worthwhile, but that travel could have taught him the same things.

The value of the broad and deep Humanist education was appreciated by some for the financial rewards that it could bring. Thomas Gainsford remarked that a fine classical education "not only elevates above the vulgar herd but leads not infrequently to positions of considerable emolument." As an aside Boris Johnson as new mayor of London was having great difficulty getting legislation through a rowdy labour opposition. He therefore announced that the rest of the meeting would be conducted in Greek!! Perhaps the classics still have a use in the modern world.

Galileo argued that truth was not to be found in the classic texts but in the modern world of science. This of course got him into serious trouble with the catholic church. His championing of Heliocentrism and the work of Copernicus was highly controversial and denied by the church as it flew in the face of the perceived sense of the Holy Scripture. The matter was investigated by the Roman Inquisition in 1615 which decided that heliocentrism was "foolish and absurd in philosophy, and formally heretical as it explicitly contradicted in many places the sense of the Holy scripture." This was reversed relatively recently (1992).

Conclusions

So, what of now, today? Should the Humanities have a significant place in the education of our doctors? Despite the controversies over the years, I believe that they should. Not in the sense of the in-depth analysis of high Latin, or in the depths of Dialecticism and Rhetoric, but to have an understanding of these subjects as made relevant to today's world. This is the challenge to make it relevant. As such it will need dedicated teachers who plan accordingly so as not to overwhelm the students in their already heavily loaded curriculum. In the words of Vittorino da Feltre in the 1650's, "promising young men must not be overworked: they must be allowed time for physical exercise and for other honest forms of leisure. More importantly they should not be 'dried out' and transformed into desiccated pedants. The teacher should not for example expect the ordinary young scholar to commit long works in Latin prose, as opposed to verse more easily learnt."[9]

So how do we do this? How do we shape the modern doctor? Especially in a world where ever more expensive medicines and therapies may become the privilege of the few? How do we show kindness, compassion and caring for our fellow humans in their hour of greatest need? We learn through the writing, painting,

9 Barlaeus, "Methodus studiorum," in Grotius 1545: 353.

music and philosophical thoughts of the greatest artists, philosophers, musicians and thinkers across the ages. Whilst writing this I am listening to Tchaikovsky's *Orestia Overture*. Everyone should know this music. It is mentally expansive, beautiful, and full of what it is to be human. Who cannot care for their fellow human when exposed to such beauty from the mind of another?

We should appreciate the Arts etc the Arts in its many forms with the encouragement to become involved in some "person" sensitive way to the extent of engendering a love of the subject and a desire to be involved in some way across a lifetime, to play and instrument, to sing in a choir, to develop the basic skills of drawing and painting. To learn to look and to see, to hear and to learn the skills of listening. To develop in as profound a way what it is to be a member of our wonderfully sensitive, caring, and innovative human being. Most profoundly to learn to love our fellow human beings in all their miraculous shapes and sizes and in all emotional states.

Then perhaps we can all regain our humanity. Our kindness. Our existentialist being relating to our fellow humans in the hours of their greatest need.

Romy Sutherland

Teaching PTSD with Film: The Case of Peter Weir's *Fearless*

For the past six years I have participated in a year-long team-taught undergraduate course for roughly one hundred and sixty Freshmen at UCLA called "Mind over Matter: The History, Science and Philosophy of the Brain." My charge in this course has been to offer students an introduction to the reading of film in order to study a series of feature films that address mental health topics. The course was conceived by Professor Scott Chandler, a distinguished UCLA neuroscientist and a specialist in brain function associated to the movement of the human body, who was keen to develop a multidisciplinary approach to the brain and the mind geared to first year college students who would receive general education credits in both the sciences and the humanities. He was responsible for teaching the electro-chemistry of the brain and for explaining brain functions involving movement of human and animal bodies. He enlisted other professors in Neuroscience, Psychology, History of Medicine and Literature to explore how their disciplines can contribute to our understanding of the mind and the brain, and as a team we created an integrated course. Since UCLA's academic year is divided into three terms (or "quarters"), Professor Chandler organized the first two quarters with lectures by the entire faculty, supported by a team of graduate student teaching assistants whose areas of expertise complemented those of the professors, and who assisted our students in writing papers on issues or topics that could be addressed from more than one disciplinary perspective such as the problem of consciousness, or the definition of pain. From the outset all faculty attend each other's lectures, and make it a point to integrate our material. The third and final term of the year, our Spring quarter, is a capstone experience, a seminar of no more than twenty students in which each of the instructors has an opportunity to offer a more in-depth study related to their own fields of expertise, as it intersects with neuroscience.

My own charge in the lecture component of the course is to contribute to this interdisciplinary project through the study of film. My initial lectures are on some key terms and concepts in film analysis (all aspects of mise-en-scène, including composition, color schemes, costume design, lighting, camera placement and movement; and the principal components of diegetic and non-diegetic sound) to ensure the students are adequately equipped to analyze the films included in the course with accuracy and precision. I also lecture on a series of films directly related to topics which other colleagues address from a range of perspectives pertaining to their respective disciplines.

In my capstone seminar I assign several recent articles by neuroscientists presenting new research on what takes place in the brain while we watch films; and

https://doi.org/10.1515/9783110788501-019

some articles by film scholars identifying the relevance of these findings for film analysis.[1] A significant component of my seminar involves training students to make a short film presenting one of the neuroscience topics they have studied during the first two quarters of the course. The pedagogical objectives of this initiative are to deepen the students' understanding of their chosen neuroscience topic, to equip them with film production skills, and to enhance their film analysis through acquisition of practical knowledge of the medium's possibilities and limitations. For this element of the course, we have access to cameras and editing equipment through UCLA's undergraduate Powell Library, and I work with UCLA technicians who specialize in assisting undergraduates with the fundamental techniques of film-making and film editing.[2] During the COVID-19 pandemic, I adjusted and redesigned this component of the course, allowing for students to produce films with their portable devices. I arranged for some training and technical support over Zoom, and made platforms available to them where they could edit their films with their group members through remote means. Each year the students' final film projects have been consistently impressive, and their course evaluations repeatedly signal that they find the Film Lab to be a particularly rewarding and useful learning experience. In fact some of the final films have so effectively illustrated certain topics covered during the first two quarters of the course that colleagues have incorporated them into future lectures for pedagogical purposes.

To give just one example, a group of students took inspiration from three different but interconnected contributions by our faculty: a lecture by our literature professor on the "Petite Madeleine" episode in Marcel Proust's *Remembrance of Things Past*, in which taste and smell usher forth an involuntary memory for the novel's protagonist and first person narrator; a lecture by our psychologist on the distinct ways short and long term memory work with our various senses, grounded on the neuroscience of memory retrieval; and my own lectures on filmic representations of subjective mental states in general and Post Traumatic Stress Disorder (PTSD) in particular. With these three elements (the literary description, the psychological explanation of memory retrieval with each of the five senses, and the discussion of techniques used by film-makers to convey or suggest internal mental states) a group of students made a short, powerful film titled "Madeleine" in which a female veteran of a recent war has an involuntary recollection of a traumatic event awoken by tastes and smells in the present that resemble tastes and smells she experienced in the theater of war. The film begins in a UCLA café in which the female protagonist takes a sip of tea and tastes a madeleine, followed by a smooth tracking shot along a wall whose muted yellow tones will blend with the color of an outdoor

1 Several of the chapters and articles I assign are from: Zacks 2015; Shimamura 2013; Nannicelli and Taberham 2014.

2 UCLA's "Studio 22," led by Vince Mitchell, has provided excellent instruction in filming, lighting, and editing to our students.

setting in which the same female actor appears tasting a madeleine, during respite from combat. This sequence is accompanied by soft rhythmic sounds subtly evoking gunfire, which gradually increase in volume. It is followed by stock footage of active combat set in a strikingly similar field, which the students found in UCLA's film library and ingeniously, and seamlessly, edited into their own footage. A reverse tracking shot then returns the protagonist from the past memory to the present café where she realizes the madeleine triggered the traumatic memories, and the realization initiates the process of addressing her PTSD. As the film concludes, the rhythmic beat of the sound track continues, suggesting that the protagonist will never be free of the memories but that, through treatment she may be able to mediate their effect on her. The experience of writing, filming and editing this piece enhanced the students' awareness and experiential understanding of filmic choices available to directors, including the application of certain conventions used in both commercial and art-house cinema to represent emotions and convey internal mental states with the use of sound, lighting, color, acting, camera movement, and editing.

The process of researching and planning the project also gave them an opportunity to strengthen a humanities perspective on topics they began studying in the science component of the course. Indeed, my seminar also involves critical thinking about science from a humanities perspective, stressing that some of the scientific research they are studying may be provisional or informed by various kinds of cultural bias. In the case of "Madeleine," for example, my students reflected on the fact that certain neurological events may take place when a traumatic memory is retrieved, but that current knowledge of neuroscience is not sufficient to determine with precision why the same experience may trigger a neurological event in one individual but not in another. This discussion was inspired by some of the lectures offered by our historian of medicine and by the topic of this essay, the representation of PTSD in film, which was the central focus of my lecture on the film *Fearless* (1994) by Peter Weir, to which I will shortly return.

The students who take my capstone seminar are self-selected as all of our students are given a range of choices including courses on more scientific or historical topics. My students wanted more in-depth engagement with the kinds of themes I addressed in my lectures for the entire group of one hundred and sixty students. It is worth noting that when we as a faculty had preliminary discussions about where the film component might be most effective in this course, we decided that my participation would be particularly suited to the advanced stage of the course, when the students are exploring higher order neurological processes with a concentration on mental illness from multidisciplinary perspectives. A film component could have been integrated in any number of other ways, of course, but this is one that has worked effectively for our objectives. I join the lecture course once our students have been studying mental illness from neurological, psychiatric, historical and literary perspectives. My role is to teach a series of films that resonate with the topics the students have been studying, including Schizophrenia and PTSD.

Just as there are literary techniques and conventions that can give a reader the illusion of having access to the internal mental states of characters – such as the stream-of-consciousness technique developed by James Joyce, Virginia Woolf and others – in my lectures in this course, I show my students that there are filmic conventions and techniques that can suggest internal mental states of characters, or the experience of certain psychological states in the viewer, and I chose films to show a variety of possibilities available to film-makers when aiming to convey subjective experiences of mental illness. I assign two films that address schizophrenia in dramatically different styles. We begin with *A Beautiful Mind* (2001) by Ron Howard to explore a skillful albeit commercially oriented treatment of the topic. We explore how Hollywood conventions and winning narrative formulas trump the realities of an individual struggling with schizophrenia, even as a popular film such as this can raise awareness of mental illness and contribute to constructive conversations about it. For a complete contrast, I then show them an intentionally non-commercial treatment of schizophrenia by the art house director David Cronenberg, *Spider* (2002) featuring a mentally challenged man who has severe difficulties expressing himself with language and suffers from schizophrenia. The eponymous protagonist goes through an arduous personal odyssey, a process whereby he recovers the painful and repressed memory that as a child he had killed his own mother, after having spent most of his life in an institution for the criminally insane. The contrast of a film like *A Beautiful Mind* with *Spider* offers our students a clear sense of the wide range of choices available to film makers, but it also makes students aware that they need to develop a critical approach to cultural objects and products such as film, even when film can play an important role in developing public awareness and stimulating conversations around public health in general and mental illness in particular.

Another topic I address in my lectures is the filmic representation of PTSD. One of the main points that our psychologist emphasizes in her lectures on PTSD is that the same traumatic event can elicit strikingly different, even opposite, responses in different individuals depending on their personal histories and circumstances. I chose to teach *Fearless* by Peter Weir, which portrays the trajectories of two individuals who survive an airplane crash in which dozens of passengers are killed. In the case of *Fearless*, my main focus is how the same catastrophic event can cause PTSD in some individuals but not in others, how PTSD can be experienced in different ways by different individuals, and the ways in which film can convey aspects of subjective experiences of PTSD. While the film is not limited to these issues and some of its concerns go beyond the experience of the condition, it does address PTSD responses to trauma directly, and this aspect of the film was the focus of my attention in the context of this course.

Before our students explore *Fearless*, they have heard lectures and read reading materials assigned by UCLA Professor Barbara Knowlton, a specialist in memory processing, and have studied the complex interactions amongst various regions of the brain in processing emotions and mental imagery. They have also studied the neurological mechanisms through which memories are encoded and recalled, and the differences in these processes between brains with and without various forms of mental illness.

When lecturing on PTSD Professor Knowlton focuses on Fear Conditioning and the encoding and retrieval of traumatic memories. The students learn the mechanisms with which the brainstem system and the amygdala system can sensitize an individual who has suffered a traumatic event to subsequent stressors and related stimuli in the future, with particular emphasis on the amygdala's role in long lasting fearful reactions (See Figure 1).

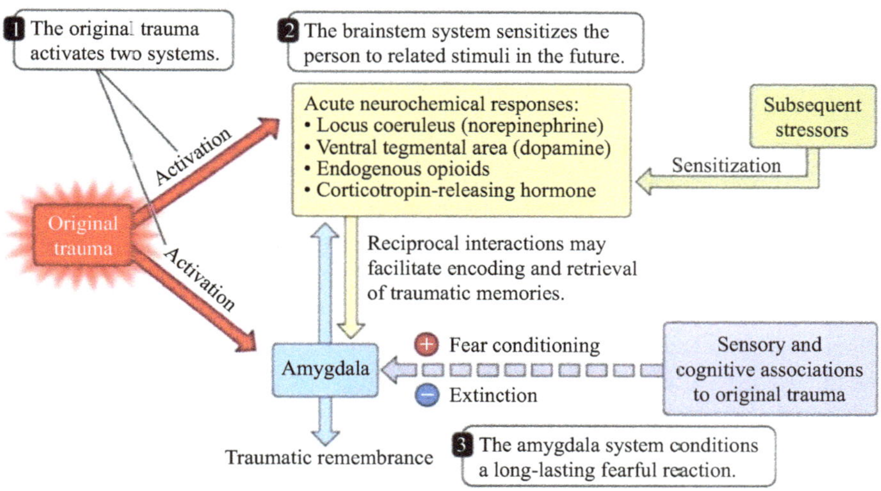

Figure 1: PTSD's neurological mechanisms, from the textbook, *The Mind's Machine. Foundations of Brain and Behavior.*

Studying the neural model of PTSD in our textbook, *The Mind's Machine. Foundations of Brain and Behavior* by Neil V. Watson and S. Marc Breedlove, highlights just how much remains to be learned about how and why PTSD can manifest itself so differently in different individuals.[3]

In my lectures I wanted to move on from the science of memory to a humanities approach, as they are becoming sensitive to phenomena such as voluntary and involuntary memory, which work differently with each of the five senses.

3 Watson and Breedlove 2016: 362.

Fearless

The first frame of *Fearless* elicits in the audience something of the disorientation and confusion the survivors of the plane crash experience in its immediate aftermath. A hand held tracking shot, filmed in slight slow-motion, follows the protagonist, Max (Jeff Bridges), as he cautiously proceeds through a corn field, grasping a boy's hand and holding a baby in his other arm: presumably his children. We, the audience, gradually see the wreckage of a plane crash for the first time through point-of-view shots from Max's perspective, only as Max, a crash survivor, sees it for the first time himself. We are surprised to hear Max's first lines, referring to the boy and baby respectively: "I'm not his father" and "I'm looking for her mother." With these opening images and lines of dialogue Weir immediately underscores parent-child relationships and establishes a pattern, namely that what appears to be the case on the surface with Max may be misleading, and warrants thoughtful examination.

The opening sequence surveys the burning wreckage of a plane's fuselage, focusing in on the responses to the scene of the film's two protagonists, Max and Carla (Rosie Perez). Max ambles through the wreckage in a state of apparent calm, demonstrating an increased threshold for anxiety, verging on numbing; while Carla is shown in an extreme state of emotional arousal as she screams uncontrollably and is physically constrained by first responders, already signaling a decreased threshold for anxiety. The sequence oscillates between point-of-view shots from both protagonists' perspectives of the carnage they are witnessing, and third person shots of each of them, revealing to the audience their respective responses. It will eventually become apparent that what each of them saw during and immediately following the accident plays a critical role in their respective cases of PTSD, and being able to acknowledge what they saw is a vital step towards their respective recoveries.

Neither Max or Carla were travelling alone on the flight. Carla was travelling with her eight-month-old son, Leonardo, and Max was travelling with his business partner and best friend, Jeff, both of whom were killed in the accident. In the immediate aftermath of the crash, Carla is distraught at the loss of Leonardo; while Max appears to be indifferent to the loss of his best friend. The narrative will eventually reveal, however, that the most challenging aspect of the ordeal for both of them to discuss is having seen their travel companions die, and acknowledging what they witnessed to each other will constitute strongly bonding moments in their relationship. The significance of what both protagonists witnessed illustrates a point our students have studied in their text book *The Mind's Machine*, "The kind of event that seems particularly likely to produce subsequent stress disorders is intense and is usually associated with witnessing abusive violence and/or death."[4]

4 Watson and Breedlove 2016: 361.

Much of the film focusses on the strong attachment Max and Carla form, which is unexpected in part because they would appear to have little in common, and their overt PTSD symptoms appear to be diametrically opposed. Max is a middle aged, upper middle class, highly educated, atheist, Caucasian male; Carla is a devout Catholic Latina woman in her twenties from a working-class background who has not had the opportunity for higher education. Max's PTSD takes the form of a feeling of elation, fearlessness, invincibility, a savior complex, reckless behavior and spending almost all his time outside his house. Carla's PTSD entails acute depression and heightened fearfulness, withdrawal, avoidance, feelings of impotence and vulnerability, and difficulty leaving her own bedroom, let alone her house. Both do share, however, a tendency to push away family and friends to whom they were previously close, and they each experience an emotional attachment to each other.

Max's form of PTSD illustrates well a point made in an article I assign the students by a psychiatrist who served on the committee that defined Post Traumatic Stress Disorder, Dr. Frank Ochberg, "Victims can appear to be indifferent, unconcerned and unharmed, when in fact they are in a state of profound post-traumatic stress."[5] The difference between the protagonists' symptoms illustrates another point in the same article: "PTSD has not only a variety of dimensions and components, but vastly different effects and implications."[6]

Max's bravado in the wake of the crash, while representing the direct opposite of Carla's behavior, is no less worrying to the psychiatrist appointed by the airline to treat the crash survivors, Dr. Pearlman (John Turturro). The doctor makes it clear that Carla and Max are the two passengers most affected by the trauma, and that he considers both of them to be at risk of suicide. Weir dexterously employs a myriad of film techniques to subtly convey that there is dissonance between Max's apparent fearlessness and an underlying fragmented psyche, haunted by the experience of being on the plane leading up to and during the crash.

My analysis of the film looks back on my colleagues' lectures in neuroscience, psychiatry, and psychology, and it adds to those perspectives by bringing to the discussion the seminal research that has been done on trauma cinema, including the work by Janet Walker, who is as keenly sensitive to issues of trauma as she is to identifying specific kinds of filmic techniques in works that portray traumatic events in affecting ways.

Given that severe trauma can explode the cohesion of consciousness, Walker has argued that the most effective works within trauma cinema figure the traumatic past as meaningful but fragmentary. In an attempt to convey characters' subjective experience of the unspeakable, such films develop an alternative to the realist representational mode, with techniques such as nonlinearity, repetition, oneiric sequences,

5 Ochberg, no date.
6 Ochberg.

symbolism, intrusive sound and flashback. Such cinematic strategies can be an effective way to intimate that the characters didn't fully perceive and process the trauma as it occurred, and that they continue to experience gaps in and resistance to memory in the wake of the traumatic event.[7] *Fearless* is a useful example of a film that incorporates the kind of filmic strategies Walker discusses, and others, to intimate aspects of subjective experiences of PTSD.

The contrast between Max's fearlessness and his underlying, masked anxiety is elegantly illustrated in two sequential shots. In the first, beautifully composed image (See Figure 2) Max is filmed from an extreme low angle atop a building, framed by an expansive blue sky and bright sunlight, with arms extended, resembling an airplane ascending.

Figure 2: Max's overt state of fearlessness.

This shot is followed, however, by an inverted image of the interior of the same corner of the building: a constricted, gray, dead end into which Max is slumped in anxiety (See Figure 3).

The juxtaposition of the two shots constitutes an eloquent visual portrayal of Max's overt presentation to the world, and his concealed emotional state, suggesting that Max's fearlessness is a coping mechanism for underlying emotional disturbance.

Throughout *Fearless*, Weir employs a range of film techniques to visually suggest Max's fragmented psyche. Immediately following the plane crash, Max flees the site of the disaster and books himself into a hotel where he takes a shower. While stepping out of the shower, he stands transfixed staring at fragments of his own body, shown to us in a series of extreme close-up, point-of-view shots from

7 Walker 2005: 809.

Figure 3: Max's concealed, underlying state of anxiety.

Max's perspective as he focusses sequentially on his feet, a scar on his torso and his face, before finally whispering skeptically "I'm not dead" (See Figure 4).

Figure 4: Point-of-view-shot from Max's perspective, struggling to conceive of himself as alive and whole.

It is as if, in the wake of the crash, he cannot conceive of himself as whole. The best he can do is acknowledge fragments of himself and struggle to assimilate something that does not make sense to him: that he is alive.

In a suspenseful scene, Max nonchalantly but intentionally walks into a busy stream of traffic, believing himself to be invulnerable. Here on the surface level of the narrative Max appears to be cheerfully exhibiting reckless behavior. Weir, however, portrays him in a series of physical fragments (See Figures 5–7).

Figures 5–7: Visual allusions to Max's counterphobic behavior and fragmented psyche.

The sequence is filmed in a succession of medium close ups on sections of Max's body as he confidently strides across the street, while cars honk and dodge out of his way to avoid hitting him. We, the audience, see a series of unconventional shots: Max's legs walking in the bottom third of a composition, with a car whizzing past obstructing the view of the rest of his body; the top of Max's face behind a car in the foreground; snippets of Max's face and body glimpsed between fast moving vehicles: a series of body fragments between silver metal bumper bars, engine fronts and car roofs, all resembling silver aircrafts. On the surface level the sequence conveys how precarious Max's deluded thinking renders him. Through unconventional framing and exaggerated lighting, however, Weir is also simultaneously visually alluding to the fragmented psyche underlying and causing the counterphobic behavior.

One of the more aesthetically sophisticated aspects of the film is the way Weir employs discreet visual strategies to suggest that Max and Carla are chronically haunted by the site of trauma. Irrespective of where they are physically, their subjective experiences in any physical locale are colored by their experiences on the plane leading up to and during the crash. Throughout the film Weir simultaneously conveys the objective "reality" (within the fiction) of the physical context in which he places his protagonists, and the protagonists' respective subjective experiences of being in that space. While maintaining a realistic register, Weir subtly suggests through several elements of mise-en-scène, including shot composition, props and lighting, that his protagonists may be in hotel rooms, homes or offices, but they *feel as if* they were still at the site of trauma, on the plane.

A group therapy meeting led by a psychiatrist for the crash survivors which Carla attends features constricted compositions and is filmed in unrealistically low-key lighting to suggest that, emotionally, the attendees continue to feel to varying degrees that they are on the plane. Several scenes featuring Max share the same visual strategies echoing the interior of the plane, which is a sophisticated visual way of implying the underlying commonalities to Max and Carla's conditions, despite their opposite surface level symptoms. When Max flees the site of the crash to a hotel, a tracking shot follows him from behind proceeding along a corridor which is unrealistically narrow and dimly lit, prefiguring a flashback we will see later in the film of Max walking down the plane's narrow, dim corridor (See Figures 8 & 9).

Figure 8: Max in a hotel corridor construed to evoke the interior of a plane.

In a later scene, once Max is reunited with his wife and son, sharing a meal in his family home, Weir defies standard composition and lighting conventions to tightly frame the three characters in unrealistically low-key lighting for a regular dinner scene (See Figure 10).

Weir also places the camera behind the protagonist, as if we were seeing the back of a passenger in the constricted space of an airplane seat, in airplane lighting. Seeing Max from behind also deprives us of access to his facial expressions, leaving us with the frustrating sense of not having enough information to make sense of his eccentric

Figure 9: Shot of a plane's interior.

Figure 10: Defying conventional composition to suggest the restricted space and dim lighting of a plane's interior.

words or behavior; while simultaneously directing our gaze to his wife's and son's confused expressions. We are hence aligned with the supporting characters' concern for the protagonist, whose emotional range has become as restricted as he was physically constricted on the plane. This scene efficiently illustrates a War Veteran's comment quoted in the Frank Ochberg article I assign our students: "Physically coming home doesn't mean being home."[8] Without any explicit commentary, these shots shed powerful light on Max's perplexing behavior of perpetually needing to flee outdoors: he is forever trying to escape the feeling of being trapped in the plane's interior.

The film technique of showing fragments of Max's body is combined with the use of dimly lit, constricted compositions, implying the interior of the plane, when Weir includes a dimly lit, extreme close-up exclusively on Max's ear, accompanied by sounds of the plane's engine, to usher us into Max's first flashback of the accident. We are given access to his memory of what took place only as Max himself is ready to re-visit it.

An intriguing question throughout the film is why Max and Carla, who on the surface appear to be so different from each other, form such a powerful and mutually healing bond at a time when they cannot connect with their own family and friends. I argue that the answer can be found in what each of them cannot discuss about their experience of the crash.

For Carla, the most traumatic element of the catastrophe is, not only that her eight-month-old infant died in the crash but that, while attempting to hold on to him as the plane was descending, she lost her grip on him and saw him fly out of her arms to his death. She is simultaneously mourning, and feeling responsible for the loss. Carla is therefore experiencing an acute form of survivor *guilt* for what she *did*: letting go of her child, but also acute *shame* for who, in her assessment, this act means she has *become*: a catastrophically failed mother and therefore an irreparably sinful individual. It is not until well into the film that she can "confess," and only to Max, how her son died, and reveal to him the debilitating degree of her attendant guilt and shame. Significantly a sensitive scene of trusting connection between the two protagonists in which Carla is, for the first time, able to consciously revisit and articulate what took place on the plane, is set in a car with mise-en-scène designed to subtly resemble the interior of an aircraft.

A detail of the accident about which Max cannot speak is whether or not he *saw* his best friend Jeff die or dead. Much is made of this point in the narrative, as whether or not he saw Jeff's dead body will have significant financial implications for both Max and Jeff's widow. The three scenes in the film in which Max displays the most heightened affect – aggressive and apparently irrational outbursts, followed by fleeing the interaction or insisting that others stop the conversation or leave the scene – are triggered by characters questioning him about whether or not he saw Jeff die or dead. At no point in the film can he actually articulate that he did see his partner's dead body. The closest he comes to revealing it is in a scene that functions as the corollary to the above-mentioned scene between Carla and Max, and is an artistic high point of the film. This scene is also set in a stationary car, in which the protagonists

8 Ochberg, no date.

occupy the front seats, while being filmed by a camera perched behind them, between the two seats, offering us a point of view analogous to that of being seated on a plane viewing the passengers in the row in front of us. The sequence, in which Max and Carla are discussing dreams, takes place in the rain and is filmed through a blue filter, imbuing the scene with an oneiric atmosphere. Despite the close friendship forged between the protagonists over the several months and multiple shared outings since the accident, Carla still doesn't know that Max was accompanied on the flight. "Were you alone on the plane?" she asks him, to which he replies "No, I was with my best friend, my partner [a pause], he was decapitated." "You saw him?" she gently enquires. Without verbally articulating anything, Max offers a barely discernable affirmative nod in reply.

I argue that an earlier scene in the film offers hints as to why Max cannot speak about having seen his partner die or dead, and why the bond between Carla and Max is so strong that it contributes significantly to their respective recoveries. Shortly after the crash, the psychiatrist, Dr. Pearlman, is so concerned that nobody can, in his words "get through to" either Max or Carla, that he arranges for the two to meet in the hope that they might connect with each other. Dr Pearlman takes Max to Carla's modest home where she lives with her husband and extended Latino family. As they are waiting outside Carla's bedroom door, Dr Pearlman explains that he is concerned that Carla won't talk and adds, "She's very Catholic, old world, filled with guilt and shame." To which Max replies, in his post-accident characteristically supercilious tone, "*I'm* filled with guilt and shame. How's that old world?" The very next words he utters, and the first he speaks to Carla on meeting her are: "When I was thirteen, I saw my father die in front of my eyes." I am suggesting that in editing together Max's unprompted and unusual revelation of being filled with guilt and shame, with his unconventional introductory greeting to Carla, Weir is subtly implying that there is a connection between Max's guilt and shame and having witnessed his father's death.[9] I am also arguing that the trauma of having *seen* his father die "in front of my eyes," as Max emphatically puts it, is triggered by having seen his partner die or dead, and it is the triggering of this former trauma that is the cause of his extreme PTSD in the aftermath of the plane crash. Both Carla and Max, therefore, have experienced a traumatic rupture in a parent-child bond (Carla the loss of her infant son; Max the loss of his father as a young adolescent), over which they had no control, but for which they experience survivor guilt and shame, in part for not having been able to prevent the respective deaths. I contend that this is the underlying point of deep connection between the two characters, which they experience even before they can articulate the critical details of what they saw during the crash. Carla's decreased threshold for anxiety and Max's increased threshold for anxiety were defensive responses to trauma,

9 In Weir's celebrated film *Witness* (1985), the director displays acute sensibility for the potential psychological impact of what a character witnesses.

which needed to be calibrated with a realistic recognition of their human vulnerabilities and strengths.

Fearless is, therefore, a particularly appropriate film for a Medical Humanities course. It is through applying attentive film analysis and an understanding of PTSD, that an audience can arrive at what is skillfully concealed beneath the surface level of the film, namely that Max's underlying trauma of having seen his father die in front of his eyes, is triggered by having seen his partner die in the crash. I argue that Max's traumatic rupture of a parent-child bond shares characteristics with Carla's traumatic rupture of a parent-child bond, and is what lies at the crux of the strong connection between the two characters, and why their close friendship ultimately leads to the beginning of a healing process for both. Furthermore, I suggest that the way Weir conceals this clue to Max's PTSD from the audience is analogous to the way Max conceals it from his own conscious mind. It is there to be uncovered in the film, just as it is there to be uncovered in Max's memories, and thus the process of actively engaging with the film has parallels to a therapeutic process.

Jorge A. Lazareff

The Humanities and Global Health: Travels with Philippa Foot and Karl Popper

> Medicine is a philosophical discipline that can be studied in any number of ways, going to medical school among them.
> Guido Ceronetti, *The Silence of the Body*, 1993.

The Project

In 2013, while visiting the Universidad Nacional de Nicaragua in Managua (UNAN-Managua), Dr. Freddy Meynard Mejía, the Dean of the Medical School, and I shared our concern about the low number of clinical research papers authored by physicians in low and middle-income countries (LMIC). Physicians in the LMIC are at the receiving end of a unidirectional flow of information from academic centers in high-income countries (HIC), even for pathologies prevalent in LMIC.

We agreed on the lack of clinical research in LMICs is an omnipresent feeling that *Roma locuta, causa finita* (Rome has spoken: the cause is over). So, what is said about a pathology in the American Medical Textbook authored by a distinguished professor who lives far away provides the interpretation of an empirical fact witnessed by a physician in Central America.

In North America, clinical medicine has embraced evidence-based medicine (EBM) as the cornerstone of a logical process for diagnosis and treatment. EBM has the aura of a novel concept, but it is not. Medicine has been based on evidence since Hippocrates and every treatment recommendation in a medical textbook is based on available evidence.

The contemporary movement of EBM relies heavily on the documentations acquired through technology, laboratory studies, and statistical analysis. The message of EBM is that there is nothing beyond the gilded data stored in a distant vault. *Roma* could be the University of California Los Angeles (UCLA) or any other Academic institution in a HIC, and the *causa finita* could be an unconditional statement such as "complete surgical excision is the treatment of choice." But, in many instances, what *Roma* recommends is not feasible at a particular time in a specific place. The physician in an LMIC feels embarrassed by the breach from what *Roma* labels as the standard of care. This sense of inferiority drives the local physician away from considering their solution as an innovation and a contribution to the body of knowledge. More importantly, it could kindle a new line of thought about a disease. If we consider that more people live in an LMIC than in a HIC, it follows that in the former, there are more patients of a given disease than in the latter; we can then agree that in LMIC, there is a wealth of untapped clinical evidence.

https://doi.org/10.1515/9783110788501-020

A way to revert this situation is to empower a rural doctor with the tools to interpret clinical evidence according to the patient's specific circumstances. With Doctor Meynard Mejía, we agreed that the essence of the project is to mutate "Roma locuta, causa finita" into "Roma locuta est, causa non est" [Rome has spoken, the cause is not yet such]. The tool we choose to achieve epistemic independence was to share the basics of analytical philosophy, the branch of philosophy that emphasizes the rules of logic to attain concise thought.

Traditionally, when contemplating the role of philosophy in medicine, scholars focus on ethics. Despite the contributions of physicians and philosophers to linking analytical philosophy and medicine. Among the physicians, we can name Edmond Murphy and Peter Rabins, and by philosophers trained as physicians as Georges Canguilhem (1904–1995) and Kenneth Schaeffer.[1] On the other hand, philosophers of science have also contributed to bridging the gap between both disciplines, most notably Carl Hempel (1905–1997) who expanded on the thought process that led Ignaz Semmelweiss (1815–1865) to elucidate the etiology and subsequent treatment of puerperal fever.[2]

The Method

The scope of Analytical Philosophy is broader than our ability to deliver it through lectures to an audience of health care graduates and undergraduates. We choose to sketch the ideas of two analytical philosophers: Karl Popper (1902–1994) and Philippa Foot (1920–2010). Karl Popper, born in Vienna, lived in many countries and had many interests from carpentry to mathematics. He is one of the salient philosophers of science, but he never formally studied philosophy; his doctorate was in psychology. Popper was a member of the Vienna Circle, a group of thinkers who proposed clarity in language and confirmation of hypothesis. At the beginning of World War II, he emigrated to New Zealand and then returned to Europe to the London School of Economics. He died in the United Kingdom in 1994. We intend to blend the core of his proposal with the specific needs of our project. Popper states that the work of scientific inquiry is to prove theories wrong, to falsify the theory. Ahead of every field of scientific inquiry lays an open landscape. He championed the concept of an open-end to any statement such as "No number of sightings of white swans can prove the theory that all swans are white. The sighting of just a black one may disprove it."[3]

1 See Murphy 1997; Rabins 2013; Canguilhem, *Critique de la raison*, 3, 1924: 125–157 (Engl. transl. Lasareff 2013: 24); Schaffner 1993.

2 Hempel 1966.

3 Popper 1963: 36.

For Popper, we challenge theories not out of animosity but driven by curiosity. The Popperian message that we deliver to our students is that to question science, is science at its best. A good representation of Popper's ideas is summarized in his eloquently titled book, *Conjectures and Refutations*.[4]

Philippa Foot was a distinguished British philosopher with academic and family ties in the United States. She was the granddaughter of President Grover Cleveland, professor of philosophy at the University of California, Los Angeles for many years. An analytical philosopher, she was concerned about the objectivity of morality. In 1967 she wrote a seminal article titled "The problem of abortion and the doctrine of the double effect" published in the Oxford Review. The doctrine of double effect refers to the "two effects that an action produces: the one aimed at, and the one foreseen but in no way desired."[5] Foot imagined the hypothetical situation that, through the work of fellow philosopher Judith Jarvis Thompson (1944–2020), became known as "The trolley dilemma." In simple terms: suppose you are the motorman of a trolley and the only thing you can control is the track the trolley will travel. If you turn the trolley to the left path, you will kill one worker, whereas you will run over five people if you veer to the right. The trolley dilemma is relevant for medical care in LMIC. In a University Hospital in a High-Income Country, all allopathic treatments are available; hence treatment is a medical decision directly inspired by the standards of Evidence-Based Medicine. The treatment priorities are different in LMIC, where there is no cardiac transplant program, no stem cell research, no state-of-the-art neonatal units and abortion is punished with a severe jail term even when the mother's life is in danger. To ponder the trolley driver's response, we must factor in many singularities that defy universal answers.

We launched the first lectures of "Introduction to Critical Thinking for Health Care Workers" in October 2014. The routine was as follows: I lectured through Webinar (WebEx) from the Academic Center in the U.S. to an audience of students gathered at an auditorium facing a widescreen.

As of March 2020, I delivered the lectures via Zoom. The original student body consisted of medical students, but later included undergraduates from nursing schools, clinical psychology, anthropology, and public health. Enrollment was voluntary. Two days before class, the students received the presentation's portable document format (pdf) and the relevant texts to that particular lecture.

A slide presentation is displayed on the screen at the auditorium at UNAN; the students see me in the top corner, and I see them on my computer.

Dialogue between students and the instructor is encouraged. We record the lecture, then upload it on a site where the students have direct access. There are no restrictions for the use of the material. For each class, there is homework. Students

---

4 Popper 1963: 36.
5 Foot 1967: 5–15.

Figure 1: Participants of the Master/Doctor in Science program gathered at the auditorium of the Medical School at the Universidad Nacional de Nicaragua, Managua, for a lecture on "Ethics and Scientific Publications". May 2022. Photo credit Dr. Manuel Pedroza Pacheco (by permission).

email theirs to the instructor. In seven years, the objective of the course remains unaltered, but the audience has changed. In the first two years, the lectures were limited to health sciences undergraduates. In 2018 UNAN-Managua created the Master/Doctor in Science Program. The critical thinking/analytical philosophy lecture series was then transferred from undergraduates to graduates enrolled in the Master/Doctor in Science Program.

Our lectures are part of an extensive curriculum covered by epidemiologists and statisticians from Nicaragua. From 2021, first-year and second-year medical and surgical residents of the 12 hospitals depending on the Minister of Health system in Managua receive six one-hour lectures on Critical Thinking/Analytical Philosophy. Till 2021, the number of graduates and undergraduates enrolled from UNAN is 876. Doctor Manuel Pedroza Pacheco, a senior UNAN-Managua faculty member and a passionate proponent of medical research, is the director of the postgraduate programs.

The Syllabus

Instead of a traditional syllabus, I structure the lectures by presenting a series of recurrent concepts. The most relevant is well conceptualized in this verse by the Portuguese poet Fernando Pessoa's (1888–1935):[6]

6 Pessoa 2006: 16. The original version is in "O Guardador de Rebanhos" 2019: 82.

> From my village, I see as much of the universe as can be seen
> from the earth,
> And so my village is as large as any town,
> For I am of the size of what I see
> And not the size of my height.
>
> Da minha aldeia vejo quanto da terra se pode ver no universo . . .
> Por iso a minha aldeia é tão grande como otra terra qualquer
> Porque eu sou do tamanho do que vejo
> E não de tamanho de minha altura

The essence is that no matter where one initiates the observation of the phenomenon, the observation will transcend. One's village can be Baltimore in the United States or Estelí in Nicaragua. From here, I expand into the universal nature of clinical observation. I tell the students that the construction of knowledge is not relevant if the patient presents in a hospital in Berlin or a remote hut. The observer is the one who rescues the circumstance from oblivion.

A phenomenon is an event that catches our attention and drives our curiosity. Our reaction to the phenomenon is dual: it is familiar, or it is not. Since the students come from different training levels and do not all have the same clinical experience, I deliver the concepts through this scenario:

> Suppose that you are walking by the window of an appliances store. Multiple TV sets are on.

In one of these the station broadcasts the image of Figure 2, a baseball game, the most popular sport in Nicaragua.

Figure 2: Image of a baseball game, probably in the United States.
Reproduced by license from Shutterstock.

In another set there is a news channel and for reasons unknown to you the TV sets are behind the thick window panel where you see the image of Figure 3.

Figure 3: Photo of a baseball game played in rural Nicaragua.
Reproduced by license from Shutterstock.

The images are different but can be recognized as baseball games. To determine that both are of the same game, we need to define the essence of the game. The observer extracts the particulars using a tool from deductive logic called Modus Ponens, which affirms the antecedent. It goes as follows:

if *p* then *q*.
p
then *q*

If I see a set of conditions that I defined as sufficient and necessary to belong to what I understand as baseball (we can call it *p*), then I am looking at a game of baseball. In the case of Figures 2 & 3, I recognize the set of premises previously defined as *p* and conclude that it is baseball (*q*). I don't find any contradiction between Figures 2 and 3. In clinical medicine, it is like looking at a pattern in the electrocardiogram of a patient with myocardial infarction and recognizing it. However, it does not have identical characteristics to the illustration on the textbook of internal medicine or even to a previous electrocardiogram.

Continuing with our hypothetical scenario of an individual strolling along the window of an appliances store where there are three TV sets, he/she, notices that on the third set a channel devoted to geography and history is displaying what we see in Figure 4.

Very likely, our subject could not identify the sport. If curious enough, he/she can start a quick search on the internet based on the elements at hand that are basically about a rider's sport, two teams, where one player carries a fleecy dead animal while two players from the other team are hustling him. The dry landscape and the physical appearance of the riders suggests that the game takes place in Central

Figure 4: Photo of a game of Buzkashi taken somewhere in Central Asia.
Reproduced by license from Shutterstock.

Asia. With those elements in hand, we can type on the search engine enough information to find that what the men are playing is Buzkashi.

For their search, the observers used familiar components such as horses or a dry landscape to start their research. They defined a set of necessary and sufficient conditions (p). But this time, the conditions reside in their memory. So, when in the future, our curious subject who investigated the name of the game displayed on the TV set of a convenience store looks at Figure 3, he/she does not need to repeat the entire search to recognizes it as Buzkashi.

The "Baseball and Buzkashi" exercise is identical to any clinical setting. In any pursuit of knowledge, we utilize induction and deduction. Induction goes from the particular to the general. On the other hand, the deduction goes from the general to the particular. The deductive method is prone to formal fallacies and the inductive to informal fallacies. We present and discuss in class both misconceptions. What we have done so far is practicing good medicine. Mastering deduction and induction is at the core of good clinical practice.[7]

To increase clinical knowledge, we engage Popper and Foot. The leitmotiv of Popper's ideas, even the political ones, spins around the concept of conjectures and refutations. In analyzing Figures 2 and 3, we discover that there are variations to the game. The data tells us that we are looking at a baseball game, but why do we accept that Figure 2A and 2B represent the same game? The students' collective task is to try to refute the assumption that it is soccer. We work on Modus Tollens or the negation of the consequent.

It runs like this:

7 Bowen 2006; Croskerry 2009.

if p then q
non-q
therefore non-p.

In Modus Ponens, we affirm the antecedent; the core of the argument is on p. In Modus Tollens is non-q, the negation of the consequent, which, risking falling in redundance, I repeat, is the essence of Popper's approach to the creation of knowledge; that of searching for the non-q.

So, if we play Modus Tollens with baseball, we can say that if (p), what we are looking at is baseball, then what we must see is 9 players from one team and the pitcher from another team, and an umpire, this is (q). But there are no 10 players: we have non-q. Therefore, we are not looking at baseball in either figure. Unless we propose a different definition of baseball and agree that both figures represent baseball. The non-p is restructuring your description of a condition.

I go on to narrate two events from the recent history of Medicine that fits Pessoa's wisdom and Popper's emphasis on conjectures and refutations:

a) In the mid-'80s, an American actor of some repute suffers the death of his wife and young child; both die of AIDS. The mother contracted the disease through a blood transfusion and he spearheads a fundraising effort to support research on a, at that time, stigmatized condition. Along the months, with some regularity, Mr. G. walks to the office of UCLA Pediatrics. Mr. G. is a lovely person. Never intrusive, never demanding, in essence, a welcome presence at UCLA. This process of Mr. G walking back and forth from the parking lot to the offices of Pediatrics does not raise the curiosity of anybody until one day, a pediatrician passes by Mr. G. and then turns around and asks point-blank, "how come you are not dead?" Mr. G. had unprotected sex with a woman with HIV that transferred it to her then-unborn child, but the husband is actively fundraising unscathed by the disease.

In Modus Tollens,

if having unprotected sexual relation (p) with a patient with AIDS
is causal of fatal infection (q),
then when I see somebody who is not sick/dead (non-q), I conclude correctly that the patient does not have AIDS.

The curious clinician is attracted to the non-q, to the Popperian refutation of the established knowledge. The traditional knowledge states that all individuals who fit the condition of p will become infected. In the Popperian metaphor, all swans are white. But you walk along the corridors of the hospital knowing that you may have seen the black swan.

b) At the beginning of the 1980s, the 4 H disease affected hemophiliacs, heroin users, homosexuals, and Haitians. Dr. Peter Piot a Belgian physician noticed that among patients who came from Africa there were as many women as men suffering from the “4H disease.” To confirm that the disease had wider reach than originally supposed, he conducted a study in Zaire. He sent his observations confirming that the disease was not limited to the “4H” to the New England Journal of Medicine, one of the most prestigious clinical journals. The journal rejected the paper because none of the female patients were Haitians, hemophiliacs, heroin addicts, or homosexuals.[8] The New England Journal of Medicine reviewers were not prepared to admit that not all swans are white. I stress to the students that some of those reviewers are the authors of the medical canon. The Lancet published it with tepid enthusiasm because the editors considered that the findings in Zaire were local.[9]

c) Also, in that decade in Perth, a relatively small Australian city, a pathologist at the Royal Perth Hospital examines biopsies of patients with stomach ulcers. Dr. Warren notices that next to the crater of the ulcer lies a structure suggestive of a bacterium. He then engages Dr. Marshall, who at the time was a medical intern, and asks him to review stomach biopsies from past years and they confirm that there seems to be an association between gastric ulcer and a bacterium. Not every sample evidenced the association, but still, the phenomenon raised the curiosity of the two inexperienced and unknown researchers.

While Dr. Warren was examining the biopsies that piqued his curiosity, hundreds of pathologists worldwide were looking at similar samples. And yet, nobody in Buenos Aires, Los Angeles, Rome or Moscow, or Managua thought about the meaning of the observation. The canon of medical knowledge was that the stomach's acidity prevented bacterial growth and previous observations of the bacterium in the stomach were dismissed as contamination.

Marshall and Warren published their seminal paper in 1984. The Lancet editors had difficulties finding reviewers who agreed that this was an important paper. Similarly, hundreds of physicians who crossed paths with Mr. G in the UCLA Center for the Health Sciences corridors did not raise the question. Again, direct observation of a phenomenon altered our understanding of a disease. The protagonists of our three stories falsified long-standing beliefs.

Another guiding principle for the lectures is Carl Sagan's (1934–1966) statement: “Science is a way of thinking and not an accumulation of knowledge,” which I present at the beginning of every class, usually after Pessoa's verses. I say that a scientist is any individual capable of formulating a question so that the answer is either a “yes” or a “no.” The physician/reporter asks how many tumors were treated at Hospital La Mascota in Managua; a physician/scientist asks if more tumors were diagnosed this year than the last. The second question allows for a yes or a no, a hypothesis, and a null hypothesis.

8 Altman 1998.
9 Piot 1984; Timberg and Halperin 2012.

The next step is the introduction of the elements necessary to validate an argument. Why compare the Atlantic Coast with the Pacific Coast and not patients from the North with patients from the South? The relation between the premises needs to be adequate and relevant to each other.

I illustrate the relevance of this progression by borrowing from Marshal and Warren. In their landmark paper in *The Lancet* they acknowledge that H.W. Steer and D.G. Colin-Jones already observed and reported bacterium in the vicinity of the ulcer in 1975.[10] Not pretending to solve this exciting puzzle of the fate of a scientific idea, I compare both papers in class. While Steer and Colin-Jones entertained the possible relation between bacterium and ulcer, Marshal and Warren went forward and set an experiment to corroborate the hypothesis.[11] The experiment was straightforward; Marshall had an abdominal endoscopy that confirmed that he did not have a gastric ulcer, then he swallowed a vial with the bacterium. A few days later he fell ill and a repeat endoscopy revealed an ulcer. He was treated with improvement of his symptoms. Marshall and Warren received the Nobel Prize in Medicine in 2005.[12] I stress that a characteristic of Marshall and Warren's short but transcendent paper is that they proposed a mechanism for developing the ulcer. The details of the mechanism are beyond the purpose of this communication.

In the first book of the Republic, Plato (428 BC–348 BC) differentiates between opinion (*doxaston*) and knowledge (*gnoston*). The latter demands a coherent explanation. Here is where I bring in the trolley dilemma. The trolley dilemma forces an explanation, a justification, a common characteristic of a sound ethical decision. And, as the course progressed along the years, this segment that we can label as dealing with ethical issues took a more prominent role.

Our approach to ethics through analytical philosophy is not original. Anne Thomson wrote Critical *Reasoning in Ethics*, demonstrating that critical reasoning focuses on giving reasons for one's beliefs and actions, on analyzing and evaluating one's own and other people's reasoning and devising and constructing better reasoning.[13] Michael Dunn and Tony Hope, in their *Medical Ethics* devote the fifth chapter to introduce the most important 'tools' of ethical reasoning, and discuss how argument, thought experiments and intuition could be combined in consideration of medical ethics.[14]

Moritz Schlick (1882–1936), the founding father of logical positivism and leader of the Vienna Circle, the bedrock of analytical philosophy until its dissolution by the Nazis in 1936, addressed ethics and analytical philosophy in *Problems of Ethics*.[15]

10 Marshall and Warren 1984.
11 Steer and Colin-Jones 1975.
12 Pinock 2005.
13 Thompson 1999.
14 Dunn and Hope 2018.
15 Schlick 1939.

Schlick affirms that ethics is a science because ethical questions have meaning and can be answered. In other words, from Foot to Popper, from medical ethics to critical thinking is a clear path: the trolley dilemma and Health Care in Central America. Throughout our course, we invite students to develop their own "trolley dilemmas." The following are cases proposed by them.

1. Two patients at an opposing distance from the hospital. One is a young man with peritonitis and the other a 73-year-old woman with spontaneous intracerebral bleeding. There is only one ambulance. We first defined the premises so they will be relevant, adequate, and appropriate. From what we know about peritonitis death, we defined the premises to septic shock and hypovolemia. The next step was to determine the availability of treatment at the referring hospital. Regarding the patient with intracerebral bleeding, the students considered the many alternatives known to control intracranial pressure.
2. A teenager born with spina bifida who has intense low back pain secondary to severe scoliosis. He requires a procedure that improves quality of life. He is on the OR schedule, but severe pain is not considered a more severe factor than an abdominal ulcer, appendicitis, and head trauma.
3. The Mesoamerican nephropathy. In the past two decades, epidemiologists observed an increased prevalence of chronic kidney disease in cane field workers in Nicaragua, Honduras, and El Salvador. The etiology is not clear; it is very likely occupational.[16] The worker confronts the dilemma of exposing himself to severe disease or finding a job for which he may not be qualified.
4. A woman who needs a blood transfusion during a surgical procedure. She is a Jehovah's Witness; her religion forbids accepting donors' blood. In Nicaragua there are not alternatives available in HIC for these cases. The patient is under anesthesia, the family members are in the waiting area. The hypothetical transfusion will be known only to the surgical team.

We discuss the issues with the students joining Popper (conjectures and refutations, modus ponens and modus tollens) and Foot (the doctrine of the double effect) in the conversation. Through these cases, we review the most common fallacies of informal logic, such as the young man being a gang member or questioning the level of knowledge of the referring physician, or the possibility of falling into a slippery slope mode, mainly when opting for devoting scarce resources to a septuagenarian or judging the intellectual strength of an individual who holds a specific religious belief. We don't lose sight of our intention, that is encouraging creation of knowledge through critical analysis of the empirical evidence.

We can't predict the outcome of this work. In class, I share with the participants the story of Henri Laborit, whose ideas are at the center of the script of Alain

16 Laws 2015.

Resnais's movie, *Mon Oncle d' Amerique* (1980), where he has a cameo appearance. Laborit's contribution to knowledge began when he was a Naval Surgeon at a French base in Tunisia. He was preoccupied with hypovolemic shock in his patients. Upon reading a research paper on the role of histamine as a vasodilator, Laborit administered an antihistaminic after surgery to his patients, but nothing relevant happened to them. Other researchers later proved that the role of histamine as a vasodilator is not applicable for preventing hypovolemic shock.

Nonetheless, Laborit reported his observations on *The Acta Chirugica Belgica* in 1949, concluding that "Antihistamines produce a euphoric quietude, our patients are calm, with restful and relaxed faces." His comment on the euphoric quietude attracted the attention of other researchers. From a compound that blocks histamines, they went to the metabolism of dopamine and finally to the synthesis of chlorpromazine and, from there, to the pharmacological treatment of mental disorders. This second-order effect establishes that every action has a consequence that ripples into another result.

The core message of our course is that the historical cases and the clinical dilemmas are not a matter of anarchic serendipity. There is a body of philosophical work that ponders the meaning of knowledge. I stress on that accumulation of data is not knowledge. Knowledge is a slow process aiming towards explaining a phenomenon. Big Data relies on correlations and while it is helpful in many circumstances its own weight suffocates creativity.[17]

Growth of the Course

Word of the work with UNAN spread to other countries. Two universities from Costa Rica enrolled in the program, and with each, we interact for one year, but somehow the concept did not grasp the attention of their authorities and we did not continue the collaboration after 2018. On the other hand, the Universidad de San Carlos (Guatemala City, Guatemala), the Universidad Cientifica del Sur (Lima, Peru) the Universidad Autonoma de Santa Ana (Santana, El Salvador) embraced the concept, and we are already planning for continuing the collaboration. From Peru we enrolled 186 undergraduate students in three consecutive series, from El Salvador 58 undergraduates, and faculty in two successive series and 25 faculty members from Guatemala and 53 undergraduates from Costa Rica.

In Nicaragua, the idea expanded and became an entity. In 2022 the course will be conducted by Nicaraguan educators. Dr. Pedroza Pacheco has organized a team of local statisticians and researchers who transmit their experience to their compatriots and share their understanding of clinical research in Nicaragua. Most likely,

17 Han 2018.

my participation will consist of videos of the lectures I delivered over the years and an introductory class.

Philippa Foot and Karl Popper are not distant Europeans but coworkers with mud in their hands. Twenty-seven researchers completed their thesis and graduated from the Master in Biomedical Research program. Looking at the list of their projects, and after reading many of them, I see that there is a determination in the researchers to apply epistemology to issues that matter to the national medical community. And this welcomed evolution rests on that, as there are facts there is also an interpretation of those facts and utilization of those facts.

I asked three of former students to assess the seminars.

One of the students from the first cohort of UNAN-Managua, Dr. Alexis Narvaez-Rojas, has a remarkable clinical research path; up to now, he has authored and co-authored twenty-three papers.

> The algorithmic thinking of the pathologies and the criticism of the processes, but the self-criticism of the methods of analysis and clinical reasoning supported by scientific evidence was an exercise that was carried out during the course, generating the bases for clinical application. The criticism of the medical problems allowed to delve into the issues and improve the techniques of searching the scientific literature, as well as the coaching of Dr. Lazareff helped the management of the databases of medical literature, six years later that coaching persists in favor of Continuous training, years later, has allowed me to continue growing in the clinical and research aspect, being evidenced by the scientific fruit of this collaboration.

Dr. Eden Chavarria, a participant in the UNAN-Managua Master/Doctorate program mentors the clinical researchers of the residents of different specialties at the Lenin Fonseca Hospital in Managua. Dr. Chavarria authored two books on the history of Medicine in Nicaragua, and two books on political history.

> I had not received a philosophy course at the Faculty of Medicine for 30 years, from the dialectical and historical materialism I approved in 1987 and 1989. Still, that absence ended with the courses in philosophy of science and critical thinking, taught by Professor Lazareff during the master's degree in biomedical research, via telematics, even during the COVID-19 pandemic. The effect has been revitalizing. In these courses, which go beyond the motivational and the thesis, I have learned about the interpretation of the scientific research process and the ability to reflect on the methods, causes, explanations, and new ways of approaching medical reality, based on the own context, to generate knowledge that contributes to finding answers to curious questions, to solving problems, and to the gradual epistemological liberation of the countries on the periphery of the planet. Science and philosophy are complementary. Every scientific act deserves a philosophical interpretation and putting that into practice, shouting cheers to science and philosophy, makes the search for critical thought more pleasant, without the uncomfortable rigidity of Rodin's The Thinker (no one can think in such a position since Borges said it). For Nicaraguans, the call has also been to publish; for what is not published does not exist and what is not published dies. Without a doubt, philosophy helps the scientist to think about the past, as history; about the present, as praxis; and as for the future, as a legacy. Thus, critical thinking inexorably leads us to a philosophical science and a scientific philosophy. I'm so sorry, Professor Hawking; but philosophy is not dead yet.

During the uprising against the government of Daniel Ortega, one of the students (class of 2017) at UNAN-Managua delivered first aid to the wounded. Elton John Rivera became a columnist for an important newspaper, now closed. In one of his columns, he praised the seminars. I asked him for a brief comment about what he got out of the lectures, and this is my translation of what he wrote.

> The critical thinking course in medicine that I received while in the fourth year of the degree, meant for me a crucial element for the understanding of health, as part of a whole, of an interrelation not only biological, but also social and environmental, I opened the door of detailed thinking for the logical analysis of things, for the comprehensive interpretation of the phenomena of cause-effect, interdependence and multidisciplinary nature of the medical sciences At the time, it was a course that enhanced my clinical understanding, but which later made more sense, beyond the hospital application; When in a context of socio-political crisis in which the population suffered and needed first aid assistance, I took to the streets to make myself available for help, my critical thinking objectified the situation and, amid a context of disproportionate violence, I understood that the character of the doctor lies in the defense of life and the lives of many were at stake, I turned my back on the system, I was pushed to an extreme contrary to the Nicaraguan government itself, for the simple fact of having decided to follow the vocation of service, opposing state violence, witnessing the havoc that it has caused thousands of people, made me a victim of reprisals policies through expulsion from the university, the interruption of my studies being in the fifth year of the race; I have received a lot of affectation as a result of my humanitarian participation in the midst of violence, but critical thinking makes me understand that I did the right thing and that is why I do not regret it. Critical thinking made me see the health sciences as a social science and both as the sciences of human life.

The experience in Latin America of bringing philosophy to a strict clinical/surgical hands-on auditorium is an endeavor worth expanding and worth reshaping by physicians and philosophers of each nation.

Postface

Alain Touwaide

Medical Humanities as a Search for Unity

A Long-term View from the History of Science

Medical Humanities have been particularly bourgeoning over the past decades, possibly inspiring a feeling of novelty, with the emergence of a new field of academic inquiry.[1] A quick glance at history reveals, however, that the Medical Humanities came to life as early as the 1940s, principally in the United States. This initial impetus was soon followed in England and, further on, in other countries, opening developments that recall in a certain way the debate of the *Two Cultures* in the early 1960s. The different designations of the field might confirm a sense of bipolarization, indeed, with the Medical Humanities at one pole and the Humanities in Medicine, for example, at the other. Whatever the reality or not of this bipolarization, the vision of the Medical Humanities is to re-humanize medicine, by reinjecting human values into a highly technologized medicine that might have lost its sense of humanity. Literature, the arts, music and any other activity put physicians in contact with the vast universe of the human experience, past or present, and remind them of the famous Latin verse of the 2nd–century BCE Latin playwriter Terentius *Nihil humani a me alienum puto* (Nothing human can be alien to me).

This search of medicine for a redefinition of its essence and, hence, of its mission, might recall the 1970s and the academic development of the fields traditionally identified as Ethnobotany and related Ethnopharmacology, both now subsumed in the Medical Ethnobiological Sciences. Contrary to what their designations might suggest, these fields go beyond the traditional knowledge and therapeutic applications of plants as they try to recover the totality of the experience of the management of health through natural resources (which are not limited to plants, but also include animals and minerals), their centuries-long knowledge, and the wisdom gained through time all of which focus on the person of the patients, the experience of their medical condition, and the tradition(s) through which they live their own body and health, or their disease.

In the same years, this attempt toward a re-centering of medicine on the human being was echoed in the medico-historical disciplines where it found its best expression in the title of a book that is a milestone in the field: *The Hippocratic Triangle*. Whereas Hippocratic medicine has traditionally been credited with the merit – which it effectively has – to have focused the physicians' attention onto the signs of the patients' disease – hence its reputation to be at the origin of clinical medicine –,

1 Being a Postface this essay does not follow the bibliographical style of the rest of the book. It includes a specific bibliography at the end, under Appendix.

https://doi.org/10.1515/9783110788501-021

a closer examination of the ancient Hippocratic texts revealed that the medicine of the Hippocratics was rather a medicine *with* the patients. According to the sub-title of the work, this *Triangle* connects the physician, the disease, and the patient. It breaks the image of the medical act limited to a linear relation between physicians and diseases, and introduces instead the treated individuals, with their personal history, their knowledge of their body, health and diseases, and their experience and perception of life. According to this triangle, the *technē* of the physician (the specialized knowledge and experience gained through practice) is no longer the unique or even the major component of the medical act; it is combined with the personal knowledge of the patient and both are inter-related in their opposition to the disease.

Ethnobotany and Ethnopharmacology allow to go further in the analysis as they introduce natural materia medica into the figure. Materia medica is not only a mediator in the relationship between the three elements of the *Hippocratic Triangle* which it does connect in effect, but it is also a vehicle that introduces into the Hippocratic triangular figure a great many components of the world of the patients through its origin: the physical environment, its construction through the society of the patients, its perception or personal projections onto it, or also bodily states and feelings generated by, or linked to, the environment and nature. The *materia medica* as it is studied by Ethnobotany and Ethnopharmacology connects the patients and the sublunary world in an interactive relationship.

This introduction of *materia medica* into the analysis has deep consequences that can be represented by a transformation of the Hippocratic triangle from a two-dimensional figure to a three-dimensional pyramid at the tip of which is health and wellbeing, at its basis the three points of the *Hippocratic Triangle*, and at its center the *materia medica*, which is connected to the four points of the pyramid. In the resulting three-dimensional figure, the *materia medica* is not a static geometric locus defined by its invariable position equidistant from the four points of the pyramid (the tip and the basis); instead, it is a point that can – and does – change position as a result of the tension between the forces in presence (the disease, the physician, and the patient) which tend to the equilibrium that is health. This equilibrium is mediated by the materia medica, which might be taken by the patients for the maintenance of health, the prevention of disease or, when necessary, the restoration of health.

Such representation of the medical process in the form of a multi-factorial, dynamic three-dimensional visualization might translate the totality of the management of health, which is not limited to medicalization *stricto sensu*, but encompasses as many components, if not all, of the patient's world. This totality might be what the Medical Humanities aims to reach.

Whatever the analysis, a question remains about the reason why medicine is currently seeking to recover its human dimension, to re-center and re-humanize itself. Going back to the debate on the *Two Cultures* – assuming that there was a clash of cultures –, it could be argued that the so-called 1910 *Flexner Report*, advocating a scientization of medicine in the United States, technologized medicine and contributed to de- or less-humanized forms of practice. However correct this might be, it could also be claimed that the transformation of therapeutics by the introduction of chemical substances best heralded by the pioneering work of Paul Ehrlich at almost the same time, undercut the traditional roots of therapeutics and isolated patients from their natural world, contrary to traditional *materia medica*. This new *materia medica*, chemical in nature, opened the way to developments that deeply transformed the art of healing. From that moment on, indeed, medicines no longer were natural substances that had been traditionally used for centuries, if not millennia, and were gradually optimized through practice and constantly repeated empirical assessment that could aptly be compared to millennia-long clinical trials. They were, instead, chemical substances, the action of which was unknown and had to be discovered by experiments, which were initially random and became later methodically organized. This transformation inverted the relation to history that had been traditionally present in remedial therapy. Until the development of pharmaco-chemistry, medicines encapsulated a practical experience that could have stretched centuries, and were in a certain way fragments of history. In pharmaco-chemistry, instead, medicines have no history at their starting point and, at least at the very beginning of the nascent pharmaco-chemistry, no known activity and no connection either with the experience of disease accumulated through the centuries; they were abstract entities the action of which still needed to be identified and subsequently understood. The objective was to connect two realms that were originally independent from each other (chemical substances on the one hand and diseases, their *histoire vécue* and the patients, on the other), whereas they had a long history of co-existence in the past. Contrary to traditionally uses of natural substances which were the result of history, pharmaco-chemical medicines had their history in front of them.

Taking the long view of history – especially of the history of sciences and medicine in the West – and bearing in mind the traditional nature of *materia medica* (particularly plants), the analysis leads as far back as Antiquity, the period improperly identified as *archaic*, Aristotle (384–322 BCE), and the Aristotelians. However distant it might seem, the Stagirite's analysis of life is relevant here as it cut with the earlier theories, which became *archaic ipso facto*, and became the theory that possibly had the longest and deepest influence in the history of human life, nature, environment, and medicine in the Western World.

In Pre-Aristotelian science, life was a continuum that connected all beings in the sublunary world. Plants, animals and humans shared the same life. According to the principle of *sympathy* that needs to be understood in its full, etymological meaning of *co-sensitivity*, they all could interact on each other through transfers of matter, energy, or abstract qualities. This was the fundamental principle of remedial therapy, which consisted in administering to a diseased individual the element(s) that could counteract or compensate for the disease. Aristotle cut this continuity and connectedness of life, and relegated plants to a lower level of life. This rupture should not be understood in Bachelard's sense, epistemic in nature, but in a much deeper sense, ontological, which can be better identified as a *fracture* of the sublunary world, separating the *human* from the *natural*, specifically the *vegetal*. This fracture deprived plants of the capacity of *participating* in the life of humans whereas this *participation* was exactly the reason why plants were used as therapeutic means, since they had the capability to restore disturbed vital functions by reintroducing essential vital elements.

However influential Aristotle's thinking and written works might have been – and they certainly were –, this theoretical vision with the hierarchical structure of life it introduced in the sublunary world, did not debase the traditional practice of therapeutics as it could have done since it disrupted the continuity of life and connectedness between all the elements of the world best represented by the use of plants as therapeutic agents. The persistence of this fundamental pre-Aristotelian principle underpinning therapeutics is best illustrated by the largest encyclopedia of therapeutics compiled in antiquity, *De materia medica* by the Greek Dioscorides (1st century CE), and also its broader Latin counterpart, the *Naturalis Historia* by Pliny (23/24–79 CE). These and other similar treatises, including their derivatives of all kinds in the following centuries, became the backbone of Western medicine for a long time (actually, until the Renaissance), with their fundamental notion of the continuity and connection between all components, humans, environment(s), nature, and the cosmos, with a totality that encompasses all the processes in the universe, from the exchange of substance between plants and human physiology to the magnetic fields of atmospheric phenomena and celestial bodies with their impact on the neuronal system or physiology of the living creatures in the terrestrial world, for example.

If the *Aristotelian fracture* did not undercut the traditional practice of therapeutic and medicine that current Ethnobotany/Ethnopharmacology searches to capture again and remained, instead, in the realm of learned natural sciences and medicine, it was nevertheless transmitted across the centuries and provided the theoretical foundations for a reification of nature and its products in a way best represented by Galen in the 2nd and early 3rd century CE, and Andreas Vesalius in the 16th century.

The case of Galen is contradictory. His major work on *materia medica*, *De simplicium medicamentorum temperamentis et facultatibus*, reduces the natural elements of the world to a set of components that it reifies – the primary matters of the world and the qualities they were credited with, air, fire, earth, water for the former, and dry, hot, cold, and moist for the latter – so as to make them compatible with the corresponding elements of the human body. Pushing this reductionism to its limit, Galen systematically omitted the description of the single materia medica (which are plants for most of them), contrary to Dioscorides' *De materia medica*. This disappearance, which could be identified as an expression of the *Aristotelian fracture*, is confirmed, though expressed in a different way, in the recently re-discovered Galenic treatise *De indolentia*, where Galen states that physicians do not need to know plants. Though possibly very attractive as an extremely achieved intellectual construction, Galen's system, which might be aptly identified as Galeno-Aristotelian, was probably not much applied in the practice of therapy in Byzantium or, at the very least, did not supplant the pre-Aristotelian concept of continuity of life in the sublunary world as a comparative study of the transmission and tradition of Galen and Dioscorides' treatises reveal, with the latter much more abundantly copied and probably also much more read and used than the former.

The reference in the field of remedial therapy remained Dioscorides' treatise, with its all-encompassing vision of the natural world. In its manuscript copies, the textual information includes the description of the plants complemented by polychrome representations of the plants that directly connect the text and nature. In the Arabic World and in the West, instead, Galen's work was more successful than in Byzantium. However, it never overshadowed Dioscorides' treatise, which was abundantly read, repeatedly translated afresh, and emulated in new, original works which expanded the range of known materia medica by introducing the species of areas not covered by Dioscorides' compilation. Later, Renaissance scientists did walk in the footsteps of their Byzantine, Arabic and medieval predecessors with their focus on Dioscorides. After they rediscovered the Greek text of *De materia medica*, read it again in its original language and mastered it, they used it as a key to describe their natural environment and also to analyze the plants from the world(s) that were discovered at that time. In so doing, they gradually separated plants from medicine, returning to plants the ontological status of which they had been deprived by Aristotle. Although plant study did not become *ipso facto* a new field of scientific investigation, it recovered nevertheless the basis it needed to be an object of study in its own right in the following decades.

Ever since the period of Western science traditionally identified as the *Scientific Revolution* – whether this Revolution did happen or not –, science has generalized the analytical procedure that consists in decomposing, divising, and reducing macrostructures to their minimal identifiable and measurable components, constantly pushing the limit of the visible and the minimal in the way of Galen in Antiquity and Vesalius in the Renaissance. Interestingly enough, however – and, ironically in a certain sense –, by a return to an anti- or, maybe better, pre-divisionist process, science in recent times has become increasingly interested in connectedness, expressed in terms of communication, inter-relation, and networks between the elements that it has persistently searched to identify and isolate in previous periods.

However paradoxical it might be considering the modern history of science, the reversal represented by this notion of network and, beyond it, of connectedness between the elements of the living world, which was broken by the *Aristotelian rupture* and further confirmed if not accentuated by Galen, might be key for interpreting Medical Humanities. Though certainly made of a dense network of components that can be identified, described, measured, and decomposed themselves in a sum of elements, life – even in a strict physico-medical reductionism – is more than the sum of the elements. It is a unique network in which the result is by no means limited to the addition of the elements. This is exactly the totality that Ethnobotany/Ethnopharmacology recovers, particularly through the study of plants as mediators that connect humans and the sublunary world with all the components and inter-relations between all forms of life created through time.

This total experience of life might be what the Medical Humanities, too, try to capture. This search is – or can be – conducted in different ways, depending on the entry points: from Medicine, it is a search to reconstitute the totality of the human life beyond a complex biological process and beyond the *Aristotelian fracture*, reconciling humans and the world rather than separating them and dividing life up to a constantly receding limit; from the Humanities, it is a search to perceive the expression of this full, deep and unitary – if not unifying – existential experience of life through its translation in works of creation (whatever the nature of such works, literary, musical, plastic or other), including by *reading* these works through the medical lens or, perhaps better, by perceiving in these works the expression of life translated – or transmuted – into art.

In spite of a possible apparent bipolarization, these two approaches to Medical Humanities are probably one and the same: perceiving and expressing the complex and multifaceted reality of the human process and experience of life in a unifying, holistic, overarching expression, returning to the allegedly *archaic*, in fact pre-Aristotelian, concept of *sympathy*, interpreted above as *co-sensitivity* but perhaps best understood as *cross-sensitivity*.

Appendix

On Medical Humanities and their initial development, see, for example Daniel M. Fox, "Who we are: the political origins of the Medical Humanities," *Theoretical Medicine* 6 (1985), pp. 327–342.

On *The Two Cultures* by Charles Percy Snow (1905–1980), see Charles Percy Snow, *The Two Cultures and the Scientific Revolution*. Cambridge: Cambridge University Press, 1959, followed by *The Two Cultures: and A Second Look. An Expanded Version of the Two Cultures and the Scientific Revolution*, Cambridge: Cambridge University Press, with several editions and reprints of both. Among the many articles about *The Two Cultures*, see for example and recently: W. Patrick McCray, "Snow's storm," *Science* 364 (6439) (3 May 2019), pp. 430–432.

On Ethnobotany/Ethnopharmacology, see for instance: Michael J. Balick and Paul Alan Cox, *Plants, People, and Culture. The Science of Ethnobotany*, 2nd Edition. Boca Raton, FL: CRC Press, 2021, and Michael Heinrich and Anna K. Jäger (eds),

Ethnopharmacology (Uppsala, London, Leiden and Amsterdam [ULLA] Series in Pharmaceutical sciences). Hoboken, NJ: Wiley, 2015, and also the following two articles (selection, chronological order): Horacio Fabrega, "The need for an ethnomedical science," *Science*, 189 (4207) (1975), pp. 969–975, and Michael Heinrich, Johanna Kufer, Marco Leonti, and Manuel Pardo-de-Santayana, "Ethnobotany and ethnopharmacology – Interdisciplinary links with the historical sciences," *Journal of Ethnopharmacology* 197 (2) (2006), pp. 157–160.

For the *Hippocratic Triangle*, see the work by Danielle Gourevitch (1941–2021): *Le triangle hippocratique dans le monde gréco-romain. Le malade, sa maladie et son médecin* (Bibliothèque des écoles françaises d'Athènes et de Rome 251). Paris and Rome: Ecole française de Rome, 1984.

On Abraham Flexner (1866–1959) and the *Flexner report*, see Abraham Flexner, *A Medical Education in the United States and Canada*. Washington, D.C.: Science and Health Publications, 1910. On the *Report*, see for example: Thomas P. Duffy, "The Flexner Report – 100 Years Later," *Yale Journal of Biology and Medicine* 84 (3) (2011), pp. 269–276.

On Paul Ehrlich (1854–1915), see Axel C. Hüntelmann, *Paul Ehrlich. Leben, Forschung, Ökonomien, Netzwerke*. Göttingen: Wallstein, 2011.

For Gaston Bachelard (1884–1962) and his concept of *epistemic rupture*, see Gaston Bachelard, *La formation de l'esprit scientifique. Contribution à une psychanalyse de la connaissance objective*. Paris: Vrin, 1938 (with multiple re-editions and reprints).

On the status of plants in ancient, medieval and Renaissance sciences and philosophy, see for example and recently: Brooke Holmes, "Pure Life: The Limits of the Vegetal Analogy in the Hippocratics and Galen," in John Z. Wee (ed.), *The Comparable Body. Analogy and Metaphor in Ancient Mesopotamian, Egyptian, and Graeco-Roman Medicine* (Studies in Ancient Medicine 49). Leiden and Boston, MA: Brill, 2017, pp. 358–386; Fabrizio Baldassarri, "In the Beginning was the Plant: The Plant-Animal Continuity in the Early Modern Reception of Galen," in Matteo Favaretti Caposampiero and Emanuela Scribano (eds), *Galen and the Early Moderns* (Archives internationales d'histoire des idées 236). Cham: Springer, 2022, pp. 55–81; Marco Sgarbi, "Renaissance Aristotelianism and the Scientific Revolution," *Physis* NS 52 (2017), pp. 329–345; and also the several essays collected in Fabrizio Baldassarri and Andreas Blank (eds), *Vegetative Powers. The Roots of Life in Ancient, Medieval and Early Modern Philosophy* (Archives internationales d'histoire des idées 234). Cham: Springer, 2021, particularly Klaus Corcilius, "Soul, Parts, and the Definition of the Vegetative Capacity in Aristotle's *De anima*" (pp. 13–34); Amber D. Carpenter, "Embodied Intelligent (?) Souls; Plants in Plato's *Timaeus*" (pp. 35–53), and Robert Vinkesteijn, "The Vegetative Soul in Galen" (pp. 55–72).

Cumulative Bibliography

Works Cited

Primary Sources

Weir, Peter, *Fearless* (Los Angeles: Warner Bros., 1994)

Secondary Sources

Briere, John and Catherine Scott, *Principles of Trauma Therapy. A Guide to Symptoms, Evaluation, and Treatment*, Los Angeles: Sage, 2015.
Caruth, Cathy, *Unclaimed Experience. Trauma, Narrative, and History*, Baltimore: Johns Hopkins Press, 2016.
Elm, Michael, Kobi Kabalek and Julia B. Köhne, eds., *The Horrors of Trauma in Cinema: Violence Void Visualization*, Newcastle: Cambridge Scholars Publishing, 2014.
Kaplan, E. Ann, Trauma Culture: *The Politics of Terror and Loss in Media and Literature*, New Jersey: Rutgers University Press, 2005.
Nannicelli, Ted and Paul Taberham, eds., *Cognitive Media Theory*, New York: Routledge, 2014.
Shimamura, Arthur P., ed., *Psychocinematics: Exploring Cognition at the Movies*. Oxford: Oxford University Press, 2013.
Walker, Janet, "The Traumatic Paradox: Documentary Films, Historical Fictions, and Cataclysmic Past Events," *Signs*, 22.4 (1997), 803–825.
Walker, Janet, *Trauma Cinema. Documenting Incest and the Holocaust* Oakland: University of California Press, 2005.
Watson, Neil V., and S. Marc Breedlove, *The Mind's Machine. Foundations of Brain Behavior* Massachusetts: Sinauer Associates, Inc., 2016.
Zacks, Jeffrey M., Flicker: *Your Brain on Movies* Oxford: Oxford University Press, 2015.

Digital Resources

Documenta Catholica Omnia, http://www.documentacatholicaomnia.eu/04z/z_1225-1274__Thomas_Aquinas__De_Ente_et_Essentia__LT.pdf.html . . . Ibid.
Ochberg, Frank, "PTSD 101 for Journalists"https://www.giftfromwithin.org/pdf/ptsd101.pdf
Sutherland, Romy, "Commanding Waves: The Films of Peter Weir," *Senses of Cinema*, Issue 34 https://www.sensesofcinema.com/2005/great-directors/weir/

Sources

Acosta 1588: José de Acosta, *Historia Natural y Moral de las Indias*: Edmundo O'Gorman, ed., (Biblioteca Americana, 38, Serie Cronistas de India). México: Fondo de Cultura Económica, 1962.
Aesopus, *Fabulae*: Donald Beecher, ed., *John Ogilby, The Fables of Aesop paraphrased in verse*. Toronto: Centre for Renaissance and Reformation Studies, 2021.
Alberto de' Zancari 1914: Alberto de' Zancari, *De cautelis medicorum* habendis, Manuel Morris, ed., *Die Schrift des Albertus de Zancariis aus Bologna. De cautelis medicorum habendis. Nach Leipziger und Pariser Handschriften*. Inaugural-Dissertation zur Erlangung der Doktorwürde in

https://doi.org/10.1515/9783110788501-022

der Medizin, Chirurgie und Geburtshilfe der Hohen medizinischen Fakultät Leipzig. Leipzig: Ferdinand Peter Nachf., 1914.

Aquinas, 2010: Thomas Aquinas, *De ente et essentia*. Turnout: Brepols Publishers, 2010.

Aristotle, *De anima*, in Keneim Foster and Silvester Humphries ed., *Aristotle's 'De anima' in the Version of William of Moerberke and the Commentary of St. Thomas Aquinas*. London: Routledge and Kegan Paul, 1951.

Aristotle, *De generatione Inimalium* ed. and transl. Peck 1953: Aristoteles, *De generatione nimalium*, Arthur Leslie Peck, ed. and transl., *Aristotle, Generation of Animals* (Aristotle XIII) (Loeb Classical Library 366). Cambridge, MA: Harvard University Press, and London: Heinemann, 1953.

Aristotle, *Nicomachean Ethics*: Harris Rackham, ed. (Aristotle XIX) (Loeb Classical Library 73). Cambridge, MA: Harvard University Press, 1933.

Arnaldo de Villanova 1504: Arnaldus de Villanova, *De cautelis medicorum*, in Balthasaris de Gabiano, ed., *Hec sunt opera Arnaldi de Villa noua nuperrime recognita ac emendata diligentique opere impressa*. Lugduni: Franciscus Fradin, 1504, ff. 215 verso – 216 verso.

Arnaldo de Villanova 1505: Arnaldus de Villanova, *De parte operativa*, in Arnaldi de Villanova, *Haec sunt opera . . . nuperrime recognita ac emendata, diligentique opere impressa . . .* Venetiis: heredes Octaviani Scoti per Bonetum Locatellum, 1505, ff. 141 verso – 149 verso.

Arnaldo de Villanova 1505: Arnaldus de Villanova, *Tractatus de amore heroyco*, in Arnaldi de Villanova, *Haec sunt opera . . . nuperrime recognita ac emendata, diligentique opere impressa . . .* Venetiis: heredes Octaviani Scoti per Bonetum Locatellum, 1505, ff. 249 recto – 250 recto (Michael R. McVaugh, ed., *De amore heroico, De dosi tyriacalium medicinarum* [*Arnaldi de Villanova Opera medica omnia* 3]. Barcelona: Universitad de Barcelona, 1985, pp. 43–54).

***Articella* 1483**: Franciscus Argilagnes, ed., *Opus praeclarum artis medicine nominatum Articella*,. Venetiis: Hermannus Liechtenstein, 1483.

Augustinus, *Confessiones*: Carolyn J.-B. Hammond, ed. and transl., *St. Augustine's Confessions*, 2 vols (Loeb Classical Library 26–27). Cambridge, MA, and London: Harvard University Press, 2014–2016.

Avicenna 1972: Avicenna (Abû-'Aly al-Husayn Ibn 'Abdallah Ibn Sînâ:), *Liber de anima seu Sextus de Naturalibus*, Simone Van Riet, ed., vol. 1. Louvain: Editions Orientalistes, 1972.

Bacon, *De erroribus medicorum*: Rogerus Bacon, *De erroribus medicorum*, Andrew George Little and Edward Theodore Withington, eds, *Opera hactenus inedita Rogeri Baconi*, Fasc. 9: *De retardatione accidentium senectutis cum aliis opusculis de rebus medicinalibus* nunc primum ediderunt. Oxford: Clarendon, 1928, pp. 150–179.

Bacon, *Advancement of Learning*: Francis Bacon, *The Twoo Bookes of Francis Bacon. Of the proficience and advancement of Learning, divine and humane*. London: [Thomas Purfoot and Thomas Creede] for Henrie Tomes, 1605 (ed. Brian Vickers, *Francis Bacon, The Major Works* [Oxford World's Classics]. Oxford: Oxford University Press, 2008, pp. 120–299.

Bacon, *Novum Organum*: Franciscus Bacon, *Instauratio magna. [Novum organum sive indicia vera de interpretatione naturae]*. Londini: Johannes Billius, 1620 Engl. transl.: Basil Montagu, *The Great Instauration of Lord Bacon, Part II: Novum Organum*, in *The Works of Francis Bacon, Lord Chancellor of England. A New Edition with a Live of the Author in three Volumes*. Philadelphia, PA: Carey and Hart, 1841, vol. 3, pp. 329–425.

Barlaeus 1645: Casparus Barlaeus, *Methodus studiorum, Praescripta Ducibus Megapolensibus, ad eorum Ephorum*, in Hugo Grotius et alii, *Dissertationes de studiis instituendis*. Amsterodami: Ludovicus Elzevirius, 1645, pp. 340–357.

Benedetti, *De medici atque egri officio* 1506: Alexander Benedictus, *de medici atque egri officio collectionum*, in Symphorien Champier, *De medicinae claris scriptoribus . . . tractatus*. [Lugduni]: Jannot De Campis, [1506 (?)], ff. xvi recto-xxvi recto.

Benedetti, *De medici atque aegri officio* 1549: Alexander Benedictus, *De medici atque aegri officio, aphorismorum sive collectionum Liber unus*, in Alexander Benedictus, *de re medica opus* *insigne & apprimè Medicinae canditatis omnibus utile*. Basileae: [Henricus Petrus], 1549, pp. 546–553.

Berengario, *Anatomia Mundini*, 1514: Jacopo Berengario da Carpi, *Anathomia Mundini noviter impressa ac per Carpum castigata*. Bononiae: Giustiniano da Rubiera, 1514.

Berengario, *Commentaria super anathomia Mundini* 1521: Iacobus Berengarius Carpensis [Jacopo Berengario da Carpi], *Carpi Commentaria cum amplissimis additionibus super Anatomia Mundini una cum textu ejusdem in pristinum & verum nitorem redacto*. Bononiae: Hieronymus de Benedictis, 1521.

Berengario, *Isagoge*, 1522: Carpus [Iacobus Carpensis], *Isagoge Breves perlucide ac uberrime in Anatomiam humani Coroporis . . .* Bononiae: Benedictus Hector, 1522 Engl. transl.: Levi Robert Lind, *Jacopo Berengario da Carpi, A Short Introduction to Anatomy (Isagogae breves)*. Translated with an Introduction and Historical Notes. Chicago, IL: University of Chicago Press, 1959.

Berengario, *Tractatus de fractura*, 1518: Jacopo Berengario da Carpi, *Tractatus de fractura calvae sive cranei*. Bononiae: Hier. de Benedictis, 1518.

Bernard of Gordon (Bernardus Gordonius), *Opus, Lilium medicinae inscriptum . . . particular II, De passionibus capitis*. Lugduni: apud G. Rovillium, 1574.

Bianchi 1748: Giambattista Bianchi, *Storia del Mostro di due Corpi, Che nacque sul Pavefe in Giugno* 1748 . . . Con alcune riflessioni. Torino: Filippo Antonio, Campana 1748.

Boaistuau 1569: Pierre Boaistuau, *Histoires prodigieuses les plus memorables qui ayent esté observées depuis la nativité de Iesus Christ, iusques à nostre siecle: extraictes de plusieurs fameux autheurs, Grecz, & Latins, sacrez & prophanes: mises en nostre langue*. Paris: Vincent Sertenas, 1560 (Engl. transl.: Edward Fenton, *Certaine Secrete wonders of Nature, containing a description of sundry strange things, seming monstrous in oeyes and judgement, bicause we are not privie to the reasons of them. Gathered out of divers learned authors as well Greeke as Latine, sacred as prophane*. London: Henry Bynneman, 1569.

Boccaccio 2017: Donald Beecher and Massimo Ciavolella, eds., and transls., *Giovanni Boccaccio, The Decameron: Selected Tales*. Peterborough: Broadview Press, 2017.

Botallo 1565/1660: Leonardus Botallus, *Commentarii duo, Alter de medici, Alter de Aegroti munere. Huic accredit admonition fungi strangulatorii*. Ludguni: Antonius Gryphius, 1565 (reproduced as *Tractatus de medici et aegri munere*, in Leonardus Botallus, *Opera omnia Medica & Chirurgica. Hac postrema editione à mendis repurgata, methodicè disposita, paragraphis distincta, notis marginalibus, & Authorum testimoniis aucta, hinc inde annotationibus illustrata, prodeunt è musaeo Joannis van Horne*. Lugduni Batavorum: Ex officina Danielis & Abrahami à Gaasbeeck, 1660, pp. 1–58.

Ital. transl.: Leonardo Caretri and Anita Bogetti Bassone, *Leonardo Botallo, I doveri del medico e del malato*. Turin: Strenna UTET, 1981).

Bright 1586: Timothy Bright, *A treatise of melancholy. Contayning the causes thereof, and reasons of the straynge effects it worketh in our minds and bodies: with the physicke cure, and spirituall consolation for such as have thereto adioyned afflicted conscience*. London: John Winde, 1586.

Browne 1904: Thomas Browne, *A true and full coppy [sic] of that which was most imperfectly and surreptitiously printed before under the name of Religio medici*. [London]: Printed for Andrew Crooke, 1645.

reproduced in *The Works of Sir Thomas Browne*, Charles Sayle, ed., 3 vols. London: Grant Richards, 1904, vol. 1, pp. 1–112; also

Thomas Browne, The Major Works, Constantinos A. Patrides, ed. London: Penguin, 1977, pp. 1–560.

Burton 1621: Robert Burton, *The Anatomy of Melancholy, What it is: With all the Kinds, Causes, Symptomes, Prognostickes, and Several Cures of it. In Three Maine Partitions with their several Sections, Members, and Subsections. Philosophically, Medicinally, Historically, Opened and Cut Up*. Oxford: John Lichfield and James Short for Henry Cripps,1621.

Burton 1927: *The Anatomy of Melancholy*, Floyd Dell and Paul Jordan-Smith, eds. New York: Tudor Publishing, 1927.

Cabello Balboa 1920: Miguel Cabello Balboa, *Historia del Peru bajo la dominacion de los Incas*. Edicion con anotaciones y concordancias con las cronicas de Indias por Horacio H. Urtega y biografia de Cabello Balboa por Carlos A. Romero. Lima: Imprenta y Libreria San Martin, 1920.

Cardano 1536: Hieronymus Cardanus, *De malo recentiorum medicorum medendi usu libellus. Eiusdem libellus de simplicium medicinarum noxa. Cum Indice eorum, quae singulis continentur capitibus*. Venetiis: [Octavianus Scotus], 1536.

Cardano ed. Sponius 1663: Carolus Sponius ed., Hieronymi Cardani, *Opera omnia tam hactenus excusa; hic tamen aucta & emendate; quam nunquam alias visa, ac primum ex Auctoris ipsius Autographis eruta*, 10 tomi. Lugduni: Ioannes Antonius Huguetan et Marcus Antonius Ravaud, 1663.

Cellini 1728: Benvenuto Cellini, *Vita di Benvenuto Cellini*. Colonia: Per Pietro Martello, 1728

Cellini, 1980: *Le opere di Benvenuto Cellini*,. Giuseppe Guido Ferrero and Enrico Carrara, eds., Torino: Unione Tipografico-Editrice Torinese, 1980.

Chabannes 1999: Ademar Chabannes, *Ademari Cabannensis Chronicon*, Pascale Bourgain, ed., Turnhout: Brepols, 1999.

Cieza de Leon (1554) 1992: Pedro Cieza de Leon, *Parte Primera de la Crónica del Perú 1554*. En Amberes: En casa de Juan Steelsio, 1554 (reprint in *Cronica del Peru, 1553*, vols 1–2. Madrid: Espasa Calpe, 1922).

Collectio Salernitana: *Collectio Salernitana ossia documenti inediti, e trattati di medicina appartenenti alla scuola medica* salernitana, raccolti ed illustrati da G. E. T. Henschel, C. Daremberg, E. S. De Renzi; premessa la storia della scuola Salvatore de Renzi, ed., 5 vols. Napoli: Filiatre-Sebezio, 1852–1859.

Crinito 1532: Pietro Crinito, *De honesta disciplina libri XXV*. Basileae: Henricus Petrus, 1532.

da Vinci 1478–1518: Leonardo da Vinci, *Notebook*, Codex Arundel 1478–1518, Arundel MS 263 (Engl. ed.: Charles D. O'Malley and J. B. de C. M. Saunders, *Leonardo da Vinci on the Human Body: The Anatomical, Physiological and Embryological Drawings of Leonardo da Vinci*. New York, NY: Gramercy Books, 2003).

de Betanzos (1880) 2004: Juan de Betanzos, *Suma y Narración de los Incas*. Madrid: Imprenta de Manuel G. Hernández 1880 (ed. Maria del Carmen Martin Rubio. Madrid: Polifemo, 2004).

de Betanzos (1987) 2015: Juan de Betanzos y el Tahuantinsuyo, *Suma y narración de los incas* [1551–1557]. Madrid: Atlas, 1987 (ed. Francisco Hernandez, Rodolfo Cerron-Palomino, *Nueva edicion de la Suma y narración de los Incas*. Lima: Fondo editorial de la Pontificia Universidad Catolica del Peru, 2015).

de Castro 1596: Rodericus de Castro, *Tractatus brevis de natura et causis pestis quae hoc anno MDXCVI Hamburgensem civitatem afflixit*. Hamburgi: Jacobus Lucius junior, 1596.

de Castro 1614: Rodericus de Castro, *Medicus Politicus, sive de officiis medico-politicis tractatus*. Hamburgi: Officina Frobeniana, 1614.

de la Hoz (1534) 1938: Pedro Sancho de la Hoz, *Relacion para Su Majestad de lo sucedico en la Conquista 1534*, Horacio H. Urteaga, ed., *Los cronistas de la conquista*. Selección, notas y concordancias (Biblioteca de cultura peruana, Prima serie, 2). Paris: Desclée de Brouwer, 1938.

de la Riva Aguero 2003: Jose de la Riva Aguero, "Epistolario," in *Obras Completas de Jose de La Riva Aguero*, vol. 28. Lima: Pontificia Universidad Catolica, 2003, pp. 387–389.

de la Vega 1609: Garcilaso de la Vega, *Primera parte de los commentarios reales: que tratan del origen de los Yncas, reyes que fueron del Peru, de su idolatria, leyes, y gouierno en paz y en guerra: de sus vidas y conquistas, y de todo lo que fue aquel imperio y su republica, antes que los españoles passaran a el. Comentarios Reales (1609)*, Lisbona: En la officina de Pedro Crasbeeck, 1609 (Buenos Aires: Emece Editores, 1945).

de Mena 1534. Cristobal de Mena, *La Conquista del Peru, llamada la Nueva Castilla*. Sevilla: Bartolomé Pérez, 1534.

del Garbo 1999: Enrico Fenzi, ed., Dino del Garbo, *La canzone d'amore di Guido Cavalcanti e i suoi antichi commenti*. Genova: il Melangolo, 1999.

Descartes 1637: René Descartes, *Discourse de la methode*. Leyde: de l'Imprimerie Ian Maire, 1637. Ian Mclean, ed. *A Discourse on the Method of Correctly Conducting One's Reason and Seeking Truth in the Science*s, Oxford: Oxford University Press, 2006.

du Laurens 1613: André du Laurens, *Second discours, auquel est traicte des maladies melancholiques, et du moyen de les guarir*, in Théophile Gelée, transl., *Toutes les oeuvres de M. André du Laurens . . . recueillies et traduites en francois* . . . Paris: [chez P. Mettayer], 1613, ff. 21 verso -1 recto.

Dubois 1539: Jacques Dubois, *Ordo et ratio in legendis Hippocratis et Galeni*. Paris: Chrétien Wechel, 1539.

Falcucci 1491: Nicolaus Falcutius, *Sermones medicinales (De febribus; De membris capitis; De membris spiritualibus; De membris naturalibus; De membris generationis; De membris exterioribus)*. Venetiis: Bernardinus Stagninus, 1490–1491.

Ferrand 2010: Jacques Ferrand, *De la maladie d'amour ou mélancolie érotique*, Donald A. Beecher and Massimo Ciavolella, eds. Paris: Éditions Garnier, 2010.

Engl. transl.: Donald A. Beecher and Massimo Ciavolella, Jacques Ferrand, *A Treatise on Lovesickness*. Syracuse: Syracuse University Press, 1990.

Ficino 2002: Marsilio Ficino, *Three Books on Life: A Critical Edition and Translation*, Carol B. Kaske, ed. Tempe, AR: Arizona Center for Medieval and Renaissance Studies, 2002.

Fracastoro 1546: Girolamo Fracastoro, *de contagione & de contagiosis morbis et eorum curatione, libri tres*. Lugduni: Guglielmus Gazeirus, 1546.

Galen, *De animi cuiuslibet peccatorum dignotione et curatione*: French transl. Vincent Barras, Terspichore Birchler, and Anne-France Morand, *L'âme et ses passions. Les passions et les erreurs de l'âme. Les facultés de l'âme suivent les tempéraments du corps* (La roue à livres 26). Paris: Belles Lettres, 1995.

Galen, *De cognoscendis curandisque animi morbis*: ed. Carl Gottlob Kühn, *Claudii Galeni Opera omnia*, vol. 5 (Medicorum graecorum opera quae exstant 5). Lipsiae: C. Cnobloch, 1823, pp. 1–57 (reprint: Hildesheim: Olms, 2001).

Galen, *De constitutione artis medicative VII*: Lat. transl. Nicolaus de Regio di Calabria, in Diomedes Bonardus de Brescia, ed., *Galieni opera*, 2 vols. Venetiae: P. Pincius, 1490, vol. 1, pp. [qq^{viii}] verso (= 111) – rr^{iii} verso (= 115).

Galen, *De cuiuslibet animi peccatorum dignotione atque medela libellus*: ed. Carl Gottlob Kühn, *Opera omnia*, vol. 5 (Medicorum graecorum opera quae exstant 5). Lipsiae: C. Cnobloch, 1823, pp. 58–103 (reprint: Hildesheim: Olms, 2001).

Galen 1529: *Galeni Pergameni libri anatomici*. Boniniae: Ioannis Baptistae Phaelli, 1529.

Galen, *Quod animi mores corporis temperamenta sequantur*: French transl. Vincent Barras, Terspichore Birchler, and Anne-France Morand, *L'âme et ses* passions. Les passions *et les erreurs de l'âme. Les facultés de l'âme suivent les tempéraments du corps* (La roue à livres 26). Paris: Belles Lettres, 1995.

Galen, *Quod optimus medicus*. Lat. transl. Erasmus: Jan Hendrik Waszink, ed., in Kazimierz Kumaniecki, Roger A. Mynors, Jan Hendrik Waszink, and Christopher Robinson, eds, *Opera* *omnia Desiderii Erasmi Roterodami recognita et adnotatione critica instructa notisque illustrata*, Ordinis primi, tomus primus. Amsterdam: North-Holland, 1969, pp. 665–669.

Galilei 1890–1909: Galileo Galilei, *Opere*, Antonio Favaro, ed. Firenze: G. Barbera, 1890–1909.

Garcilaso de la Vega (1609) 1945: Garcilaso de la Vega, *Primera parte de los commentarios reales: que tratan del origen de los Yncas, reyes que fueron del Peru, de su idolatria, leyes, y gouierno en paz y en guerra: de sus vidas y conquistas, y de todo lo que fue aquel imperio y su republica, antes que los españoles passaran a el.* Lisboa: En la oficina de Pedro Crasbeek, 1609.

Reprint: *Comentarios Reales (1609)*, 2 vols. Buenos Aires: Emece, 1945.

Glissenti 1596: Fabio Glissenti, *Discorsi morali contra il dispiacer del morire, detto Anatophilia*. In Venetia: Domenico Farri, 1596.

Gonzales de Holguin 1975: D. Gonzales de Holguin, *Gramatica y arte de la nueva lengua general de todo el Peru llamada lengua qquichua o lengua del Inca (1607)*. Cabildo Valdivia-Georgetown: Franz Wolf, Heppenheim, 1975.

Guaman Poma de Ayala 1615 (1993): Felipe Guaman Poma de Ayala, *El primer nueva corónica y buen gobierno* [1615–1616], Ms. Copenhague, Det Kongelige Bibliotek (Royal Library), Gammel Kongelig Samling (GKS), 2232, 4°, ed. Franklin Pease, *Guaman Poma, Nueva Cronica y Buen Gobierno 1615*. Lima: Fondo de Cultura Economica, 1993.

Haly Abbas 1492: Haly Abbas (Ali ibn al ʿAbbas al-Majusi), *Liber medicinae qui dicitur Regalis dispositio quem ex arabico in latinam transtulit facultatem Stephanus Antiochenus*. Venetiis: Bernardus ricius. impensa . . . Joannis Domenici de Nigro, 1492.

Harvey 1577: Gabriel Harvey, *Ciceronianus*. London: Bynneman, 1577.

Gabriel Harvey's Ciceronianus, Harold S. Wilson and Clarence A. Forest, eds. Lincoln: The University, 1945.

Hippocrates, 1944: *Corpus Hippocraticum, Epidemics*, W. D. Smith, ed. Leiden: Brill, 1944.

Hippocrates 1923/2004: Hippocrates, *De Prisca medicina*: William H. S. Jones, ed. and transl., *Hippocrates*, vol. 1 (Loeb Classical Library 147). London: William Heinemann, and Cambridge, MA: Harvard University Press, 1923, pp. 1–64 (reprint 2004).

***Homer Iliad 1966**: Homer, Iliad, Harold Bloom, ed.* New York: Chelsea Publishers, 1996.

Ibn Hazm: Arthur John Arberry, transl., *Ibn Hazm, The Ring of the Dove: A Treatise on the Art and Practice of Arab Love*. London: Luzac Oriental, 1994.

Ingrassia 1544: Ioannes Philippus Ingrassia, *Iatropologia, Liber, quo multa adversus barbaros medicos disputantur, collegiique modus ostenditur ac multae quaestiones tam physicae quam chirurgicae discutiuntur: eiusdem Quaestio, quae Capitis Vulneribus ac Phrenitidi medicamenta convenient*. Venetiis: Ioannes Gryphius, 1544.

Isidore of Seville 2006: Isidore of Seville, *The Etymologies*, Stephen A. Barney, W.J. Lewis, J.A. Beach, Oliver Berghof, and Muriel Hall, eds. Cambridge: Cambridge University Press, 2006.

Kircher 1650: Athanasius Kircher, *Musurgia universalis sive ars magna consoni et dissoni in X. libros digesta*. Romae: Ex typographia Haeredum Francisci Corbelletti, 1650.

Kramer and Sprenger 1928: Heinrich Kramer and James Sprenger, *The Malleus Maleficarum*, Montague Summers, transl. London: John Rodker, 1928; (reprint: New York: NY: Dover Publications, 1977).

Liceti 1616: Fortunio Liceti, *De monstrorum natura, caussis et differentiis libri duo*. Patavii: Apud Paulum Frambottum, 1616.

Liceti 1616: Fortunio Liceti, *De perfecta constitutione hominis in utero liber unus*. Patavii: Apud Paulum Frambottum, 1616.

Liceti 1618: Fortunio Liceti, *De spontaneo viventium ortu*. Vicentiae: Domenico Amadio, Francesco Bolzetta, 1618.

Mann 1936: Thomas Mann, "The Dilettante," in *Stories of Three Decades*, H. T. Lowe-Porter, transl. New York, NY: The Modern Library, 1936.
Mann 1936: Thomas Mann, *Little Herr Friedemann*, in *Stories of Three Decades*, H. T. Lowe-Porter, transl. New York, NY: The Modern Library, 1936.
Mann 1992: Thomas Mann, *Confessions of Felix Krull, Confidence Man*. Denver Lindsley, transl. New York, NY: Vintage International, 1992. Original title: Bekenntrisse des Hochstaplers Felix Krull. Der Memoiren, erster Teil. First published in German in 1954 and in English in 1955.
Mann 1993: Thomas Mann, *Buddenbrooks: The Decline of a Family*, John E. Woods, transl. New York, NY: Alfred A. Knopf, 1994.
First ed. published as *Buddenbrooks: Verfall einer* Familie. Berlin: S. Fischer, 1901.
Mann 1996: Nicholas Mann, "The Origins of Humanism," in J. Kraye, ed., *The Cambridge Companion to Renaissance Humanism*. Cambridge: Cambridge University Press, 1986.
Mann 1995 and 2005: Thomas Mann, *The Magic Mountain*. John E. Woods, transl. New York, NY: Alfred A. Knopf, 1995, (reprint 2005).
Mann 1997: Thomas Mann, *Doctor Faustus: The Life of the German Composer Adrian Leverkühn as Told by a Friend*, John E. Woods, transl. New York, NY: Vintage International, 1997.
Milton 2005: John Milton, *Paradise Lost*, David Scott Kastan, ed. Indianapolis, IN.: Hackett Publishing, 2005.
Montaigne 2003a: Michel De Montaigne, "Of A Monstrous Child," in Michel de Montaigne, *The Complete Works: Essays, Travel Journal, Letters*, Donald M. Frame, ed. and transl. New York, NY: Everyman's Library, 2003.
Montaigne 2003b: Michel de Montaigne, *The Complete Works: Essays, Travel Journal, Letters*, Donald M. Frame, ed. and transl. New York, NY: Everyman's Library, 2003.
Nicholas of Lyra 1519: Nicholas of Lyra, *Postilla seu Expositio literalis & moralis Nicolai de Lira ordinis minorum super epistolas: & euangelia quadragesimalia*. Venetiis: sumptibus Laurentij Lorij Salodiensis: per Alexandrum de Bindonis, 1519.
Novalis 1997: Novalis, *Philosophical Writings*, (Margaret Mahony Stoljar, ed. and transl.). New York, NY: State University of New York Press, 1997.
Ondegardo (1571) 1940: Juan Polo de Ondegardo: "Relaciones de los fundamentos acerca del notable daño que resulta de no guardar a los indios sus fueros 1571," in *Colección de documentos inéditos relativos al descubrimiento, conquista y colonización de las posesiones españoles en América y Occeanía*, sacados bajo la direccion de Francisco de Cárdenas y Espejo, Joaquín Francisco Pacheco, Luis Torres de Mendoza, vol. 17. Madrid: Real academia de la historia, 1872, pp. 7–100 (*Revista Histórica* 13 [1940], pp. 125–196).
Ortelius 1570: Abraham Ortelius, *Theatrum orbis terrarum*. Antverpiae: Apud Aegid. Coppenium Diesth, 1570.
Paré 1573: Paré, Ambroise, *Deux livres de chirurgie, I. De la generation de l'homme, et maniere d'extraire les enfans hors du ventre de la mere, ensemble ce qu'il faut faire pour la faire mieux, et plus tost accoucher, avec la cure de plusieurs maladies qui luy peuvent survenir. 2. Des monstres tant terrestres que marins, avec leurs portrait. Plus un petit traité des plaies faites aux parties nerveuses*. Paris: André Wechel, 1573.
Critical edition Jean Céard, *Ambroise Paré, Des monstres et prodigies*. Édition critique et commentée (Travaux d'Humanisme et Renaissance 115). Genève: Librairie Droz, 1971.
Engl. transl.: Janis L. Pallister, *Ambroise Pare, On Monsters and Marvels*. Chicago, IL: University of Chicago Press, 1982.
Ital. transl.: Massimo Ciavolella, *Ambroise Paré. Mostri e prodigi* (Omikron 54) Rome: Salerno Editrice, 1996.

Petrarca 2016: Francesco Petrarca, *De secreto conflictu curarum mearum*, Nicolas Mann, ed. and transl., *My secret book* (The I Tatti Renaissance Library 72). Cambridge, MA, and London: Harvard University Press, 2016.
Pietro d'Abano 1472: Petrus de Abano, *Conciliator differentiarum philosophorum et precipue medicorum*. Mantue: Thomas Septem Castrensis et Joannes Burster, 1472.
Pizarro 1571: Pedro Pizarro, *Relacion del descubrimiento y la conquista del Peru y su estado antiguo 1571*. Peru: Fondo Editorial, 1978.
Pliny the Elder 1997: Pliny the Elder, *Naturalis Historia* 35.6, in Julius von Scholsser ed. *Histoire du protrait en cire*. Paris: Macula, 1997.
Pomponazzi 2011: Pietro Pomponazzi, *De incantationibus*, Vittoria Perrone, ed. Florence: Compagni, 2011.
Ramazzini 1745: Bernardino Ramazzini, *Le malattie degli artefici*. In Venezia: appresso Domenico Occhi, 1745.
Riolan 1605: Jean Riolan, *De monstro nato Lutetiae anno 1605. Disputatio philosophica*. Parisiis: Apud Olivarium Varennaeum,1605.
Roger Bacon 1928: Rogerus Bacon, *De erroribus medicorum*, Andrew George Little and Edward Theodore Withington, eds. *Opera hactenus inedita Rogeri Baconi*, Fasc. 9: *De retardatione accidentium senectutis cum aliis opusculis de rebus medicinalibus nunc primum ediderunt*. Oxford: Clarendon, 1928, pp. 150–179.
Rüff 1580: Jacob Rüff, *De conceptu et generatione hominis*. Francoforti a Moenum: apud Georgium Corvinum, 1580.
Sancho de Leon 1977: Pedro Sancho de Leon, *Historia de la Conquista del Peru 1534*, Phillip A. Means, transl., *An Account of the Conquest of Peru*. Boston: Longwood Press, 1977.
Sarmiento de Gamboa 2019: Pedro Sarmiento de Gamboa, *Historia de los Incas, 1572/1965 (Segunda parte de la historia general, llamada indica)*, Madrid: Biblioteca de Autores Españoles, *Ediciones Atlas* vol. 135 (2019), pp. 193–279.
Shakespeare 1970: William Shakespeare, *The Two Noble Kinsmen*, G.R. Proudfoot, ed. London: Edward Arnold, 1970.
Siccus 1551: Ioannes Antonius Siccus, *De optimo medico caput primum libri primi de antiqua medicina eiusdem*. Venetiis: Iuntae, 1551.
Tolstoy 2016: Leo Tolstoy, *The Death of Ivan Ilyich* (Penguin Classics). New York: Penguin-Random House, 2016.
Varchi 1858: Benedetto Varchi, *De la generazione de' mostri*, in *Opere di Benedetto Varchi*. Trieste: Sezione Letterario-Artistica del Lloyd Austriaco, 1858.
Vasari G. 1832–1838: *Le opere di Giorgio Vasari. Pittore e Architetto Aretino*. Parte seconda. Firenze: David Passigli e Soci, 1832–1838.
Weinrich 1595: Martin Weinrich, *De ortu monstrorum commentarius in quo essentia, differentiae, causae et affecitones explicantur*. Wroclaw: Sumptibus Heinrici Osthusii, 1595.
Woolf 1926: Virginia Woolf, *On Being Ill*, *The New Criterion*, 4.1 (January 1926), pp. 32–45.
Zerbi 1495: Gabriel Zerbus, *Opus perutile de cautelis medicorum*. [Venetiis: Christophorus de Pensis, 1495].
Germ. transl. et comm. Mariacarla Gadebusch Bondio, Manuel Förg and Christian Kaiser, *Gabriele Zerbi: Über die Kautelen der Ärzte / "De cautelis medicorum" (ca. 1495)*. Stuttgart: Franz Steiner Verlag, 2019.

Secondary Literature

Adams 2000: John Adams, "Novanglus, Thoughts on Government, Defense of the Constitution," in George Carey, ed., *The Political Writings of John Adams*. Washington D.C.: Regnery Publishing, 2000.

Adolphs 2001: Ralph Adolphs, "The Neurobiology of Social Cognition," *Current Opinion in Neurobiology* 11 (2001), pp. 231–239.

Agamben 1998: Giorgio Agamben, *Homo Sacer: Sovereign Power and Bare Life*. Stanford, CA: Stanford University Press, 1998.

Agrimi and Crisciani 1988: Jole Agrimi and Chiara Crisciani, *Edocere medicos: medicina scolastica nei secoli XIII-XV* (Istituto Italiano per gli Studi Filosofici: Hippocratica Civitas 2). Naples: Guerini, 1988.

Agrimi and Crisciani 1990: Jole Agrimi and Chiara Crisciani, "Per una ricerca su *experimentum-experientia*: riflessione epistemologica e tradizione medica (secoli XIII-XV)," in Pietro Janni and Innocenzo Mazzini, eds, *Presenza del lessico greco e latino nelle lingue contemporanee*. Macerata: Università degli studi di Macerata, 1990, pp. 9–49.

Alexander 1995: B. Alexander, "Bartonellosis in Equador," *American Journal of Tropical Medicine and Hygiene* 52 (4) (1995), pp. 354–359.

Allison, Pezzia, and Gerstein 1974: Marvin J. Allison, Alejandro Pezzia, Enrique Gersten, et al., "A case of Carrion's disease associated with human sacrifice from the Huari culture of southern Peru," *American Journal of Physical Anthropology* 41 (1974), pp. 295–300.

Aloisi 2020: Silvia Aloisi, "Galileo's Lost Tooth, Finger Go on Show in Florence," *Reuters* June 8 (2010) https://www.reuters.com/article/us-italy-galileo-odd/galileo-lost-tooth-fingers-go-on-show-in-florence-idUSTRE6573R520100608Accessed March 15, 2020.

Alston 1944: Mary Niven Alston, "Attitudes of the Church towards Dissection before 1500," *Bulletin of the History of Medicine* 16 (1944), pp. 221–238.

Altman 1998: Lawrence K. Altman, "Battle-Scarred Veteran is General in Global War on Aids," *The New York Times*, July 21, 199. https://archive.nytimes.com/www.nytimes.com/library/national/science/072198sci-docworld.2.jpg.html Accessed March 15 2022.

Ammaniti and Gallese 2014: Massimo Ammaniti and Gallese Vittorio, *The Birth of Intersubjectivity. Psychodynamics, Neurobiology, and the Self*. New York, NY: Norton & Company, 2014.

Amsterdam Guild of Surgeons 1631–1731; Rembrandt, "The Anatomy Lesson of Dr. Nicolaes Pulp" 1632.

Andreason 1987: Nancy Anderson, "Genius and Madness," *American Journal of Psychiatry* 144 (10) (1987), pp. 1288–1292.

Antoine 2001: E. Antoine, "Ex voto," in André Vauchez, ed., *Encyclopedia of the Middle Ages*. Cambridge: James Clarke & Co. 2001 (e-reference edition). http://www.oxfordreference.com Accessed 25 July 2009.

Archivio di Stato di Firenze: *Archivio Mediceo avanti il Principato*, Inventario, vol. 1. Filza II, Pandulfus Sigismundus de Malatestis. Arimini (December 5, 1463, n. 492). Rome: Ministero dell'Interno, 1966, f. 31.

Argyris 1982: Chris Argyris, *Reasoning, Learning, and Action: Individual and Organizational*. San Francisco, CA: Jossey-Bass, 1982.

Argyris and Schön 1974: Chris Argyris, and Donald A. Schön, *Theory in Practice: Increasing Professional Effectiveness. San Francisco:* Jossey-Bass, 1974.

Arias-Stella 1987: Javier Arias-Stella, "Verruga Peruana Mimicking Malignant Neoplasms," *American Journal of Dermatopathology* 9 (1987), pp. 279–384.

Ariès 1981: Philippe Ariès, *The Hour of Our Death*. New York, NY: Vintage, 1981.

Arieti 1999: Stefano Arieti, "Un medico tra Bologna e le corti padane alle soglie del Rinascimento scientifico: Jacopo Berengario da Carpi," in Euride Fregni, ed., *Archivi, territori, poteri in area Estense (secc. XVI-XVIII)*. Rome: Bulzoni, 1999, pp. 1–8.

Armaroli 1995: Maurizio Armaroli, *Le Cere Anatomiche Bolognesi del Settecento*, Università degli Studi Bologna, Accademia delle Scienze, Catalogue of the exhibition 1981. Bologna: Clueb, 1995.

Arnold and Ford-Jones 2000: Sandra R. Arnold and E. Lee Ford-Jones, "Congenital Syphilis: A Guide to Diagnosis and Management," *Pediatrics & Child Health* 5 (8) (2000), pp. 463-469.

Aron 1955: Raymond Aron, *L'opium des intellectuels*. Paris: Calmann-Lévy, 1955.

Arrese Estrada, Maquiña, and Perrard 1992: Jorge Arrese Estrada, Cesar Maguiña, and Gerald Pierrard, "La verruga peruana pasado, presente y future," *Piel* 7 (1992), pp. 350–353.

Arrizabalaga 2009: Jon Arrizabalaga, "Medical Ideals in the Sephardic Diaspora: Rodrigo de Castro's Portrait of the Perfect Physicians in Early Seventeenth-Century Hamburg," *Medical History* 53. 29 (2009), pp. 107–124.

Arrizabalaga, Henderson, and French 1997: Jon Arrizabalaga, John Henderson, and Roger French, *The Great Pox: The French Disease in Renaissance Europe*. New Haven, CT: Yale University Press, 1997.

Arroyo Erazo 1938: Cesar Arroyo Erazo, "La Bartonellosis, epidemia olvidada con una letalidad sin precedents en el Departamento de Nariño," *Revista medicina* (1938) DOI: http://www.revistamedicina.net. Accessed March 5, 2022.

Ascari and Crespi 1964: Tiziano Ascari and Mario Crespi, "Barigazzi, Iacopo (Iacopo Berengario, Iacopo da Carpi)," in Alberto M. Ghisalberti, ed., *Dizionario biografico degli Italiani* vol. 6. Rome: Istituto della Enciclopedia Italiana, 1964, pp. 360–364.

Augé 2014: Marc Augé, *Une ethnologie de soi: Le temps sans âge*. Paris: Seuil, 2014.

Austin 1962: John L. Austin, *How to Do Things with Words*. Cambridge, MA: Harvard University Press, 1962.

Avendaño 1908: Luis Avendaño, "La epidemia de tifus exantematico de 1525 (Huayna Capac)" in "Dictamen sobre la tesis de Julio C. Tello: La antiguedad de la sifilis en el antiguo Peru," *Anales de la Universidad Mayor de San Marcos*. Lima, 1908, pp. 2–212. DOI: https://hdl.handle.net/20.500.12672/1455. Accessed March 5, 2022.

Azzaroli 1997: Maria Luisa Azzaroli, "La Specola. The zoological museum of Florence University," in Azzaroli et al. eds, *La ceroplastica nella scienza e nell'arte. Atti del I Congresso Internazionale*. Florence: Leo S. Olschki, 1977, pp. 1–22.

Baille 2017: Yves Baille, *Les monstres chez Ambroise Paré: un regard rétrospectif*. Marseille: Presses Universitaires de Provence, 2017.

Balbuena 2014: Francisco Balbuena, "The Pioneering Work of Marguerite Sechehaye into the Psychotherapy of Psychosis: A Critical Review," *Swiss Archives of Neurology and Psychiatry* 165 (5) (2014), pp.167–174.

Baker 1996: Jeffrey P. Baker, *The Machine in the Nursery: Incubator Technology and the Origins of Newborn Intensive Care*. Baltimore, MD: Johns Hopkins University Press, 1996.

Balatri 1755: Filippo Balatri, *Testamento (burlesco) o sia ultima volontà di Filippo Balatri nativo Alfeo*. Manuscript Munich, Bayerische Staatsbibliothek, Cod. It. 329.

Baldi 1824: Bernardino Baldi, *Vita e fatti di Federigo di Montefeltro Duca di Urbino*, vol. 3. Rome: Perego Salvioni, 1824.

Balint 1957: Michael Balint, *The Doctor, His Patient and the Illness*. London: Tavistock Publications, 1957.

Ballesteros Gaibros 1963: Manuel Ballesteros Gaibros, "Descubrimiento y Conquista del Peru," in *Historia de America y de los Pueblos Americanos*, vol 9. Barcelona: Salvat, 1963, p. 62.

Ballestriero 2007: Roberta Ballestriero, "The art of ceroplastics. Clemente Susini and the collection of the anatomical wax model of the University of Cagliari," in Alessandro Riva, ed., *Flesh & Wax. The Clemente Susini's Anatomical Models in the University of Cagliari*. Nuoro: Ilisso, 2007, pp. 35–45.

Ballestriero and Edwards 2017: Roberta Ballestriero and William Edwards, "Wax and Science," *Bee Farmer Magazine*, Special edition 2017, pp. 9–13.

Ballestriero and Richardson 2014: Roberta Ballestriero and Ruth Richardson, *Joseph Towne at the Gordon Museum*. Uckfield: Pureprint, 2014.

Barbour 2013: Reid Barbour, *Sir Thomas Browne: A Life*. Oxford: Oxford University Press, 2013.

Baron-Cohen 2003: Simon Baron-Cohen, *The Essential Difference: The Truth about the Male and Female Brain*. New York, NY: Basic Books, 2003.

Baron-Cohen, Golan, Wheelwright and Hill 2004: Simon Baron-Cohen, Ofer Golan, Sally Wheelwright, and Jacqueline J. Hill. *Mind Reading: The Interactive Guide to Emotions*. London: Jessica Kingsley, 2004.

Barthes 2017: Roland Barthes, *Non si riesce mai a parlare di ciò che si ama*. Milan: Mimesis, 2017.

Bartolini 2015: Donatella Bartolini, "On the Borders: Surgeons and Their Activities in the Venetian State (1540–1640)," *Medical History* 59 (1) (2015), pp. 83–100.

Baruffaldi 1760: Girolamo Baruffaldi, *La mammana istruita per validamente amminiftrare il S. Sagramento del Battesimo in cafo di neceffità alle creature nafcenti*. Trento: Francesco Michele Battisti, 1760.

Baskins 2012: Cristelle Baskins, "Lost in Translation: Portraits of Sitti Maani Gioerida della Valle in Baroque Rome," *Early Modern Women: An Interdisciplinary Journal* 7 (2012), pp. 241–260.

Bates 2005: Alan W. Bates, *Emblematic Monsters: Unnatural Conceptions and Deformed Births in Early Modern Europe* (Clio Medica 77) Leiden: Brill, 2005.

Bates, Bleakley and Goodman 2014: Victoria Bates, Alan Bleakley, and Sam Goodman, eds, *Medicine, Health and the Arts: Approaches to the Medical Humanities*. Abingdon: Routledge, 2014.

Bauer 2018: Brian S. Bauer, *Cuzco Antiguo. Tierra Natal de los Incas* (2nd ed.) Cuzco: Impresiones Lezama, 2018.

Bauman 2008: Zygmunt Bauman, *Does ethic have a chance in a world of consumers*? Cambridge MA: Harvard University Press, 2008.

It. transl. G. Bettini, Le sfide dell'etica. Milan: Feltrinelli, 1996.

Barzellotti 1818: Giacomo Barzellotti, *Questioni di medicina legale secondo lo spirito delle leggi civili e penali veglianti nei governi d'Italia* vol. 1. Pisa: Prosperi, 1818.

BBC News: BBC News, "Syphilis and Gonorrhea Up by One-Fifth." https://www.bbc.com/news/health-44368741. 5 June 2018.Accessed October 15, 2018.

Beauchamp and Childress 2001: Tom L. Beauchamp and James F. Childress, *Principles of Biomedical Ethics*. New York, NY: Oxford University Press, 2001.

Bechah and Capo and Mege and Raoult 2008: Yassina Bechah, Christian Capo, Jean-Louis Mege, and Didier Raoult, "Epidemic Typhus," *Lancet Infectious Diseases* 8 (2008), pp. 417–426.

Beck 2003: C. H. Beck, *Mittellateinisches Wörterbuch bis zum ausgehenden 13. Jahrhundert*, vols. 3 and 4. Munich: Bayerische Akademie der Wissenschaften, 2003.

Belting 1994: Hans Belting, *Likeness and* Presence, transl. Edmund Jephcotton. Chicago. IL: University of Chicago Press, 1994.

Bénat 2006: Louise Bénat, "*Los hombres son como nosotros, que de otra manera bestias y monstruos serían* . . . Los monstruos en la historiografía colonial spañola de la primera mitad del siglo XVI," in Eddy Stols, Werner Thomas, and Johan Verberckmoes, eds, *Naturalia,*

Mirabilia & Monstruosa en los Imperios Ibéricos (siglos XV-XIX). Leuven: Leuven University Press, 2006, pp. 235–254.

Benedetti 2008: Fabrizio Benedetti, *Placebo Effects: Understanding the Mechanisms in Health and Disease*. Oxford: Oxford University Press, 2008.

Benedetti 2013: Fabrizio Benedetti, *Il caso di G. L. La medicina narrativa e le dinamiche nascoste della mente*. Rome: Carocci, 2013.

Benedetti 2011: Fabrizio Benedetti. *The Patient's Brain. The Neuroscience Behind the Doctor–Patient Relationship*. Oxford: Oxford University Press, 2010.

Benitez-Bribiesca 1999: Luis Benitez-Bribiesca et al. "Dendritic Spine Pathology in Infants with Severe Protein-Calorie Malnutrition," *Pediatrics* 104 (2) (1999), e21.

Bennet 2005: Jim Bennet, "Museums and the History of Science: Practitioner's Postscript," *Isis* 96 (2005), pp. 602–608.

Bentley 1985: James Bentley, *Restless Bones: The Story of Relics*. London: Constable, 1985.

Beretta, Conforti and Mazzarello 2016: Marco Beretta, Maria Conforti, and Paolo Mazzarello eds., *Savant Relics: Brains and Remains of Scientists*. Sagamore Beach, MA: Watson Publishing International, 2016.

Berriot-Salvadore and Mironneau 2003: Évelyn Berriot-Salvadore and Paul Mironneau, eds., *Ambroise Paré (1510–1590), Practique et écriture de la science à la Renaissance*. Paris: Honoré Champion, 2003.

Bert 2003: Giorgio Bert, *Parole di medici, parole di pazienti: counselling e narrativa in medicina*. Rome: Il Pensiero Scientifico, 2003.

Bert 2007: Giorgio Bert, *Medicina narrativa. Storie e parole nella relazione di cura*. Rome: Il Pensiero Scientifico, 2007.

Bianconi 1942: Luigi Bianconi, *Viaggio in Levante di Pietro Della Valle*. Florence: Sansoni, 1942.

Bianucci 2016: Raffaella Bianucci et al., "'From Father to Son': Early Onset Gout in Guidobaldo I Da Montefeltro, Duke of Urbino (1472–1508)," *European Journal of Internal Medicine* 36 (2016), pp. e28–e30.

Bianucci 2018: Raffaella Bianucci et al., "Michelangelo Buonarroti (1475–1564) Had the Deconditioning Syndrome While Painting the Sistine Chapel Ceiling," *Medical Hypotheses* 113 (2018), pp. 13–14.

Biasin 1975: Gian-Paolo Biasin, *Literary Diseases: Theme and Metaphor in the Italian Novel*. Austin, TX: University of Texas Press, 1975.

Billig 2008: Michael Billig, *The Hidden Roots of Critical Psychology*. London: Sage 2008.

Blanton 1919: Smiley Blanton. "Mental and Nervous Changes in the Children of the Volksschulen of Trier, Germany" *Mental Hygiene* 3 (1919), pp. 343–386.

Bleakley 2014: Alan Bleakley, "Towards a Critical Medical Humanities," in Victoria Bates, Alan Bleakley, and Sam Goodman, eds., *Medicine, Health and the Arts: Approaches to the Medical Humanities*. Abingdon: Routledge, 2014, pp. 17–26.

Bleuler 1987: Paul Eugen Bleuler, *Dementia Praecox or the Group of Schizophrenias* (transl. Joseph Zinkin). New York, NY: International Universities Press, 1987.

Blum 2002: Carol Blum, *Strength in Numbers: Population, Reproduction, and Power in Eighteenth-Century France*. Baltimore, MD: Johns Hopkins University Press, 2002.

Blumenfeld-Kosinski 1991: Renate Blumenfeld-Kosinski, *Not of Woman Born: Representations of Caesarean Birth in Medieval and Renaissance Culture*. Ithaca, NY: Cornell University Press, 1991.

Bobbio 1955: Norberto Bobbio, *Politica e cultura*. Turin: Einaudi, 1955.

Bobbio 2010: Marco Bobbio, *Il malato immaginato. I rischi di una medicina senza limiti*. Turin: Einaudi, 2010.

Bocchi and Ceruti 2007: Gianluca Bocchi and Mauro Ceruti, eds., *La sfida della complessità*. Milan: Bruno Mondadori, 2007.
Bodei 1979a: Remo Bodei, "Comprendere, modificarsi. Modelli e prospettive di razionalità trasformatrice," in Aldo Gargani, ed., *Crisi della ragione*. Turin: Einaudi, 1979, pp. 197–240.
Bodei 1979b: Remo Bodei, "Filosofia," in *La cultura del 900. Filosofia – linguistica e semiotica – storiografia*. pp. 15–232. Milan: Gulliver, 1979, pp. 15–232.
Bodei 1991: Remo Bodei, *Geometria delle passioni. Paura, speranza e felicità: filosofia e uso politico*, Collana *Campi del sapere*. Milan: Feltrinelli, 1991.
Bodei 1998: Remo Bodei, *Il noi diviso. Ethos e idee nell'Italia repubblicana*. Turin: Einaudi, 1998.
Bodei 2000: Remo Bodei, *Le logiche del delirio. Ragione, affetti, follia*. Bari-Rome: Laterza, 2000.
Bodei 2002: Remo Bodei, *Destini personali. L'età della colonizzazione delle coscienze*. Milan: Feltrinelli, 2002.
Bodei 2005: Remo Bodei, *Una scintilla di fuoco. Invito alla filosofia*. Bologna: Zanichelli, 2005.
Bodei 2006: Remo Bodei, *Piramidi di tempo. Storie e teoria del déjà vu*. Bologna: Il Mulino, 2006, pp. 61–62.
Bodei 2008: Remo Bodei, *Paesaggi sublimi. Gli uomini davanti alla natura selvaggia. Milan:* Bompiani, 2008.
Bodei 2009: Remo Bodei, *La vita delle* cose. Rome-Bari: Laterza, 2009.
Bodei 2013: Remo Bodei, *Immaginare altre vite. Realtà, progetti*. Milan: Feltrinelli, 2013.
Bodei 2014a: Remo Bodei, *La civetta e la talpa. Sistema ed epoca in Hegel*. Bologna: Il Mulino, 2014.
Bodei 2014b: Remo Bodei, *Generazioni. Età della vita, età delle cose*. Rome-Bari: Laterza, 2014.
Bodei 2015a: Remo Bodei, *La filosofia del Novecento (e oltre)*. Milan: Feltrinelli, 2015.
Bodei 2015b: Remo Bodei, *Ordo amoris. Conflits terrestres et bonheurs célestes*. Paris: Les Belles Lettres, 2015.
Bodei 2016a: Remo Bodei, *Limite, Parole Controtempo*. Bologna: Il Mulino, 2016.
Bodei 2016b: Remo Bodei, *Scomposizioni. Forme dell'individuo moderno*. Bologna: Il Mulino, 2016.
Bodei 2017a: Remo Bodei, *Le forme del bello*. Bologna: Il Mulino, 2017.
Bodei, Giorello, Marzano and Veca 2017b: Remo Bodei, Giulio Giorello, Michela Marzano, and Salvatore Veca, *Le virtù cardinali. Prudenza, Temperanza, Fortezza, Giustizia*. Bari-Rome: Laterza, 2017.
Bodei 2019: Remo Bodei, *Dominio e sottomissione. Schiavi, animali, macchine, Intelligenza Artificiale*. Bologna: Il Mulino, 2019.
Bollas 2016: Christopher Bollas, *When the Sun Bursts: The Enigma of Schizophrenia*. New Haven, CT: Yale University Press, 2016.
Bolzoni 2002: Lina Bolzoni, *La rete delle immagini. Predicazioni in volgare dalle origini a Bernardino da* Siena. Turin: Einaudi, 2002.
Bonnar 1865: G. L. Bonnar, "A Critical Inquiry Regarding Superfoetation: With Cases," *Edinburgh Medical Journal* 10 (7) (1865), pp. 582–598.
Borgna 2017a: Eugenio Borgna, *L'ascolto gentile*. Turin: Einaudi 2017.
Borgna 2017b: Eugenio Borgna, *Le parole che ci salvano*. Turin: Einaudi, 2017.
Borgna 2020: Eugenio Borgna, *Speranza e disperazione*. Turin: Einaudi, 2020.
Botallo 1981: Leonardo Botallo, *I doveri del medico e del malato*, Leonardo Caretri and Anita Bogetti Fassone, eds. Turin: Strenna UTET, 1981.
Bourgery and Jacob 2005: Jean Baptiste Marc Bourgery and Nicholas Henri Jacob, *Atlas of Human Anatomy and Surgery (1830–1850)*, Jean-Marie Le Minor and Henry Sick, eds. Los Angeles, CA: Taschen, 2005.
Bowen 2006: J. Bowen, "Educational Strategies to Promote Clinical Diagnostic Reasoning," *New England Journal of Medicine* 355 (21) (2006), pp. 2217–2223.

Brachter 2005: Mirja Martha Brachter, *Ibn al-Jazzar's Zad al-Musafir and Costantinus Africanus' Latin Version Viaticum Peregrinantis*: *A Comparative Study* (Master's thesis, International Islamic University, Malaysia, 2005). Selangor: International Islamic University, 2005.
Bravo Guerreira 1977: Maria C. Bravo Guerreira: "La Muerte de Huayna Capac. 1530: Precisiones Cronologicas," *Revista de Indias*, 37 (1977), pp. 1–22.
Briere and Scott 2015: John Briere and Catherine Scott, *Principles of Trauma Therapy. A Guide to Symptoms, Evaluation, and Treatment*. Los Angeles, CA: Sage, 2015.
Brody 1987: Howard Brody, *Stories of Sickness*. New Haven: Yale University Press, 1987.
Brooks 1993: Francis Brooks, "Revising the Conquest of Mexico: Smallpox, Sources and Populations," *Journal of Interdisciplinary History* 24 (1) (1993), pp. 1–29.
Brown 1988: Peter Brown, *The Body and Society: Men, Women and Sexual Renunciation in Early Christianity*. New York: Columbia University Press, 1988.
Brown and Pollitt 1996: J. Larry Brown and Ernesto Pollitt, "Malnutrition, Poverty and Intellectual Development," *Scientific American*, vol. 274, no. 2 (February 1996), pp. 38–43.
Brown 1981: Elizabeth A. R. Brown, "Death and the Human Body in the Later Middle Ages: The Legislation of Boniface VIII and the Division of the Corpse," *Viator* 12 (1981), pp. 221–270.
Brown 1966: Roy Brown. "Organ Weight in Malnutrition with Special Reference to Brain Weight," *Developmental Medicine and Child Neurology* 8 (1966), pp. 512–522.
Bruner 1990: Jerome Bruner, *Acts of Meaning*. Cambridge, MA: Harvard University Press, 1990.
Bruner 2002: Jerome Bruner, *Making Stories: Law, Literature, Life*. New York, NY: Farrar Straus & Giroux, 2002.
Bucci 2006: Roberto Bucci, ed., *Manuale di medical humanities*. Rome: Zadigroma, 2006.
Burton 2007: June K. Burton, *Napoleon and the Woman Question: Discourses of the Other Sex in French Education, Medicine, and Medical* Law *1799–1815*. Lubbock, TX: Texas Tech University Press, 2007.
Cabello Balboa 1576–1586: Miguel Cabello Balboa, *Historia del Peru bajo la dominacion de los Incas. Edicion con anotaciones y concordancias con las cronicas de Indias por Horacio H. Urtega y biografia de Cabello Balboa por Carlos A. Romero*, original 1576–1586. Lima: Imprenta y Libreria San Martin, 1920.
Caceres, Rodriguez-Taifur, Bravo, and Maguiña 1996: Hector Caceres, Juan Rodriguez-Tafur, Francisco Bravo, Ciro Maguiña, "Apuntes epidemiologicos de la enfermedad de Carrion," *Boletin Societad Peruana Dermatologia* 17 (1996), pp. 5–9.
Cagnetta 1998: François Cagnetta, "Gaetano Giulio Zummo," in Paolo Giansiracusa, ed., *Gaetano Giulio Zumbo*. Milan: Fabbri, 1998, pp. 61–69.
Camera Apostolica, *Introitus et exitus*, annuum 1468, in Archivio Segreto Vaticano, regesto 471, f. 137v.
Canalis and Ciavolella 2018: Rinaldo F. Canalis and Massimo Ciavolella, eds, *Andreas Vesalius and the Fabrica in the Age of Printing. Art, Anatomy, and Printing in the Italian Renaissance* (Cursor Mundi 33). Turnhout: Brepols, 2018.
Cangiamila 1751: Francesco Emanuele Cangiamila, *Embriologia Sacra, ovvero dell'Uffizio de' Sacerdoti, Medici, e Superiori, circa l'Eterna Salute de' Bambini racchiusi nell'Utero*. Milan: Giuseppe Cairoli, 1751.
Canguilhem 1991: Georges Canguilhem, *Le normal et le pathologique*. Paris: Presses Universitaires de France, 1966.
Engl. trans. Carolyn R. Fawcwtt, *The Normal and the Pathological*. New York, NY. Zone Books, 1991.
Canguilhem 2008: Georges Canguilhelm, *La connaissance de la vie*. Paris: Vrin, 1952.

Engl. transl. Stephanos Geroulanos and Daniela Ginsburg, *Knowledge of Life*. New York, NY: Fordham University Press, 2008)

Capecelatro 1817: Giuseppe Capecelatro, *Lettera pastorale o sia istruzione canonica sul battesimo cristiano*. Naples: Porcelli, 1817.

Cardoso and Gomes 2008: Hugo Cardoso and J. E. A. Gomes, "Trends in adult Stature of Peoples Who Inhabited the Modern Portuguese Territory from the Mesolithic to the Late 20th Century," *International Journal of Osteoarchaeology* 19 (6) (2008), pp. 711–725.

Carlin 2005: Claire Carlin. ed., *Imagining Contagion in Early Modern Europe*. Basingstoke, NH: Palgrave Macmillan, 2005.

Carlino 1999: Andrea Carlino, *Books of the Body: Anatomical Ritual and Renaissance Learning*, transl. Anne C. Tedeschi. Chicago, IL: The University of Chicago Press, 1999.

Carrier 2002: David Carrier, "Why Were There No Public Art Museums in Renaissance Italy?," *Notes in the History of Art* 22 (1) (2002), pp. 44–50.

Carruthers 2008: Mary Carruthers, *The Book of Memory: A Study of Memory in Medieval Culture*. Cambridge: Cambridge University Press, 2008.

Caruth 2016: Cathy Caruth, *Unclaimed Experience. Trauma, Narrative, and History*. Baltimore, MD: Johns Hopkins Press, 2016.

Carver 1996: Raymond Carver, 'What the Doctor Said', in *All of Us: Collected Poems*. London: Harvill Press, 1996, p. 113.

Casper 1998: Monica J. Casper, *The Making of the Unborn Patient: A Social Anatomy of Fetal Surgery*. New Brunswick, NJ: Rutgers University Press, 1998.

Cassidy, Werner, Rourke, Zubernis, and Balaraman 2003: Kimberly Wright Cassidy, Rebecca Stetson Werner, Mary Rourke, Lynn S. Zubernis, and Geetha Balaraman, "The Relationship between Psychological Understanding and Positive Social Behavior," *Social Development* 12 (2) (2003), pp. 198–221.

Castaldi 1947: Luigi Castaldi, *Francesco Boi 1767–1860, primo cattedratico di Anatomia Umana a Cagliari e le cere anatomiche fiorentine di Clemente Susini*. Florence: Leo S. Olschki, 1947.

Cavazza 1982: Silvano Cavazza, "La doppia morte: Resurrezione e battesimo in un rito del Seicento," *Quaderni Storici* 17 (2) (1982), pp. 551–82.

Céard 1996: Jean Céard, *La nature et les prodiges: l'insolite au XVIe siècle*. Geneva: Droz, 1996.

Ceccio 1978: Joseph Ceccio, ed., *Medicine in Literature*. New York, NY: Longman, 1978.

Centers for Disease Control and Prevention (CDC) 2018: Centers for Disease Control and Prevention (CDC), "History of Smallpox." www.cdc.gov/smallpox/history/history.html Accessed October 15 2018.

Ceronetti 1993: Guido Ceronetti, *The Silence of the Body*. New York, NY: Farrar Strauss Giroux, 1993.

Cerron-Palomino 2010: Rodolfo Cerron-Palomino, "El contacto inicial quechua-castellano: la conquista del Peru en dos palabras," *Lexis* 34 (2) (2010), pp. 369–381.

Cerron-Palomino 2015: Rodolfo Cerron-Palomino, "Betanzos como quechuista," in Francisco Hernandez and Rodolfo Palomino, eds. *Juan de Betanzos y el Tawantinsuyo. Nueva Edicion de la Suma y Narracion del Peru*. Lima: Fondo Editorial Pontificia Universidad Catolica del Peru, 2015, pp. 29–39.

Chamberland 2009: Celeste Catherine Chamberland, "Honor, Brotherhood, and the Corporate Ethos of London's Barber-Surgeons' Company, 1570–1640," *Journal of the History of Medicine and Allied Sciences* 64 (3) (2009), pp. 300–332.

Charon 2001: Rita Charon, "Narrative Medicine: A Model for Empathy, Reflection, Profession, and Trust," *Journal of the American Medical Association* 286 (2001), pp. 1897–1902.

Charon 2006a: Rita Charon, *Honoring the Stories of Illness*. Oxford: Oxford University Press, 2006.

Charon 2006b: Rita Charon, *Narrative Medicine*. Oxford: Oxford University Press, 2006.

Charon 2017: Rita Charon et al., *The Principles and Practice of Narrative Medicine*. Oxford: Oxford University Press, 2017.

Charon and Montello 2002: Rita Charon and Martha Montello, eds, *Stories Matter: The Role of Narrative in Medical Ethics and Psychoanalysis and Narrative Medicine*. New York, NY: Routledge, 2002.

Cheng 2012: Sandra Cheng, "The Cult of the Monstrous: Caricature, Physiognomy, and Monsters in Early Modern Italy," *Preternature: Critical and Historical Studies on the Preternatural* 1 (2) (2012), pp. 197–231.

Chilvers 2003: Ian Chilvers, *Oxford Concise Dictionary of Art & Artists*. Oxford: Oxford University Press, 2003, pp. 197, 204.

Chiozzi 1817: "Chiozzi Fratelli e Sorelle c. Casazza Carlo ed Antonio," *Giurisprudenza Pratica secondo la Legislazione Austriaca attivata nel Regno Lombardo-Veneto ossia Collezione di Decisioni Sentenze e Decreti in Materia* Civile, *Commerciale, Criminale e di Diritto Pubblico*, 4 (1817), pp. 265–283.

Chomsky 1969: Noam Chomsky, *American Power and the New Mandarins*. New York, NY: Pantheon, 1969.

Christoffersen-Deb 2012: Astrid Christoffersen-Deb, "Viability: A Cultural Calculus of Personhood at the Beginnings of Life," *Medical Anthropology Quarterly* 26 (4) (2012), pp. 575–94.

Christopoulos 2021: John Christopoulos, *Abortion in Early Modern Italy*. Cambridge, MA: Harvard University Press, 2021.

Ciancio 2017: Cristina Ciancio, *Il momento della morte come evento giuridico: Definire, tutelare, gestire fra Ottocento e primo Novecento*. Bologna: University of Bologna Press, 2017.

Ciavolella 1976: Massimo Ciavolella, *La 'malattia d'amore' dall'Antichità al Medioevo*. Rome: Bulzoni, 1976.

Clark and Crawford 1994: Michael P. Clark and Catherine Crawford, eds, *Legal Medicine in History*. Cambridge: Cambridge University Press, 1994.

Clarke and Dewhurst 1972: Edwin Clarke and Kenneth Dewhurst, *An Illustrated History of Brain Function*. Berkeley, CA: University of California Press, 1972.

Clover, Sanford, and Butterwick 2013: Darlene E. Clover, Kathy Sanford, and Shauna Butterwick, eds, *Aesthetic Practices and Adult Education*. London: Routledge, 2013.

Cobo 2018: Bernabé Cobo, *Historia del Nuevo Mundo*. New York, NY: Wentworth Press, 2018.

Conforti 2008: Maria Conforti, "Chirurghi, mammane, ciarlatani. Pratica medica e controllo delle professioni," in Antonio Clericuzio and Germana Ernst, eds, *Il Rinascimento italiano e l'Europa*. vol. 5: *Le scienze*. Treviso: Angelo Colla Editore, 2008, pp.323–340.

Conlon 2013: Joseph M. Conlon, "The Historical Impact of Epidemic Typhus," 2013. https://www.montana.edu/historybug/documents/TYPHUS-Conlon.pdf Accessed March 5, 2022.

Contarino 1569: Luigi Contarino, *La nobiltà di Napoli in Dialogo*. Naples: Giuseppe Cacchii, 1569.

Corradi 1877: Alfonso Corradi, *Dell'ostetricia in Italia dalla metà dello scorso secolo fino al presente*. Pisa: Gamberini e Parmeggiani, 1877.

Costa 1987: Alessandro Costa, *Binswanger. Il mondo come* Progetto. Rome: Studium, 1987.

Costello 1956: Harry T. Costello, "Recollections of Royce's Seminar on Comparative Methodology," *The Journal of Philosophy* Feb 2, 1956, Volume 53, Issue 3 (February 1956), pp. 72–76.

Couliano 1984: Ioan P. Couliano, *Eros et magie à la Renaissance*. Paris: Flammarion, 1984. Engl. transl. Margaret Cook, *Eros and Magic in the Renaissance*. Chicago, IL: Chicago University Press, 1987).

Couser 2004: Thomas G. Couser, *Vulnerable Subjects: Ethics and Life Writing*. Ithaca, NY: Cornell University Press, 2004.

Crawford 2015: Julie Crawford, *Marvelous Protestantism: Monstrous Births in Post-Reformation England*. Baltimore, MD: Johns Hopkins University Press, 2005.

Crisciani 2005: Chiara Crisciani, "Histories, Stories, *Exempla*, and Anecdotes: Michele Savonarola from Latin to Vernacular," in Gianna Pomata and Nancy Siraisi, eds, *Historia: Empiricism and Erudition in Early Modern Europe*. Cambridge, MA: MIT Press, 2005, pp. 297–324.
Croskerry 2009: Pat Croskerry, "Clinical Cognition and Diagnostic Error: Applications of a Dual Process Model for Reasoning," *Advances in Health Sciences Education* 14 (2009), pp. 27–35.
Cusumano 2012: Nicola Cusumano, "Ricerche sulla teratologia in Sicilia (Secoli XVI-XVIII)," *Studi Storici* 53 (2012), pp. 855–881.
dal Prete 2015: Ivano dal Prete, "Cultures and Politics of Preformationism in Eighteenth-Century Italy," in Raymond Stephanson and Darren N. Wagner, eds, *The Secrets of Generation: Reproduction in the Long Eighteenth Century*. Toronto: University of Toronto Press, 2015, pp. 59–78.
Dalton, Allison, and Pezia 1976: Harry P. Dalton, Marvin J. Allison, and Alejandro Pezia, "The Documentation of Communicable Diseases in Peruvian Mummies," *Medical College of Virginia Quarterly* 12 (1976), pp. 43–48.
Damasio 1994: Antonio Damasio, *Descartes' Error: Emotion, Reason, and the Human Brain*. New York, NY: Avon Books, 1994.
Damasio 1999: Antonio Damasio, *The Feeling of What Happens: Body and Emotion in the Making of Consciousness*. New York, NY: Harcourt Brace, 1999.
Dannenfeldt 1985: Karl Dannenfeldt, "Egyptian Mumia: The Sixteenth Century Experience and Debate," *Sixteenth Century Journal* 16 (2) (1985), pp. 163–180.
Daston 2015: Lorraine Daston, "Epistemic Images," in Alina Payne, ed., *Vision and its Instruments*. University Park, PA: Pennsylvania State University Press, 2015, pp.13–55.
Daston and Galison 2007: Lorraine Daston and Peter Galison, *Objectivity*. New York, NY: Zone Books, 2007.
Daston and Lubeck 2011: Lorrain Daston and Elizabeth Lubeck, eds, *Histories of Scientific Observation*. Chicago, IL: The University of Chicago Press, 2011.
Daston and Park 2001: Lorraine Daston and Katharine Park, *Wonders and the Order of Nature: 1150–1750*. New York, NY: Zone Books, 1998 (second edition 2008).
Davies 2011: Brian Davies, *Thomas Aquinas on God and Evil*. New York, NY: Oxford University Press, 2011.
Davis 2019: Dana-Ain Davis, *Reproductive Injustice: Racism, Pregnancy, and Premature Birth*. New York, NY: New York University Press, 2019.
Daza and Barrera 2001: Luis B. Daza and Juan Barrera, "Historia y leyenda acerca de los Incas enterrados en el hospital San Andres de Lima," *Revista de Neuropsiquiatria* 64 (2001), pp. 18–35.
De Carolis 2001: Stefano De Carolis, "Le malattie dei Malatesta," *Bollettino dell'Ordine dei Medici Chirurghi e degli Odontoiatri della Provincia di Rimini* 2 (2) (2001), pp. 39–42.
De Carolis 2002: Stefano De Carolis, "Le ricognizioni dei resti mortali di Sigismondo Pandolfo Malatesta: un'anamnesi retrospettiva," *Penelope. Arte Storia e Archeologia* 1 (2002), pp. 29–39.
De Carolis and Galassi 2017: Stefano De Carolis and Francesco M. Galassi, "Le Malattie di Sigismondo," in Manlio Masini, ed., *Sigismondo Pandolfo Malatesta. Signore di Rimini*. Rimini: Panozzo Editore, 2017, pp. 79–82.
de Gutiérrez-Mahoney and Schechter 1972: Carlos Guillermo de Gutiérrez-Mahoney and Mannie M. Schechter, "The Myth of the *rete mirabile* in Man," *Neuroradiology* 4 (1972), pp. 141–158.
De Moulin 1988: Daniel De Moulin, *A History of Surgery. With Emphasis on the Netherlands*. Dordrecht: Martinus Nijhoff Publishers, 1988
de Renzi 1853: Salvatore de Renzi, *Collectio Salernitana*, vol. 2. Naples: Filiatre-Sebezio, 1853.

Del Soldato 2017: Eva Del Soldato, "Saving the Philosopher's Soul: The *De pietati Aristotelis* by Fortunio Liceti," *Journal of the History of Ideas* 78 (4) (2017), pp. 531–547.

Dewey 1996: Larry A. Hickman, ed., *John Dewey, The Collected Works of John Dewey, 1882–1953: The Electronic Edition*. Charlottesville, VA: InteLex Corp., 1996.

Dickinson 1961: Emily Dickinson, *The Complete Poems*. Boston, MA, New York, NY, and London: Little-Brown and Company, 1961.

Diethelm 1971: Oskar Diethelm, *Dissertations of Psychiatric Interest*. Basel: Karger, 1971.

Di Pietro 1971: Pericle Di Pietro, "Contributo alla biografia di Berengario da Carpi," *Deputazione di Storia Patria delle Antiche Province Modenesi* 10 (6) (1971), pp. 34–44.

Dingwall 1995: Helen M. Dingwall, *Physicians, Surgeons and Apothecaries: Medicine in Seventeenth Century Edinburgh*. East Linton: Tuckwell Press, 1995.

Dominguez Favra 2008: Nicanor Dominguez Favra, "Betanzos y los Quipucamayos en la epoca de Vaca de Castro (Cuzco 1543)," *Revista Andina* 46 (2008), pp. 155–192.

Dominguez Favra 2010: Nicanor Dominguez Favra, "El Licenciado Polo: Fuente del discurso sobre la descendencia y gobierno de los Ingas (1602–1608)," *Revista Historica* (PUCP) 34 (2010), 131–144.

Dostoevsky 1990: Fyodor Dostoevsky, *The Brothers Karamazov: A Novel in Four Parts with Epilogue*, transl. Richard Pevear and Larissa Volokhonsky. New York, NY: Farrar, Strauss and Giroux, 1990.

Drusini and Rippa Bonati 2006: Andrea Drusini and Maurizio Rippa Bonati, "Le fragili ossa del poeta: Lo studio antropologico dei resti del Petrarca," in Valeria Finucci, ed., *Petrarca, canoni, esemplarità*. Rome: Bulzoni, 2006, pp. 327–346.

Dubow 2010: Sara Dubow, *Ourselves Unborn: A History of the Fetus in Modern America*. Oxford: Oxford University Press, 2010.

Duden 1993a: Barbara Duden, *Disembodying Women: Perspectives on Pregnancy and the Unborn*. Cambridge, MA: Harvard University Press, 1993.

Duden 1993b: Barbara Duden, *The Woman beneath the Skin: A Doctor's Patients in Eighteenth-Century Germany*. Cambridge, MA: Harvard University Press, 1998.

Dumbar-Temple 1940: Ella Dumbar-Temple, "La Descendencia de Huayna Capac," *Revista Historica* 13 (1940), pp. 31–77.

Dunn and Hope 2018: Michael Dunn and Tony Hope, *Medical Ethics*. A Very Short Introduction. Oxford: Oxford University Press, 2018.

Duran 1944: Carlos Martinez Duran, "Histopatologia de las lesiones del tifo exantematico-primera investigacion hecha en Guatemala," *Boletin de la Oficina Sanitaria Panamericana* 23 (9) (1944), pp. 791–795.

Duviols 1979: P. Duviols, "Datation, paternité et idéologie de la declaration de les Quipucamayos a Vaca de Castro." *Essais publié en homage à la mémoire de Marcel Bataillon (1895 – 1977)*. Paris: Foundation Singer-Polignac, 1979, pp. 583–591.

Edgerton 1985: Samuel Y. Edgerton Jr, *Pictures and Punishment: Art and Criminal Prosecution During the Florentine Renaissance*. Ithaca, NY: Cornell University Press, 1985.

Ekman and Friesen 1976: Paul Ekman and Wallace V. Friesen, *Pictures of Facial Affect*. Palo Alto, CA: Consulting Psychologists Press, 1976.

Ekman 1999: Paul Ekman, *Handbook of Cognition and Emotion*. Sussex: John Wiley & Sons, 1999.

Elias and **Norbert Elias 1982**: *Ueber die Einsamkeit der Sterbenden in unseren Tagen*. Frankfurt: Suhrkamp, 1982.

Elkins 1995: James Elkins, "Art History and Images That Are Not Art," *Art Bulletin* 77 (4) (1995), pp. 553–571.

Elm 2015: Michael Elm, Kobi Kabalek, and Julia B. Köhne, eds, *The Horrors of Trauma in Cinema: Violence Void Visualization*. Newcastle: Cambridge Scholars Publishing, 2015.

Engels 2009: Friedrich Engels, *The Condition of the Working Class in England*. Oxford: Oxford University Press, 2009.

Eurich 2008: S. Amanda Eurich, "Between the Living and the Dead: Preserving Confessional Identity and Community in Early Modern France," in Michael J. Halvorson and Karen J. Spierling, eds, *Defining Community in Early Modern Europe*. London and New York, NY: Routledge, 2008, pp. 43–61.

Fabbri 2017: Paolo Fabbri, *L'efficacia semiotica. Risposte e repliche*. Milan: Mimesis, 2017.

Fabbri 2021: Paolo Fabbri, *Biglietti d'invito per una semiotica marcata*, Gianfranco Marrone, ed., Milan: Bompiani, 2021.

Faietti and Scaglietti Kelescian 1995: Marzia Faietti and Daniela Scaglietti Kelescian, *Amico Aspertini*. Modena: Artioli, 1995.

Fanti 2001: Mario Fanti, *Confraternite e città a Bologna nel medioevo e in età moderna*. Rome: Herder, 2001.

Farrington 1995: Ian Farrington, "The mummy, palace and estate of Inca Huayna Capac at Quispeguanca Tawantinsuyu," *An International Journal of Inka Studies* 1 (1995), pp. 55–65.

Favaro 1928: Giuseppe Favaro, *Gabriele Falloppia modenese (1523–1562), Studio biografico*. Modena: Tip. Immacolata Concezione, 1928.

Fenner, Henderson, Arita, Jezek, Ladnyi, and Danilovihch 1988: Frank Fenner, Donald A. Henderson, Isao Arita, Zdenek Jezek, Ivan Ladnyi, and Ivan Danilovich, "Smallpox and Its Eradication," *World Health Organization*, 1988. https://apps.who.int/iris/handle/10665/39485.Accessed February 22, 2022.

Fernández de Navarrete 1964: Martín Fernández de Navarrete, *Colección de viajes y descubrimientos que hicieron por mar los españoles desde fines del siglo XV*. 3 vols. Madrid: Atlas, 1964.

Ferrari 1987: Giovanna Ferrari, "Public Anatomy Lessons and the Carnival: The Anatomy Theater of Bologna," *Past and Present* 117 (1987), pp. 50–106.

Filippini 1995: Nadia Maria Filippini, *La nascita straordinaria: Tra madre e figlio la rivoluzione del taglio cesareo (sec. XVIII-XIX)*. Milan: Franco Angeli, 1995.

Ferrari 1996: Giovanna Ferrari, *L'esperienza del passato: Alessandro Benedetti filologo e medico umanista*, in *Biblioteca di Nuncius, studi e testi XXII*. Florence: Leo S. Olschki, 1996.

Findlen 1990: Paula Findlen, "Jokes of Nature and Jokes of Knowledge: The Playfulness of Scientific Discourse in Early Modern Europe," *Renaissance Quarterly* 43 (1990), pp. 292–331.

Findlen 1996: Paula Findlen, *Possessing Nature: Museums, Collecting, and Scientific Culture in Early Modern Italy*. Berkeley, CA: University of California Press, 1996.

Findlen 2019: Paula Findlen, "The Museum's Renaissance Revisited: Histories, Objects, Exhibits," *I Tatti Studies in the Italian Renaissance* 22 (2) (2019), pp. 295–310.

Finucci 2001: Valeria Finucci, "Maternal Imagination and Monstrous Birth: Tasso's *Gerusalemme liberata*," in Valeria Finucci and Kevin Brownlee, eds, *Generation and Degeneration: Tropes of Reproduction in Literature and History from Antiquity through Early Modern Europe*. Durham, NC: Duke University Press, 2001, pp. 41–80.

Finucci 2003: Valeria Finucci, "The Masquerade of Manhood: The Paradox of the Castrato." In Valeria Finucci, *The Manly Masquerade: Masculinity, Paternity and Castration in the Italian Renaissance*. Durham, NC: Duke University Press, 2003, pp. 226–280.

Finucci 2015a: Valeria Finucci, *The Prince's Body: Vincenzo Gonzaga and Renaissance Medicine*. Cambridge, MA: Harvard University Press, 2015.

Finucci 2015b: Valeria Finucci, "Thinking through Death: The Politics of the Corpse," *Journal of Medieval and Early Modern Studies* 45 (1) (2015), pp. 1–6.

Finucci and Brownlee 2001: Valeria Finucci and Kevin Brownlee, eds, *Generation and Degeneration: Tropes of Reproduction in Literature and History from Antiquity through Early Modern Europe*. Durham, NC: Duke University Press, 2001.

Flavey 2008: Kathleen Falvey, "Scaffold and Stage: Comforting Rituals and Dramatic Traditions in Late Medieval and Renaissance Italy," in Nicholas Terpstra, ed., *The Art of Executing Well : Rituals of Execution in Renaissance Italy*. Kirksville, MO: Truman State University Press, 2008, pp. 13–30.

Flint 1984: Valerie I. J Flint, "Monsters and the Antipodes in the Early Middle Ages and Enlightenment," *Viator: Medieval and Renaissance Studies* 15 (1984), pp. 65–80.

Fodéré 1822: François-Emmanuel Fodéré, "Riflessioni del Professore Barovero sull'articolo *viabilité*, scritta dal Professore Fodéré, ed inserito nel vol. 57 del *Dictionnaire des Sciences Médicales*," *Repertorio Medico Chirurgo* 29 (1822), pp. 191–214.

Fontes da Costa 2004: Palmira Fontes da Costa, "The Medical Understanding of Monstrous Births at the Royal Society of London during the First Half of the Eighteenth Century," *History and Philosophy of the Life Sciences* 26 (2) (2004), pp. 157–75.

Foot 1967: Philippa Foot, "The Problem of Abortion and the Doctrine of the Double Effect," *Oxford Review* 5 (1967), pp. 5–15.

Forbes 1985: Thomas R. Forbes, *Surgeons at the Bailey: English Forensic Medicine to 1878*. New Haven, CT: Yale University Press, 1985.

Fornaciari 2000: Antonio Fornaciari, "L'esplorazione della tomba di Federico II da Montefeltro: notizie preliminari," *Archeologia Postmedievale* 4 (2000), pp. 211–218.

Fornaciari 2018: Antonio Fornaciari et al., "Gout in Duke Federico of Montefeltro (1422–1482), A New Pearl of the Italian Renaissance," *Clinical and Experimental Rheumatology* 36 (1) (2018), pp. 15–20.

Fortuna 2019: Stefania Fortuna, "Editions and Translations of Galen from 1490 to 1540," in Petros Bouras-Vallianatos and Barbara Zipser, eds, *Brill's Companion to the Reception of Galen*. Leiden: Brill, 2019, pp. 437–452.

Foscolo 1979: Ugo Foscolo, "Ai Sepolcri," in Mario Pazzaglia (ed), *Letteratura italiana. Testi e critica con lineamenti di storia letteraria*. Bologna: Zanichelli, 1979.

Foucault 1963: Michel Foucault, *Naissance de la clinique: une archéologie du regard médical*. Paris: PUF, 1963.

Foucault 1994a: Michel Foucault, *Poteri e strategie. L'assoggettamento dei corpi e l'elemento sfuggente,* Pierre Dalla Vigna, ed., Milan: Mimesis, 1994.

Foucault 1994b: Michel Foucault, *The Birth of the Clinic*, Alan Sheridan-Smith, transl. New York, NY: Vintage Books, 1994.

Foucault 2001: Michel Foucault, *Biopolitica e liberalismo: detti e scritti su potere ed etica, 1975–1984*, Ottavio Marzocca, ed., Milan: Medusa, 2001.

Foucault 2003: Michel Foucault, *Society Must be Defended: Lectures at the Collège de France, 1975–76*, transl. David Macey. New York, NY: Picador, 2003.

Foucault 2008: Michel Foucault, *Le gouvernement de soi et des autres: Cours au Collège de France (1982–1983)*, Frédéric Gros, François Ewald and Alessandro Fontana, eds., Paris: Gallimard; Seuil, 2008.

Foucault 2017: Michel Foucault, *Il filosofo militante. Archivio Foucault. Vol. 2: Interventi, colloqui, interviste. 1971–1977*, Alessandro Dal Lago, ed. Milan: Feltrinelli, 2017.

Gall 1825: France Joseph Gall, *Sur les fonctions du cerveau et sur celles de chacune de ses parties*. 6 vols. Paris: Baillière, 1825.

Frank 1807: Johann Peter Frank, *Sistema completo di polizia medica*, vols 2 and 3. Milan: Pirotta e Maspero, 1807.

Frank 2002: Joseph Frank, *Dostoevsky: The Mantle of the Prophet*, 1871–1881. Princeton, NJ: Princeton University Press, 2002.
Frank 1985: Arthur W. Frank, *The Wounded Storyteller*. Chicago, IL: Chicago University Press, 1985.
Frazer 1967: James G. Frazer, *The Golden Bough*. London: Macmillan, 1967 (abridged from the publication of 1922).
Freeman 2001: Walter J. Freeman, *How Brains Make Up Their Minds*. New York, NY: Columbia University Press, 2001.
French 1985: Roger K. French, "Berengario da Carpi and the Use of Commentary in Anatomical Teaching," in Roger K. French and Iain M. Lonie, eds., *The Medical Renaissance of the Sixteenth Century*. Cambridge: Cambridge University Press, 1985, pp. 42–74.
French 1993: Roger K. French, "The Medical Ethics of Gabriele de Zerbi," in Andrew Wear, Johanna Geyer-Kordesch, and Roger Kenneth French, eds., *Doctors and Ethics. The Earlier Historical Setting of Professional Ethics*. Amsterdam: Rodopi, 1993, pp. 72–97.
French 1999: Roger K. French, *Dissection and Vivisection in the European Renaissance, Renaissance Quarterly*, Vol 53, Issue 3. Aldershot: Ashgate, 1999.
Freud 1978: Sigmund Freud, *Civilization and its Discontents*, in *The Standard Edition of the Complete Psychological Works of Sigmund Freud*. London: Hogarth Press, 1953–1974 [1978 reprint], vol. 21 (1955), pp. 64–145.
Frischmann and Selinger 2019: Brett Frischmann and Evan Selinger, eds., *Re-Engineering Humanity*. Cambridge: Cambridge University Press, 2019.
Frith and Johnstone 2003: Chris Frith and Eve Johnstone. *Schizophrenia: A Very Short Introduction*. Oxford: Oxford University Press, 2003.
Fulcheri 2002: Ezio Fulcheri, "Fortunio Liceti: un punto di svolta negli studi sui mostri e l'inizio della moderna teratologia," *Pathologica* 94 (2002), pp. 263–268.
Gadebusch Bondio 2005: Mariacarla Gadebusch Bondio, "Leonardo Botallo, *De via sanguinis* – edizione, traduzione ed analisi di un testo riscoperto nel '600," *Medicina nei secoli* 17 (3) (2005), pp. 663–694.
Gadebusch Bondio 2012: Mariacarla Gadebusch Bondio, "Vom Ringen der Medizin um eine Fehlbarkeitskultur. Epistemologische und ethische Reflexionen." in Mariacarla Gadebusch Bondio and Agostino Paravicini Bagliani, eds, *Errors and Mistakes: A Cultural History of Fallibility* (Micrologus' Library 49). Florence: SISMEL-Edizioni del Galluzzo, 2012, pp. 291–311.
Gadebusch Bondio 2014: Mariacarla Gadebusch Bondio, ed., *Medical Ethics. Premodern Negotiations between Medicine and Philosophy* (Aurora. Schriften der Villa Vigoni 2). Stuttgart: Franz Steiner Verlag, 2014.
Gadebusch Bondio and Förg 2020: Mariacarla Gadebusch Bondio and Katharina-Louise Förg, "Als Arzt politisch handeln. Roderigo de Castros Medicus-politicus zwischen Anspruch, Ideal und Praxis," in Mariacarla Gadebusch Bondio, Manuel Förg, and Christian Kaiser, eds, *Menschennatur in Zeiten des Umbruchs – Das Ideal des politischen Arztes in der frühen Neuzeit*. Berlin: De Gruyter, 2020, pp. 83–114.
Gaedke 2017: Andrew Gaedke, *Modernism and the Machinery of Madness*. Cambridge: Cambridge University Press, 2017.
Galassi and Borghi 2015: Francesco M. Galassi and Claudio Borghi, "A Brief History of Uric Acid: From Gout to Cardiovascular Risk Factor," *European Journal of Internal Medicine* 26 (5) (2015), p. 373.
Galassi 2017: Francesco Maria Galassi, *Un mondo senza vaccini? La vera storia*. Rome: Gruppo CIV Edizioni, 2017.
Galassi 2018: Francesco M. Galassi et al., "Boccaccio e la paleopatologia," *Heliotropia* 15 (2018), pp. 267–280.

Galassi and Varotto 2021: Francesco M. Galassi and Elena Varotto, "The Alleged Shoes of Michelangelo Buonarroti: Anthropometrical Considerations," *Anthropologie* (Czech Republic) 59 (1) (2021), pp. 97–99.

Galen 1821–1833: Carl Gottlob Kühn, ed., *Claudii Galeni Opera omnia*. Leipzig: Knobloch, 1821–1833, 20 tomes in 22 volumes (reprint: Hildesheim: Olms, 2001).

Gallagher 1987: Janet Gallagher, "Prenatal Invasions and Interventions: What's Wrong with Fetal Rights," *Harvard Women's Law Journal* 10 (1987), pp. 9–58.

Gallerani 2012: Manuela Gallerani, *L'impegno lieve. Il razionalismo critico e l'ideale estetico*. Naples: Loffredo, 2012.

Gallerani 2015: Manuela Gallerani, *Prossimità inattuale. Un contributo alla filosofia dell'educazione problematicista*. Milan: Franco Angeli, 2015.

Gallerani 2016: Manuela Gallerani, "Narrative Medicine and Stories of Illness: Caring for the Sick Requires Active Listening," in Laura Formenti and Linden West, (eds.), *Stories that make a difference. Exploring the collective, social and political potential of narratives in adult education research*. Lecce: Pensa MultiMedia, 2016, pp. 341–47.

Gallerani 2020: Manuela Gallerani, "The Responsibility in Speaking of Care," in *Studium Educationis* 21 (3) (2020), pp. 19–31.

Gallerani 2021a: Manuela Gallerani, "Disease in Art and Art(ist) in Disease: Reflections on Paradigmatic Works by Leonardo da Vinci and Michelangelo Buonarroti," in Rinaldo F. Canalis and Massimo Ciavolella eds., *Disease and Disability in Medieval and Early Modern* Art *and Literature*, (Cursor Mundi 38). Turnhout: Brepols, 2021, pp. 229–261.

Gallerani 2021b: Manuela Gallerani, "Images, Thoughts and Words on Disease in the Works of Michelangelo," in Cristopher Craig, Enrico Fongaro, Raffaele Milani, and James Tink, eds., *Images, Philosophy, Communication: Aesthetics and Thought in Japan and the World*. Milan: Mimesis International, 2021, pp. 93–109.

Gallese 2003: Vittorio Gallese, "La molteplice natura delle relazioni interpersonali: ricerca di un comune meccanismo neurofisiologico," *Networks* 1 (2003), pp. 24–47.

Gallese and Goldman 1998: Vittorio Gallese and Alvin I. Goldman, "Mirror neurons and the simulation theory of mind-reading," *Trends in Cognitive Sciences* 12 (1998), pp. 493–501.

Gallese, Fadiga, Forgassi, Rizzolatti 1996: Vittorio Gallese, Luciano Fadiga, Leonardo Fogassi and Giacomo Rizzolatti, "Action recognition in the premotor cortex," *Brain* 119 (1996), pp. 593–609.

Gallese, Keysers, and Rizzolati 2004: Vittorio Gallese, Christian Keysers, and Giacomo Rizzolatti, "A unifying view of the basis of social Cognition," *Trends in Cognitive Sciences* 8 (2004), pp. 396–403.

Galluzzi 1998: Paolo Galluzzi, "The Sepulchers of Galileo: The 'Living' Remains of a Hero of Science," in Peter Machamer, ed., *The Cambridge Companion to Galileo*. Cambridge: Cambridge University Press, 1998, pp. 417–447.

Ganguilhem, 1924: Georges Ganguilhem, *Critique de la raison appliquee et mentale*. Paris: Flammarion, 1924.

Engl. transl.: Jorge Lazaref, *A Criticism of Mental and Applied Reason*. Berkeley, CA: World of Knowledge, 2013.

Garcia-Caceres and Garcia 1991: Uriel Garcia-Caceres and Fernando U. Garcia, "Bartonellosis. An Immunodepressive Disease and the Life of Daniel Alcides Carrion," *Pathology Patterns* v. 95 (4), Suppl. 1 (1991), pp. 58–66.

Garcia-Caceres 1998: Uriel Garcia-Caceres, "Historiografia de la enfermedad de Carrion. Analisis histografico de la iconografia de la brucellosis humana. 1a parte," *Folia Dermatologica del Peru* 9 (1998), pp. 47–54.

Garin 1974: Eugenio Garin, *Intellettuali del XX secolo*. Rome: Editori Riuniti, 1974.
Gelfand 1980: Toby Gelfand, *Professionalizing Modern Medicine: Paris Surgeons and Medical Science and Institutions in the 18th Century*. Westport, CT: Greenwood Press, 1980.
Genoways and Andrei 2009: Hugh Howard Genoways and Mary Ann Andrei, *Museum Origins: Readings in Early Museum History and Philosophy*. London: Routledge, 2009.
Gentilcore 2006: David Gentilcore, *Medical Charlatanism in Early Modern Italy*. Oxford: Oxford University Press, 2006.
Ghadessi 2013: Touba Ghadessi, "Lords and Monsters: Visible Emblems of Rule," *I Tatti Studies in the Italian Renaissance* 16 (1–2) (2013), pp. 491–523.
Giansiracusa 1991: Paolo Giansiracusa, ed., *Vanitas Vanitatum. Studi sulla ceroplastica di Gaetano e Giulio Zumbo*. Siracusa, Arnaldo Lombardi, 1991.
Giazotto 1988: Remo Giazotto, *Il grande viaggio di Pietro Della Valle, il "Pellegrino" (1612–26), La Turchia, la Persia, l'India con il ritorno a Roma*. Rome: Edizioni Torre d'Orfeo, 1988.
Giannecchini and Moggi-Cecchi 2008: Monica Giannecchini and Jacopo Moggi-Cecchi, "Stature in Archaeological Samples from Central Italy: Methodological Issues and Diachronic Changes," *American Journal of Physical Anthropology* 135 (2008), pp. 284–292.
Gilles-Chikhaoui 2010: Audrey Gilles-Chikhaoui, "Circulation des substances, circulation des discours, circulation des textes: l'importance du plaisir dans l'acte générationnel chez Ambroise Paré," *Les Chantiers de la Création* 3 (2010). https://journals.openedition.org/lcc/299?lang=frAccessed March 5 2022.
Gilligan 1982: Carol Gilligan, *In a Different Voice. Psychological Theory and Women's Development*. Cambridge, MA: Harvard University Press, 1982.
Ginsberg 1999: Bob Rosenthal, Peter Hale, and Bill Morgan, (eds.), *Allen Ginsberg, Death & Fame: Poems 1993–1997*. New York, NY: Harper Collins, 1999, pp. 98–99.
Ginsberg 2006: Allen Ginsberg, *Collected Poems, 1947–1997*. New York, NY: Harper Collins, 2006, p. 1130.
Ginzburg 1986: Carlo Ginzburg, *Miti emblemi spie. Morfologia e storia*. Turin: Einaudi, 1986.
Gnudi Teach and Webster 1950: Marta Gnudi Teach and Jerome Pierce Webster, *The Life and Times of Gaspare Tagliacozzi, Surgeon of Bologna, 1545–1599*. New York, NY: H. Reichner, 1950.
Goffman 1963: Ervin Goffman, *Stigma: Notes on the management of Spoiled Identify*. Englewood Cliffs, NJ: Prentice-Hall, 1963
Goffman 1986: Ervin Goffman, *Frame Analysis: An Essay on the Organization of Experience*. Boston, MA: Northeastern University Press 1986.
Goleman 2006: Daniel Goleman, *Emotional Intelligence*. New York, NY: Bantam Books, 2006.
Goleman 2007: Daniel Goleman, *Social Intelligence: The New Science of Human Relationships*. London: Random, 2007.
Gonzales, Maguiña, Heras, Conde-Salazar 2007: R. Cesar Gonzales, V. Ciro Maguiña, M. Felipe Heras, and Luis Conde-Salazar, "Bartonellosis: Fiebre de La Oroya o Verruga Peruana. Enfermedad Ocupacional?," *Medicina y Seguridad del Trabajo* 53 (2007), pp. 35–41.
Good 1994: Byron J. Good, *Medicine, Rationality, and Experience: An Anthropological Perspective*. Cambridge: Cambridge University Press, 1994.
Gooday 2008: Graeme Gooday, "Placing or Replacing the Laboratory in the History of Science?" *Isis* 99 (4) (2008), pp. 783–795.
Goodwin 2020: Michelle Goodwin, *Policing the Womb: Invisible Women and the Criminalization of Motherhood*. Cambridge: Cambridge University Press, 2020.
Gordon White 1991: David Gordon White, *Myths of the Dog Man*. Chicago, IL: University of Chicago Press, 1991.
Gottschall 2012: Jonathan Gottschall, *The Storytelling Animal: How Stories Make Us Human*. Boston, MA: Mariner Books, 2012.

GOV.UK 2016: Public Health England, "PHE reminds Londoners to practice safe sex as rates of syphilis are 3 times higher in the capital than anywhere else in England. www.gov.uk/government/news/phe-publishes-report-on-syphilisAccessed October 15 2018.
Grafton 1979: Anthony Grafton, "Rhetoric, Philology, and Egyptomania in the 1570's: J.J. Scaliger's Invective against M. Guilandinus's Papyrus," *Journal of the Warburg and Courtauld Institutes* 42 (1979), pp. 167–194.
Grafton 1996: Anthony Grafton, *The New Science and the Tradition of Humanism*, in Jill Craye, ed., *The Cambridge Companion to Renaissance Humanism*. Cambridge: University Press, 1996.
Grafton, Shelford, and Siraisi 1992: Anthony Grafton, April Shelford, and Nancy Siraisi, *New Worlds, Ancient Texts: The Power of Tradition and the Shock of Discovery*. Cambridge, MA: Harvard University Press, 1992.
Gramsci 1949: Antonio Gramsci, *Gli intellettuali e l'organizzazione della cultura*. Turin: Einaudi, 1949.
Green 2002: Christopher D. Green, "Where Did the Ventricular Localization of Mental Faculties Come From?," *Journal of History of the Behavioral Sciences* 39 (2) (2002), 131–142.
Greenblatt 1991: Stephen Greenblatt, *Marvelous Possessions: The Marvels of the New World*. Chicago, IL: University of Chicago Press, 1991.
Greenhalgh 1999: Trisha Greenhalgh, "Narrative Based Medicine in an Evidence-Based World," *British Medical Journal* January 1999, pp. 318–23.
Grendler 1999: Paul Grendler, "The University of Bologna, the City, and the Papacy," *Renaissance Studies* 13 (4) (1999), pp. 475–485.
Guerra 1988: Francisco Guerra, "The Earliest American Epidemic," *Social Science History* 12 (3) (1988), pp. 305–325.
Guerrini 2012: Anita Guerrini, "The Value of a Dead Body," in Helen Deutsch and Mary Terrall, eds, *Vital Matters*. Toronto: University of Toronto Press, 2012, pp. 246–264.
Guerrini 2015: Anita Guerrini, *The Courtiers' Anatomists: Animals and Humans in Louis XIV's Paris*. Chicago, IL: University of Chicago Press, 2015.
Guillen 1983: Edmundo Guillen, "El enigma de las momias Inca," *Boletin de Lima* 28 (1983), pp. 29–42.
Guy 1844: William A. Guy, *Principles of Forensic Medicine*. London: Renshaw, 1844.
Halpern 2004: Susan P. Halpern, *The Etiquette of Illness: What to say when you can't find the word*. London: Bloomsbury, 2004.
Hampe 1982: Teodoro Hampe, "Las Momias de los Incas en Lima," *Revista del Museo Nacional* 46 (1982), pp. 405–418.
Han 2018: Byung-Chul Han, *The Expulsion of the Other: Society, Perception and Communication Today*. Cambridge: Polity Press, 2018.
Hanke 1959: Lewis Hanke, *Aristotle and the American Indians: A Study of Race Prejudice in the Modern World*. Chicago, IL: Henry Regnery Company, 1959.
Harari 2017: Yuval Noam Harari, *Homo Deus: A Brief History of Tomorrow*. New York, NY: Harper Collins, 2017.
Harper, Zuckerman, Kingston, and Armelagos 2011: Kristin N. Harper, Molly K. Zuckerman, Megan L. Harper, John D. Kingston, and George J. Armelagos, "The Origin and Antiquity of Syphilis Revisited: An Appraisal of Old World and Pre-Columbian Evidence for Treponemal Infection" *American Journal of Physical Anthropology*, Supplement 143 (53) (2011), pp. 99–133.
Harrington 2019: Ann Harrington, *Mind Fixers: Psychiatry's Troubled Search for the Biology of Mental Illness*. New York, NY: Norton 2019.
Harrison 2003: Robert Pogue Harrison, *The Dominion of the Dead*. Chicago, IL: University of Chicago Press, 2003.

Heckers 2000: Stephan Heckers, "Neural Models of Schizophrenia," *Dialogues in Clinical Neuroscience* 2 (3) (2000), pp. 267–79.
Heidegger 2009: Martin Heidegger, *History of the Concept of Time: Prolegomena*. Bloomington, IN: Indiana University Press 2009.
Hempel 1996: Carl Hempel, *Philosophy of Natural Science* Englewood Cliffs: Prentice Hall, 1996.
Henrich 2020: Joseph Henrich, *The Weirdest People in the World: How the West Became Psychologically Peculiar and Particularly Prosperous*. London: Allen Lane, Penguin Random House, 2020.
Hernandez 1993: Max Hernandez, *Memoria del Bien Perdido: Conflicto, Identidad y Nostalgia en la Obra del Inca Garcilaso de la Vega*. Lima: Instituto de Estudios Peruanos, 1993.
Hernandez 2016: Max Hernandez, "The incomplete journey of Inca Garcilaso de la Vega," *El Correo del Peru*, 28 (2016), pp. 1–7.
Higden 1865–1886: Churchill Babington, ed., *Ranulph Higden, Polychronicon Ranulphi Higden monachi cestrensis, together with English translation of John Trevisa and an Unknown Writer of the Fifteenth Century*, 9 vols. London: Longman & Co., 1865–1886.
Hippocrates: *Ancient medicine*, in Hippocrates, *Works*, William H. S. Jones. ed. and transl. Cambridge: Harvard University Press, 2004.
Hirai 2002: Hiroshi Hirai, "Concepts of Seeds and Nature in the Work of Marsilio Ficino," in Michael J.B. Allen and Valery Rees, eds, *Marsilio* Ficino: *His Theology, His Philosophy, His Legacy*, Leiden: Brill, 2002, pp.257–284.
Hirai 2006: Hiro Hirai, "Âme de la terre: Génération spontanée et origine de la vie: Fortunio Liceti critique de Marsile Ficin," *Bruniana & Campanelliana* 12.2 (2006), p. 451–469.
Hirai 2011: Hiroshi Hirai, *Medical Humanism and Natural Philosophy* (History of Science and Medicine Library, 26; Medieval and Early Modern Science, 17). Leiden: Brill, 2011.
Holmes 2014: Frederick Holmes: "Typhus on the Eastern Front," In *Medicine in the First World War," University of Kansas Magazine* (2014) https://www.kumc.edu/school-of-medicine/academics/departments/history-and-philosophy-of-medicine/archives/wwi/essays/medicine/typhus-on-the-eastern-front.html/ Accessed March 7, 2022.
Howard and Goldschmidt 2018: Jacqueline Howard and Debra Goldschmidt, "US measles outbreak is largest since disease was declared eliminated in 2000," 25 April 2019. www.cnn.comhttps://edition.cnn.com/2019/04/24/health/measles-outbreak-record-us-bn/index.html. Accessed October 15 October 2018.
Huet 1993: Marie Hélène Huet, *Monstrous Imagination*. Cambridge, MA: Harvard University Press, 1993.
Huet 2004: Marie Hélène Huet, "Monstrous Medicine," in Laura Lunger Knoppers and Joan B. Landes, eds, *Monstrous Bodies/Political Monstrosities in Early Modern Europe*. Ithaca, NY: Cornell University Press, 2004, pp. 127–147.
Huygens 2018: Constantin Huygens, *Memoires*. Sydney: Wentworth Press, 2018.
Hazm 1994: Arthur John Arberry, transl., Ibn Hazm, The Ring of the Dove: A Treatise on the Art and Practice of Arab Love. Engl. transl. Arthur John Arberry,. London: Luzac Oriental, 1994.
Impey and MacGregor 1985: Oliver Impey and Arthur MacGregor, *The Origins of Museums: The Cabinet of Curiosities in Sixteenth and Seventeenth Century Europe*. Oxford: Oxford University Press, 1985.
Jackson 1996: Mark Jackson, *Newborn Child Murder: Women, Illegitimacy and the Courts in Eighteenth-century England*. Manchester: Manchester University Press, 1996.
Jacquart 2012: Danielle Jacquart, "De la faillibilité de l'art médical aux erreurs du praticien au début du XIVe siècle. Une imperceptible margei," in Mariacarla Gadebusch Bondio and Agostino Paravicini Bagliani, eds, *Errors and Mistakes: A Cultural History of Fallibility* (Micrologus' Library 49). Florence: SISMEL-Edizioni del Galluzzo, 2012, pp. 129–146.

Jacquart and Thommaset 1985: Danielle Jacquart and Claude Thommaset, "L'amour 'héroique' à travers le traité d'Arnaud de Villeneuve," in Jean Céard, Pierre Naudin and Michel Simonin, eds, *La folie et le* corps. Paris: Presses de l'École normale supérieure, 1985, pp. 143–158.

James 1902: William James, The *Varieties of Religious Experience*: *A Study in Human Nature*. New York, NY: Longmans, Green and Co., 1902.

James 1949: William James, *PRAGMATISM*: *A New Name for Some Old Ways of Thinking*. New York, NY: Longmans, Green and Co., 1949.

James 1981: William James, *The Principles of Psychology*. Cambridge, MA: Harvard University Press, 1981.

Jensen 2007: Meg Jensen, "Tradition and Revelation. Moments of Being in Virginia Woolf's Major Novels," in Morag Shiach, ed., *The Cambridge Companion to the Modernist Novel*. Cambridge: Cambridge University Press, 2007, pp. 112–125.

Jimenez de la Espada 1892: Marcos Jimenez de la Espada, *Una Antigualla Peruana*, Madrid: M. Gines Hernández, 1892.

Jonas 1985: Hans Jonas, *Technik, Medizin und Ethik: Zur Praxis des Prinzips Verantwortung*. Frankfurt am Main: Insel Verlag, 1985.

Jonas 1979: Hans Jonas: Das Prinzip Verantwortung: Versuch einer Ethik für die technologische Zivilisation. Frankfurt am Main: Insel Verlag, 1979.

Jonson 2001: Helen Ostovich, ed: *Ben Jonson, Every Man out of his Humour*. Manchester: Manchester University Press, 2001.

Kaiser and Schenkel 2019: Christian Kaiser and Peter M. Schenkel, "Pietro d'Abano über die Bedeutung der theoretischen Wissenschaften für den Arzt (mit einer kritischen Edition und Übersetzung der *Differentia prima* des *Conciliator)*," in Christian Kaiser, Leo Maier and Oliver Maximilian Schrader, eds, *Die nackte Wahrheit und ihre Schleier. Weisheit und Philosophie in Mittelalter und Früher Neuzeit – Studien zum Gedenken an Thomas Ricklin*. Münster: Aschendorff, 2019, pp. 191–247.

Kandel 2019: Eric Kandel, *The Disordered Mind*. New York, NY: Farrar, Straus, and Giroux, 2019.

Kannen 2012: Victoria Kannen, "My body speaks to them': Instructor Reflections on the Complexities of Power and Social Embodiments," *Teaching in Higher Education* 17 (6) (2012), pp. 637–648.

Kantorowicz 1957: Ernst H. Kantorowicz, *The Prince's Two Bodies: A Study in Mediaeval Political Theology*. Princeton, NJ: Princeton University Press, 1957.

Kaplan 2005: E. Ann Kaplan, *Trauma Culture: The Politics of Terror and Loss in Media and Literature*. New Brunswick, NJ: Rutgers University Press, 2005.

Kaplan and Berman 2010: Stephen J. Kaplan and Marc G. Berman, "Directed Attention as a Common Resource for Executive Functioning and Self-Regulation," *Perspectives on Psychological Science* 5 (1) (2010), pp. 43–57.

Kaufman and Morgan 2005: Sharon Kaufman and Lynn Morgan, "The Anthropology of the Beginnings and Ends of Life," *Annual Review of Anthropology* 34 (2005), pp. 317–341.

Kellenbenz 1958: Hermann Kellenbenz, "Sephardim an der unteren Elbe. Ihre wirtschaftliche und politische Bedeutung vom Ende des 16. bis zum Beginn des 18. Jahrhunderts," *Vierteljahrschrift für Sozial- und Wirtschaftsgeschichte*, 40 (1958), pp. 323–331.

Kemp 2010: Martin Kemp, "Style and Non-Style in Anatomical Illustration: From Renaissance Humanism to Henry Grey," *Journal of Anatomy* 216 (2) (2010), pp. 192–208.

Kenney 1992: Sally Jane Kenney, *For Whose Protection? Reproductive Hazards and Exclusionary Policies in the United States and Britain*. Ann Arbor, MI: University of Michigan Press, 1992.

Kirakofe and Maar 2002: James B. Kirakofe and John S. Maar, "The Great Andean Epidemic of 1524–5: Smallpox or Bartonellosis?" An unpublished paper presented at symposium "Disease and Disaster in Pre-Columbian and Colonial America." Sponsored by the *Interamerican Institute for Advanced Studies in Cultural History Conference*, Washington, D.C. 2002.

Kirakofe and Maar 2008: James B. Kirakofe and John S. Maar, "Marching to Disaster: The Catastrophic Convergence of Inca Imperial Policy, Sand Flies, and el Niño in the 1524 Andean Epidemic," in Daniel H. Sandweiss and Jeffrey Quilter, eds, *El Niño. Catastrophism, and Culture Change in Ancient America*. Washington D.C.: Dumbarton Oaks Research Library and Collection and Harvard University Press, 2008, pp. 145–186.

Klestinec 2011: Cynthia Klestinec, *Theaters of Anatomy: Students, Teachers, and Traditions of Dissection in Renaissance Venice*. Baltimore, MD: Johns Hopkins University Press, 2011.

Knapton 2016: Sarah Knapton, "Leonardo da Vinci Paintings analysed for DNA to Solve Grave Mystery," *Telegraph*, May 5, 2016. https://www.telegraph.co.uk/science/2016/05/05/leonardo-da-vinci-paintings-analysed-for-dna-to-solve-grave-myst/ Accessed June 20, 2021

Knoppers and Landes 2004: Laura Lunger Knoppers and Joan B. Landes, eds, *Monstrous Bodies/ Political Monstrosities in Early Modern Europe*. Ithaca, NY: Cornell University Press, 2004.

Kontje 2006: Todd Kontje, "Exotic Heimat: Province, Nation, and Empire in Tho'as Mann's *Buddenbrooks*," *German Studies Review* 29 (3) (2006), pp. 495–514.

Korpiola and Lahtinen, 2015: Mia Korpiola and Anu Lahtinen, "Cultures of Death and Dying in Medieval and Early Modern Europe: An Introduction," in Mia Korpiola and Anu Lahtinen (eds), *Cultures of Death and Dying in Medieval and Early Modern Europe*. Helsinki: Helsinki Collegium for Advanced Studies, 2015, pp. 1–31.

Kosmin 2021: Jennifer Kosmin, *Gender, Authority, and Midwifery in Early Modern Italy: Contested Deliveries*. Abingdon and New York, NY: Routledge, 2021.

Kolb 1984: David Kolb, *Experiential Learning: Experience as the Source of Learning and Development*. Englewood Cliffs, NJ: Prentice-Hall, 1984.

Koudounaris 2011: Paul Koudounaris, *The Empire of Death: A Cultural History of Ossuaries and Charnel Houses*. New York, NY: Hudson and Thames, 2011.

Krell 2007: David Farrell Krell, "Two Apothecaries: Novalis and Derrida," *Studies in Romanticism* 46 (3) (2007), pp. 289–309.

Kristeva, Moro, Odemark and Engebretsen 2018: Julia Kristeva, Marie Rose Moro, John Odemark, and Eivind Engebretsen, "Cultural Crossings of Care: An Appeal to the Medical Humanities," *Medical Humanities* 44 (2018), pp. 55–58.

Kusukawa 2011: Sachiko Kusukawa, *Picturing the Book of Nature: Image, Text, and Argument in Sixteenth-Century Human Anatomy and Medical Botany*. Chicago, IL: University of Chicago Press, 2011.

Lacan 1991: Jacques Lacan, *Le séminaire, Livre VIII. Le transfert*. Paris: Seuil, 1991.

Laing 1959: R. D. Laing, *The Divided Self: An Existential Study in Sanity and Madness*. Harmondsworth: Pelican, 1959.

Lanfranco 1995: Cesar Lofranco, "Notas biograficas de Daniel A. Carrion 1885–1995. Centenario de su inmolacion," Diagnostico 16 (1995), pp. 88–96.

Lakoff and Johnson 1980: George Lakoff and Mark Johnson, *Metaphors We Live By*. Chicago, IL: University of Chicago Press, 1980.

Lakoff and Johnston 1999: George Lakoff and Mark L. Johnson, *Philosophy in the Flesh: The Embodied Mind and Its Challenge to Western Thought*. New York, NY: Basic Books, 1999.

Lange 2016: Steven Lange, "Is Medical Humanism a Humanism?" *Training: The Agora of the Medical Student Community* 2 (2016), p. 2.

Lantos 2001: John D. Lantos, *The Lazarus Case: Life-and-Death Issues in Neonatal Intensive Care*. Baltimore, MD: Johns Hopkins University Press, 2001.

Lanza, Azzaroli Puccetti and Poggesi 1979: Benedetto Lanza, Maria Luisa Azzaroli Puccetti, and Marta Poggesi, *Le Cere Anatomiche della Specola*. Florence: Arnaud, 1979.
Laqueur 2015. Thomas F. Laqueur, *The Work of the Dead: A Cultural History of Mortal Remains*. Princeton, NJ: Princeton University Press, 2015.
Lastres 1956: Juan Lastres, "Encefalopatías de la verruga peruana," *Anales de la Facultad de Medicina* 39 (1956), pp. 73–104.
Launer 2002: John Launer, *Narrative-Based Primary care: A Practical Guide*. Abingdon: Radcliffe Medical Press, 2002.
Laurenza 2003: Domenica Laurenza, *La Ricerca dell'armonia. Rappresentazioni anatomiche nel Rinascimento* (Biblioteca di Nuncius, Studi e testi 47). Florence: Leo S. Olschki, 2003.
Lavoreira 1901: Daniel Eduardo Lavoreira, "El arte de curar entre los antiguos peruanos," *Anales de la Universidad Nacional Mayor de San Marcos* 29 (1901), pp. 159–255.
Lawrence 2012: Randee Lipson Lawrence, "Intuitive Knowing and Embodied Consciousness," *New Directions for Adult and Continuing Education* 134 (2012), pp. 5–13.
Laws 2015: Rebecca Laws et al., "Changes in Kidney Function Among Nicaraguan Sugarcane Workers." *International Journal Occupational Environmental Health* 21 (3) (2015), pp. 241–250.
Lazzarini 2011: Elena Lazzarini, "Wonderful Creatures: Early Modern Perceptions of Deformed Bodies," *Oxford Art Journal* 34 (3) (2011), pp. 415–431.
Lazzeri 2016a: Davide Lazzeri et al., "Osteoarthritis in the Hands of Michelangelo Buonarroti," *Journal of the Royal Society of Medicine* 109 (5) (2016), pp. 180–183.
Lazzeri 2016b: Davide Lazzeri et al., "The Mystery of Michelangelo Buonarroti's Goiter," *Rambam Maimonides Medical Journal* 7 (1) (2016), pp. 1–5.
Leavy 2020: Patricia Levy, *Method Meets Art. Arts-Based Research Practice*. New York, NY: Guilford Press, 2020.
Leder 1992: Drew Leder, *The Body in Medical Thought and Practice*. Boston: Kluwer Academic Publishers, 1992.
Leeming 2019: William Leeming, "Fortunio Liceti's Strategic Use of *Lusus Naturae* in *De Monstris* (1634–1665) and the Self-Assured Semiology of Naturalized Early Modern Science," in Justyna Jajszczok and Aleksandra Musia, eds., *The Body in History, Culture, and the Arts*. London: Routledge, 2019, pp. 11–26.
Legacey 2019: Erin-Marie Legacey, *Making Space for the Dead: Catacombs, Cemeteries, and the Reimagining of Paris, 1780–1830*. Ithaca, NY: Cornell University Press, 2019.
Lehmann and Stroux 2003: Paul Lehmann and Johannes Stroux, *Mittellateinisches Wörterbuch bis zum ausgehenden 13. Jahrhundert*, ed. Bayerische Akademie der Wissenschaften. Munich: Beck, 2003.
Leon 1951: Luis Leon, *Relacion cronologica del tifus exantematico en el Ecuador*. Quito: Imprenta de la Universidad, 1951.
Leone and Privitera 2004: Salvino Leone and Salvatore Privitera, eds., *Nuovo Dizionario di Bioetica*. Rome-Acireale: Città Nuova, 2004.
Lévinas 1980: Emmanuel Lévinas, *Totalità e Infinito. Saggio sull'esteriorità*. Milan: Jaca Book, 1980.
Lévinas 1991: Emmanuel Lévinas, *Entre Nous. Essais sur le penser-à-l'autre*. Paris: Grasset, 1991.
Lévi-Strauss 1962: Claude Lévi-Strauss, *Le pensée sauvage*. Paris: Plon, 1962.
Lewis 1967: Aubrey Julian Lewis, "The Psychopathology of Insight," in *Inquiries in Psychiatry: Clinical and Social Investigations*. New York, NY: Routledge, 1967, pp. 16–29.
Lewis 2016: Margaret Brannan Lewis, *Infanticide and Abortion in Early Modern Germany*. London and New York, NY: Routledge, 2016.
Linden 1999: David E. J. Linden, "Gabriele Zerbi's *De cautelis medicorum* and the Tradition of Medical Prudence," *Bulletin of the History of Medicine* 73(1) (1999), pp. 19–37.

Lingiardi and McWilliams 2017: Vittorio Lingiardi and Nancy McWilliams, eds., *Psychodynamic Diagnostic Manual. PDM-2*. New York, NY: Guilford Press, 2017.
Lingiardi 2018: Vittorio Lingiardi, *Diagnosi e destino*. Turin: Einaudi, 2018.
Lindquist 2018: Sherry C.M. and Asa Simon Mittman, *Medieval Monsters: Terrors, Aliens, Wonders*. New York, NY: The Morgan Library and Museum, 2018.
Lippi 2017: Lippi, Donatella, "Case study of Berengario da Carpi and Lorenzo de' Medici," *British Journal of Surgery* 104 (2017), pp. 1–5.
Long 2006: Kathleen Long, *Hermaphrodites in Renaissance Europe: Women and Gender in the Early Modern World*. London: Ashgate, 2006.
Long 2011: Pamela O. Long, *Artisan/Practitioners and the Rise of the New Sciences, 1400–1600*. Corvallis, OR: Oregon State University Press, 2011.
Long 2015: Pamela O. Long, "Trading Zones in Early Modern Europe," *Isis* 106 (4) (2015), pp. 840–847.
Lopez-de-Guimaraes, Vera-Guzmán, Menacho-López, Avila-Polo, Loarte-López (2017): D. Lopez-de-Guimaraes, J. Vera-Guzmán, Julio Menacho-López, Francisco Ávila-Polo, and Carlos Loarte-López, "Neurobartonelosis: 17 casos observados en Huaraz," *Acta medica peruana* 21 (1) (2017), pp. 8–15.
Lorenzi 2017: Rossella Lorenzi, "Leonardo da Vinci 'Relics' Discovered that Can Potentially Provide His DNA," *Seeker*, April 19, 2017. https://www.seeker.com/culture/history/leonardo-da-vinci-relics-discovered-that-can-potentially-provide-his-dna Accessed June 10, 2021.
Lovell 1992: George W. Lovell "Heavy Shadows and Black Night: Disease and Depopulation in Colonial Spanish America," *Annals of the Association of Medical Cartographers* 8 (23) (1992), pp. 426–443.
Lowes 1914: John Livingston Lowes, "The Lovers' Maladye of Heroes," *Modern Philology* 11 (1914), pp. 481–546.
Ludwig 1992: Arnold M. Ludwig, "Creative Achievement and Psychopathology: Comparison Among Professions," *American Journal of Psychotherapy* 46 (3) (1992), pp. 330–56.
Lummus 2017: David Lummus, "Placing Petrarch's Legacy: The Politics of Petrarch's Tomb and Boccaccio's Last Letter," *Renaissance Quarterly* 70 (2017), pp. 435–473.
Luppi 2010: Anna Luppi, "Per una breve *Historia Monstruorum*. Ragioni di una favolosa passione del Cinquecento," in Andrea Carlino, Roberto P. Ciardi, Anna Luppi, Annamaria Petroli Tofani, *L'anatomia tra arte e medicina: Lo studio del corpo nel tardo Rinascimento*. Milan: Silvana Editoriale, 2010, pp. 51–59.
Lüthy and Newman 1997: Christopher Lüthy and William R. Newman, "The Fate of Hylomorphism: Matter and Form in Early Modern Science," *Early Science and Medicine* 2 (1997), pp. 215–352.
Mack 2003: Peter Mack, "Humanist Rhetoric and Dialectic," in Jill Kraye, ed., *The Cambridge Companion to Renaissance Humanism*. Cambridge: Cambridge University Press, 2003, pp. 82–99.
Maclean 2002: Ian Maclean, *Logic, Signs, and Nature in the Renaissance: The Case of Learned Medicine*. Cambridge: University Press, 2002.
Maguiña and Gottuzzo 2001: Cesar Maguiña and E. Gotuzzo, "Bartonellosis new and old," *Infection Diseases Clinics of North America* 14 (2001), pp. 1–22.
Maguiña and Maguiña 1984: Cesar Maguiña Vargas and Teófilo Maguiña Cueva, "Nuevos aportes sobre la historia de la verruga peruana o enfermedad de Carrion," *Diagnostico* 13 (1984), pp. 47–51.
Maguiña Vargas, Peña, and Ponce 2008: Ciro Maguiña Vargas, Renzo Peña, Carlos Ponce, et al., "Manifestaciones clinicas y complicaciones de la fase aguda de bartonelosis o fiebre de la Oroya en pacientes atendidos en el Hospital Nacional Cayetano Heredia," *Revista Medica Herediana*, 19 (2008), pp. 1–6.

Makunda, Moghbeli, Rizzo, Niepold and DeLisser 2019: Neha Makunda, Nazanin Moghbeli, Adam Rizzo, Suzannah Niepold, and Horace M. DeLisser, "Visual Art Instruction in Medical Education: A Narrative Review," Medical Education. Online 24.1 (December 2019). URL https://pubmed.ncbi.nlm.nih.gov/30810510/Accessed March 12, 2022

Mann 2017: Sarah Mann, "Artful Medicine," Fienberg School of Medicine, Northwestern University, Second Annual Conference, 2018. Focusing on Arts, Humanities to Develop Well-Rounded Physicians. Center for Bioethics and Medical Humanities, Association of American Medical Colleges. AAMC News, August 15, 2017.

Manning 1992: Rita C. Manning, *Speaking from the Heart. A Feminist Perspective on Ethics*. Lanham, MD: Rowman & Littlefield, 1992.

Mari et al 2006: Francesco Mari et al., "The Mysterious Death of Francesco I de' Medici and Bianca Capello: An Arsenic Murder?" *British Medical Journal* 333 (7582) (2006), pp. 1299–1301.

Marks 2021: Jay W. Marks, "Medical Definition of Hippocratic Oath," Medicine Net, 2021. https://www.medicinenet.com/hippocratic_oath/definition.htm Accessed February 22, 2022

Marpurg 1755: Friedrich Wilhelm Marpurg, *Historisch-Kritische Beyträge zur Aufnahme der Musik*. Notenbeisp: Schütz, 1755.

Marshall and Warren 1984: Barry Marshall and Robin Warren, "Unidentified curved bacilli in the stomach of patients with gastritis and peptic ulceration," *Lancet* 323 (1984), pp. 1311–1315.

Martin, Huck, and Hutton 1988: Luther H. Martin, Huck Gutman and Patrick H. Hutton eds., *Technologies of the Self: A Seminar with Michel Foucault*. Amherst: University of Massachusetts Press, 1988.

Martini 2021: Mariano Martini et al., "Michelangelo Buonarroti's Alleged Goiter: Anatomical Fact or Literary Misinterpretation?," *Endocrine Practice* 27 (12) (2021), pp. 1264–1265.

Martinotti 1911: Giovanni Martinotti, *L'insegnamento dell'anatomia in Bologna prima del secolo XIX*. Bologna: Azzoguidi, 1911.

Martinotti 1923: Giovanni Martinotti, "Testamento di M. Jacopo Barigazzi, o Berengario da Carpi, lettore di Chirurgia nello studio di Bologna," *Rivista di Storia delle scienze mediche e naturali* 14 (3–4) (1923), pp. 1–11.

Mattheson 1739: Johann Mattheson, *Der Vollkommene Capellmeister*. Hamburg: Christian Herold, 1739.

Maturana and Varela 1987: Humberto R. Maturana, and Francisco J. Varela 1987, *The Tree of Knowledge: The Biological Roots of Human Understanding*. Boston: Shambhala Publications, 1987.

Maurette 2018: Pablo Maurette, "The Organ of Organs: Vesalius and the Wonders of the Human Hand," *Journal of Medieval and Early Modern Studies* 48 (1) (2018), pp. 105–124.

McCaa 1994: Robert McCaa, "Spanish and Nahuatl views on smallpox and demographic catastrophe in the conquest of Mexico," *Journal of Interdisciplinary History* 25 (3) (1995), pp. 397–431.

McCaa, Nimlos and Hampe-Martinez 2004: Robert McCaa, Aleta Nimlos, and Teodoro Hampe-Martinez, "Why blame smallpox? The death of Inca Huayna Capac and the demographic destruction of Tawantinsuyo (Ancient Peru)," *Journal of Interdisciplinary History* 34 (4) (2004), pp. 1–38. www.hist.umn.edu/~rmccaa/aha 2004 Accessed February 10, 2022.

McClive 2016: Cathy McClive, *Menstruation and Procreation in Early Modern France*. London: Routledge, 2016.

McVaugh 1970: Michael R. McVaugh, "Arnald of Villanova," in Charles C. Gillispsie, ed., *Dictionary of Scientific Biography* vol. 1. New York, NY: Charles Scribner's Sons, 1970, pp. 289–291.

Méhul 1996: Etienne Nicholas Méhul, *Stratonice*, Opera comique in one act, 1792. Libretto by François Benoit Hoffman. CD Capella Coloniensis, William Christie, dir. Paris: Erato, 1996.

Melchiorre 2000: Virgilio Melchiorre, *I luoghi del comprendere* (Metafisica e storia della metafisica, 21). Milan: Vita e Pensiero, 2000.

Meli 1821: Domenico Meli, *Delle proprietà vitali dell'utero gravido e de' parti che avvengono dopo la morte della pregnante*. Milan: A.S. Brambilla, 1821.
Meredith 1999: Stephen C. Meredith, "Mortal Illness on the Magic Mountain," in Stephen D. Dowden, ed., *A Companion to Thomas Mann's Magic Mountain*. Columbia SC: Camden House, 1999.
Merrill and West 2009: Barbara Merrill and Linden West, *Using Biographical Methods in Social Research*. London: Sage, 2009.
Middelkoop, Noble, Wadum and Boos 1998: Norbert Middelkoop, Petria Noble, Morgen Wadum, and Ben Broos. *Rembrandt under the Scalpel: The Anatomy Lesson of Dr. Nicolaes Tulp Dissected*. The Hague: Mauritshuis, 1998.
Mieli 1921: Aldo Mieli, *Gli scienziati italiani*. vol. 1, part 1. Rome: Nardecchia, 1921.
Minkowski 1923: Eugène Minkowski, "Étude psycologique et analyse phénoménologique d'un cas de mélancolie schizophrénique," *Journal de psychologie normale et pathologique*, 20 (10) (1923), pp. 543–560.
Minkowski 1967: Eugène Minkowski, "Studio psicologico e analisi fenomenologica di un caso di melancolia schizofrenica (1923)," in Eugène Minkowski, Viktor Emil von Gebsattel, and Erwin W. Straus, eds., *Antropologia e psicopatologia*, Danilo Cargnello, ed., Milan: Bompiani, 1967, pp. 15–40.
Minkowski 1968: Eugène Minkowski, *Le temps vécu. Etudes phénoménologiques et psychopathologiques*. Neuchâtel: Delachaux et Nestlé, 1968.
Minnick, Anderson, Lima, Battisti, Lawyer, and Birties 2014: Michael F. Minnick, Burt E. Anderson, Amorce Lima, James M. Battisti, Phillip G. Lawyer, and Richard J. Birties, "Oroya Fever and Verruga Peruana: Batonellosis Unique to South America," *PloS Neglected Tropical Diseases* 8 (7) (2014), pp. 1–39.
Mishkin and Ungerleider 1982: M. Mishkin and L. G. Ungerleider, "Contribution of Striate Inputs to the Visuospatial Functions of Parieto-Preoccipital Cortex in Monkeys," *Behavioural Brain Research* 6 (1) (1982), pp. 57–77.
Montes-Santiago 2013: Julio Montes-Santiago, "The Lead-poisoned Genius: Saturnism in Famous Artists Across Five Centuries," *Progress in Brain Research* 203 (2013), pp. 223–240.
Monti 2000: Maria Teresa Monti, "Epigenesis of the Monstrous Form and Preformistic Genetics (Lémery-Winslow-Haller)," *Early Science and Medicine* 5 (1) (2000), pp. 3–32.
Morin 1970: Edgar Morin, *L'homme et la mort*. Paris: Seuil, 1970.
Morin 1999: Edgar Morin, *L'intelligence de la complexité*. Paris: L'Harmattan, 1999.
Moskwitz and Heim 2011: Andrew Moskwitz and Gerhard Heim, "Eugen Bleuler's *Dementia Praecox or the Group of Schizophrenias* (1911), A Centenary Appreciation and Reconsideration," *Schizophrenia Bulletin* 37 (3) (2011), pp. 471–479.
Mozelius 2017: Mozelius et al., "Motivating Factors and Tangential Learning for Knowledge Acquisition in Educational Games," *The Electronic Journal of E-Learning* 15(4) (2017).URL https://eric.ed.gov/?id=EJ1154724 Accessed March 13, 2022.
Mucillo 1994: Maria Mucillo,"Falcucci, Niccolò," in *Dizionario Biografico degli Italiani*, vol. 44. Rome: Instituto della Enciclopedia Italiana, (1994), pp. 401–404.
Mukunda et al 2019: Neha Mukinda, *Visual Art Instruction in Medical Education: A Narrative Review*. Published Online, November 2019. https://pubmed.ncbi.nlm.nih.gov/30810510/. Accessed February 5, 2022.
Muir 2005: Edward Muir, *Ritual in Early Modern Europe*. Cambridge: Cambridge University Press, 2005.
Muñoz-Sanchez 2006: Agustin Muñoz-Sanchez, "La gripe de Cristobal Colon. Hipótesis sobre una catástrofe ecológica," *Enfermedades Infecciosas y Microbiología Clínica* 24 (5) (2006), pp. 326–334.

Murphy 1997: Edmond Murphy, *The Logic of Medicine*. Baltimore, MD, and London: John Hopkins University Press, 1997.

Nadel and Camaioni 1993: Jacqueline Nadel and Luigia Camaioni, eds, *New Perspectives in Early Communicative Development*. London: Routledge, 1993.

Naghshineh et al 2008: Sheila Naghshineh, "Formal Art Observation Training Improves Medical Students' Visual Diagnostic Skills," in *Journal of General Internal Medicine* 23 (7) (2008), pp. 991–997.

Nancy 2017: Jean-Luc Nancy, *La custodia del senso. Necessità e resistenza della poesia*. Bologna: EDB, 2017.

Naso 1983: Irma Naso, *Medici e strutture sanitarie nella società tardo-medioevale* (Studi e ricerche storiche 18). Milan: Franco Angeli, 1983.

Natali and Gamba 1993–94: Arturo Natali and Antonio Gamba, "Per la storia della quinta vertebra lombare di Galileo Galilei conservata all'Università di Padova," in *Atti dell'Istituto Veneto di Scienze, Lettere ed Arti. Classe di scienze fisiche, matematiche e naturali*, 152 (1993–1994), pp. 63–85.

Nechepurenko 2021: Ivan Nechepurenko, "Joseph Brodsky Slept Here: The Great Poet's Cranky Neighbor Couldn't Care Less," *New York Times*, September 12, 2021, https://www.nytimes.com/2021/09/12/world/europe/joseph-brodsky-museum-russia.html?searchResultPosition=1 Accessed November 22 2021

Newman, Danziger, and Cohen 1987: Ruth G. Newman, Marie A. Danziger, Mark Cohen, (eds.), *Communication in Business Today*. Washington, D.C.: Heath and Company, 1987.

Niccoli 1980: Ottavia Niccoli, "'Menstruum quasi Monstruum': Parti Mostruosi e Tabù Mestruale nel '500," *Quaderni Storici* 15 (44) (1980), pp. 402–428.

Noddings 2013: Nel Noddings, *Caring. A Relational Approach to Ethics and Moral Education*. Berkeley, CA: University of California Press, 2013.

Northoff 2019: Georg Northoff, *La neurofilosofia e la mente sana. Imparare dal cervello malato*. Milan: Raffaello Cortina, 2019.

Nussbaum 2011: Martha Nussbaum, *Creating Capabilities: The Human Development Approach*. Cambridge, MA: Harvard University Press, 2011.

Nussbaum 2016: Martha C. Nussbaum, *Not for Profit: Why Democracy Needs the Humanities*. Princeton, NJ: Princeton University Press, 2016.

Nussbaum and Sen 1993: Martha Nussbaum and Amartya Sen, eds., *The Quality of Life*. Oxford: Oxford University Press, 1993.

Nutton 1988: Vivian Nutton, *From Democedes to Harvey: Studies in the History of Medicine from the Greeks to the Renaissance*. London: Variorum Reprints, 1988.

Nutton 1999: Vivian Nutton, "Representation and Memory in Renaissance Anatomical Illustration," in Fabrizio Meroi and Claudio Pogliano, eds., *Immagini per conoscere: dal Rinascimento alla rivoluzione scientifica*. Atti della giornata di Studio (Firenze, Palazzo Strozzi, 29 ottobre 1999) (Istituto nazionale di studi sul Rinascimento. Atti di convegni 21). Florence: Leo S. Olschki, 1999, pp. 61–80.

Nutton 2019: Vivian Nutton, "Renaissance Galenism, 1540–1640: Flexibility or an Increasing Irrelevance?," in Petros Bouras-Vallianatos and Barbara Zipser, eds., *Brill's Companion to the Reception of Galen*. Leiden: Brill, 2019, pp. 472–486.

Obladen 2011: Michael Obladen, "Historical Notes on Viability. Part 1: Measures of Viability," *Journal of Perinatal Medicine* 39 (2011), pp. 563–569.

Ochberg 2004: Frank Ochberg, "PTSD 101 for Journalists," *Giftfromwithin.org*, March 24, 2009. DOI: https://www.giftfromwithin.org/pdf/ptsd101.pdf Accessed March 10, 2022.

O'Gorman 1961: Edmundo O'Gorman, *The Invention of America: An Enquiry into the Historical Nature of the New World and the Meaning of Its History*. Bloomington, IN: Indiana University Press, 1961.

Orletti 2000: Franca Orletti, *La conversazione diseguale. Potere e interazione*. Rome: Carocci, 2000.

Orletti and Iovino 2018: Franca Orletti and Rossella Iovino, *Il parlar chiaro nella comunicazione medica. Tra etica e linguistica*. Rome: Carocci, 2018.

Ovaska 2016: Anna Ovaska, "Minds in Crises: Literary Representations of Psychotic Experiences," in Elizabeth Kovach, Ansgar Nünning and Imke Polland, eds., *Literature and Crises: Conceptual Explorations and Literary Negotiations*. Trier: Wissenschaftlicher Verlag Trier, 2016, pp. 159–73.

Paganelli 2013: Leonardo Paganelli, "Ezra Pound in Rimini," *Linguistics and Literature Studies* 1 (1) (2013), pp. 43–45.

Palisca 1985: Claude Palisca, *Humanism in Italian Renaissance Musical Thought*. New Haven, CT: Yale University Press, 1985.

Palmer 1979: Richard Palmer, "Physicians and Surgeons in Sixteenth-Century Venice," *Medical History* 23 (4) (1979), pp. 451–460.

Palmer 1985: Richard Palmer, "Medical Botany in Northern Italy in the Renaissance," *Journal of the Royal Society of Medicine* 78 (1985), pp. 149–157.

Pamo Reyna 1993: Oscar Pamo Reyna, "Algunas consideraciones medicas sobre las cronicas del Peru," *Medicina Herediana* 4 (1) (1993), pp. 33–47.

Pancino 1984: Claudia Pancino, *Il Bambino e l'acqua sporca: Storia dell'assistenza al parto dalle mammane alle ostetriche (secoli XVI-XIX)*. Milan: Franco Angeli, 1984.

Panksepp 1998: Jaak Panksepp, *Affective Neuroscience: The Foundations of Human and Animal Emotion*. New York, NY: Oxford University Press, 1998.

Paré 1982: Ambroise Paré, *Discours d'Ambroise Paré avec une table des plus notables matières continues esdits discours*. Paris: Gabriel Buon, 1581.

Engl. transl.: Ambroise Paré, *The Apologie and Treatises of Ambroise Paré, Containing the Voyages with many of His Writings upon Surgery*, Geoffrey Keynes, ed. London: Falcon, 1952.

Park 1985: Katharine Park, *Doctors and Medicine in Early Renaissance Florence*. Princeton NJ: Princeton University Press, 1985.

Park 1993: Katharine Park, "The Criminal and the Saintly Body: Autopsy and Dissection in Renaissance Italy," *Renaissance Quarterly* 47 91) (1993), pp. 1–33.

Park 1995: Katharine Park, "The Life of the Corpse: Division and Dissection in Late Medieval Europe," *Journal of the History of Medicine and Allied Sciences* 50 (1) (1995), pp. 111–132.

Park 1999: Katharine Park, "Natural Particulars: Medical Epistemology, Practice, and the Literature of Healing Springs," in Anthony Grafton and Nancy Siraisi, eds., *Natural Particulars: Nature and the Disciplines in Renaissance Europe*. Cambridge, MA: MIT Press, 1999, pp. 347–368.

Park 2022: Sowon S. Park, "Deixis and Dissociation: On the Adaptive Power of Dissociated 'I's," *Modern Fiction Studies* 68 (4) (2022, forthcoming).

Pastore 1998: Alessandro Pastore, *Il Medico in tribunale: La perizia medica nella procedura penale d'Antico Regime (secoli XVI-XVIII)*. Bellinzona: Casagrande, 1998.

Patron 1894: Pablo Patron, "La enfermedad mortal de Huayna Capac," *La Cronica Medica* 11.131 (1894), pp. 179–183.

Pavillon 1789: Nicholas Pavillon, *Istruzioni teologiche e morali per amministrare, e rilevare i sacramenti divise in due parti*, vol. 1. Siena: Luigi e Benedetto Bindi, 1789.

Panzanelli 2008: Roberta Panzanelli, *Ephemeral Bodies: Wax Sculpture and the Human Figure*. Los Angeles: Getty Research Institute, 2008.

Pease 1988: Franklin Pease, "Notas sobre una nueva edicion de la Suma y memorias de los Incas," *Historica* 12 (1988), pp. 183–193.

Pelling 1998: Margaret Pelling, *The Common Lot: Sickness, Occupations and the Urban Poor in Early Modern England*. London: Longman, 1998.

Pernick 1996: Martin S. Pernick, *The Black Stork: Eugenics and the "Death" of "Effective" Babies in American Medicine and Motion Pictures since 1915*. Oxford: Oxford University Press, 1996.

Perocco 1997: Daria Perocco, *Viaggiare e raccontare. Narrazione di viaggio ed esperienza tra Cinque e Seicento*. Alessandria: Edizioni dell'Orso, 1997.

Pesce 1955: Hugo Pesce, "Lepra en el Peru Precolombino," *Anales la Facultad de Medicina, Universidad Nacional Mayor de San Marcos* 38 (1955), pp. 48–64.

Pesenti 1978: Tiziana Pesenti, "'Professores chirurgie,' 'medici ciroici' e 'barbitonsores' a Padova nell'età di Leonardo Buffi da Bertapaglia († dopo il 1448)," *Quaderni per la Storia dell'Università di Padova* 11 (1978), pp. 1–38.

Pessoa 2006: Fernando Pessoa, *A Little Larger than the Entire Universe: Selected Poems*, Richard Zenith, transl. New York: Penguin, 2006, p. 16.

Pessoa 2019: "O Guardador de Rebanhos" in *Un corazón de nadie. Antología Poética*. Barcelona: Galaxia Gutenberg, 2019, p. 82.

Peters and Biesta 2008: Michael A. Peters and Gert Biesta, eds, *Derrida, Deconstruction, and the Politics of Pedagogy*. Frankfurt: Peter Lang, 2008.

***Peterson 2010**: Mary Peterson and Elizabeth Salvagio, "Figure-ground Perception," Scholarpedia* 5 (4) (2010), p. 4320.

Pinock 2005: Stephen Pinock, "Nobel Prize Winners Robin Warren and Barry Marshall," *Lancet* 366, Issue 9495 (2005), pp. 366, 1429.

Piot 1984: Pete Piot, et al., "Acquired Immunodeficiency Syndrome in a Heterosexual Population in Zaire," *Lancet* 324 (1984), pp. 65–69.

Piva 1991: Gino Piva, *Manuale pratico di tecnica pittorica*. Milan: Ultico Hoepli, 1991.

Pivano 2005: Fernanda Pivano, *Pagine Americane. Narrativa e poesia 1943–2005*. Milan: Frassinelli, 2005.

Plomin 2018: Robert Plomin, *Blueprint: How DNA Makes Us Who We Are*. Cambridge, MA: MIT Press, 2018.

Pogliano 2015: Claudio Pogliano, *Senso lato: Il tatto e la cultura occidentale*. Rome: Carocci, 2015.

Polo 1913: Jose Toribio Polo, *Apuntes sobre las epidemias del Perú*. Lima: F. Barrionuevo, 1913.

Poma 2014: Roberto Poma, "La médicine morale de Luigi Luisini. Le premier traitement médico-philosophique des passions de l'âme," in Mariacarla Gadebusch Bondio, ed., *Medical Ethics. Premodern Negotiations between Medicine and Philosophy* (Aurora. Schriften der Villa Vigoni 2). Stuttgart: Franz Steiner Verlag, 2014, pp.161–174.

Popper 1963: Karl Popper, *Conjectures and Refutations*. New York, NY: Routledge,1963.

Porras Barrenechea 1986: Raúl Porras Barrenechea, *Los Cronistas del Peru (1528 – 1650) y otros ensayos* (Biblioteca Clasicos del Peru). Lima: Editorial e Imprenta DESA, 1986.

Porter and Rousseau 1999: Roy Porter and George Sebastian Rousseau, *Gout. The Patrician Malady*. New Haven,CT: Yale University Press, 1999.

Poska 1998: Allyson M. Poska, *Regulating the People: The Catholic Reformation in Seventeenth Century Spain*. Series Cultures, Beliefs & Traditions: Medieval and Modern People 5. Leiden: Brill, 1998.

Powers 2017: Ron Powers, *No One Cares About Crazy People: The Chaos and Heartbreak of Mental Health in America*. New York, NY: Hachette Books, 2017.

Preti 2004: Cesare. Preti, s.v. "Ingrassia, Giovanni Filippo," *Dizionario Biografico degli Italiani*, vol. 62. Rome : Instituto della Enciclopedia Italiana (2004) (digital edition) https://www.treccani.it/enciclopedia/giovanni-filippo-ingrassia_%28Dizionario-Biografico%29/Accessed December 22, 2021.

Procacci 1965: Ugo Procacci, *La Casa Buonarroti a Firenze*. Milan: Electa, 1965.
Prosperi 2005: Adriano Prosperi, *Dare l'anima: Storia di un infanticidio*. Turin: Einaudi, 2005.
Prosperi 2013: Adriano Prosperi, *Delitto e perdono: La pena di morte nell'orizzonte mentale dell'Europa cristiana, XIV-XVIII secolo*. Turin: Einaudi, 2013.
Pundel 1969: J. Paul Pundel, *Histoire de l'opérationé: Etude historique de la césarienne dans la médecine, l'art et la littérature, les religions et la legislation*. Bruxelles: Presses Académiques Européennes, 1969.
Putti 1937: Vittorio Putti, *Berengario da Carpi: saggio biografico e bibliografico, seguito dalla traduzione del* "De fractura calvae sive cranei." Bologna: Cappelli, 1937.
Quinlan 2009: Sean M. Quinlan, "Monstrous Births and Medical Networks: Debates over Forensic Evidence, Generation Theory, and Obstetrical Authority in France, ca. 1780–1815," *Early Science and Medicine* 14 (5) (2009), pp. 599–629.
Rabins 2013: Peter Rabins, *The why of Things*. New York, NY: Columbia University Press, 2013.
Raimondi 1903: Carlo Raimondi, "Una lettera inedita di P. A. Mattioli a Gabriele Falloppio, " *Bullettino Senese di Storia Patria* 10 (1903), pp. 286–289.
Raimondi 1906: Carlo Raimondi, "Lettere di P. A. Mattioli ad Ulisse Aldrovandi," *Bullettino Senese di Storia Patria* 13 (1906), pp. 21–185.
Raoult, Reed, and Dittmar 2008: Didier Raoult, David Reed, Katharina Dittmar, et al., "Molecular Identification of Lice in Pre-Columbian Mummies," *The Journal of Infectious Diseases* 197 (4) (2008), pp. 535–543.
Recalcati 2011: Massimo Recalcati, *Il miracolo della forma*. Milan: Bruno Mondadori, 2011.
Recalcati 2016: Massimo Recalcati, *Il mistero delle cose*. Milan: Feltrinelli, 2016.
Redfield Jamison 1993: Kay Redfield Jamison, *Touched with Fire: Manic-depressive Illness and the Artistic Temperament*. New York, NY: The Free Press, 1993.
Regalado de Hurtado 2015: Liliana Regalado de Hurtado, "La expansion y el proceso sucesorio de los Incas segun Betanzos," in Francisco Hernandez and Rodolfo Cerron, eds., *Juan de Betanzos y el Tahuantinsuyo, Nueva edicion de la Suma y narracion de los Incas*, Lima: Fondo Editorial, Pontificia Universidad Catolica del Peru, 2015.
Reid and Linden 2014: Hazel Reid and Linden West, eds, *Constructing Narratives of Continuity and Change: A Transdisciplinary Approach to Researching Lives*. London: Routledge, 2014.
Ricci 1998: Giovanni Ricci, *Il principe e la morte: Corpo, cuore, effigie nel Rinascimento*. Bologna: Il Mulino, 1998.
Ricciardi 1884: E. Ricciardi, "Francesco Cortese," *Giornale di medicina militare* 32 (1) (1884), pp. 3–21.
Ricœur 2007: Paul Ricœur, *Etica e morale*. Brescia: Morcelliana, 2007.
Riddle 1997: John M. Riddle, *Eve's Herbs: A History of Contraception and Abortion in the West*. Cambridge, MA: Harvard University Press, 1997.
Riedel 2005: Stefan Riedel, "*Edward Jenner and the history of smallpox and vaccination*," *Proceedings (Baylor University. Medical Center)*, 2005 Jan; 18.1, pp. 21–25. https://www.ncbi.nlm.nih.gov/pmc/articles/PMC1200696/ Accessed April 20 2019.
Rinieri 1600: Valerio Rinieri, "Diarii delle cose più notabili seguite nella città di Bologna 1520 fino al 1613," Biblioteca Universitaria di Bologna, MS 434, vol. 4, c. 2v (1600), in Valeria Finucci, *The Prince's Body: Vincenzo Gonzaga and Renaissance Medicine*. Cambridge, MA: Harvard University Press, 2015, pp. 92 and 201.
Riva 2007: Alessandro Riva, ed., *Flesh & Wax: The Clemente Susini's Anatomical Models in the University of Cagliari*. Nuoro: Ilisso Edizioni, 2007.
Riva, Conti, Solinas and Loy 2010: Alessandro Riva, Gabriele Conti, Paola Solinas, and Francesco Loy, "The Evolution of Anatomical Illustration and Wax Modelling in Italy, from the 16th to Early 19th Centuries," *Journal of Anatomy* 216 (2010), pp. 209–222.

Rizzolatti and Craighero 2004: Giacomo Rizzolatti and Laila Craighero, "The Mirror-Neuron System," *Annual Review of Neuroscience* 27 (1) (2004), pp. 169–192.
Rizzolatti and Gnoli 2016: Giacomo Rizzolatti and Antonio Gnoli, *In te mi specchio. Per una scienza dell'empatia*. Milan: Rizzoli, 2016.
Rizzolatti and Sinigaglia 2006: Giacomo Rizzolatti and Corrado Sinigaglia, *So quel che fai. Il cervello che agisce e i neuroni specchio*. Milan: Raffaello Cortina, 2006.
Rizzolatti and Sinigaglia 2019: Giacomo Rizzolati and Corrado Sinigaglia, *Specchi nel cervello. Come comprendiamo gli altri dall'interno*. Milan: Raffaello Cortina, 2019.
Roberts 1997: Dorothy Roberts, *Killing the Black Body: Race, Reproduction, and the Meaning of Liberty*. New York, NY: Vintage, 1997.
Rogers 1951: Carl R. Rogers, *Client-Centred Therapy: Its Current Practice, Implications, and Theory*. Boston, MA: Houghton Mifflin, 1951.
Rogers 1966: Carl. R. Rogers, "Client-centered Therapy," in Silvano Arieti, ed., *American Handbook of Psychiatry*, vol. 3. New York, NY: Basic Books, 1966: pp. 183–200.
Rogers 1963: Jacques Rogers, *Les sciences de la vie dans la pensée française du XVIIIe siècle: la génération des animaux de Descartes à l'Encyclopédie*. Paris: Armand Colin, 1963.
Romano 2019: Cesare Romano, "Freud, Morelli e la nascita del paradigma indiziario in psicoanalisi," *Psicoterapia e scienze umane* 53 (2) (2019), pp. 261–280.
Rossi 1998: Sergio Rossi, ed., *Scienza e miracoli nell'arte del 600: Alle origini della medicina moderna*. Milan: Electa, 1998.
Rostworowski 2001: Maria Rostworowsky, *Obras Completas I. Pachacutec Inca Yupanki*. Lima: Instituto de Estudios Peruanos, 2001.
Rostworowski 2014: Maria Rostworowsky, *Obras Completas VIII. Historia del Tawantinsuyu*. Lima: Instituto de Estudios Peruanos, 2014.
Roth 1991: Philip Roth, *Patrimony. A True Story*. New York, NY: Simon & Schuster, 1991.
Roth 2000: Rachel Roth, *Making Women Pay: The Hidden Costs of Fetal Rights*. Ithaca, NY: Cornell University Press, 2000.
Rowe 1945: John H. Rowe, "Absolute Chronology in the Andes Area," *American Antiquity*, 10 (3) (1945), pp. 233–284.
Rowe 1978: John H. Rowe, "La Fecha de la Muerte de Wayna Qhapaq," *Historica* 2 (1978), pp. 83–88.
Rühli 2016: Frank Rühli et al., "Palaeopathology: Current Challenges and Medical Impact," *Clinical Anatomy* 29 (7) (2016), pp. 816–822.
Russell 1917: Bertrand Russell, *Mysticism and Logic and Other Essays*. Watford: Taylor Garnett Evans & Co., 1917, p. 30.
Saettone-León 2004: Arturo Saettone-León, "Verruga peruana," *Dermatologia Peruana* 14 (2) (2004), pp. 121–134.
Sahlins 2008: Marshall Sahlins, *The Western Illusion of Human Nature*. Chicago, IL: Prickly Paradigm Press, 2008.
Sarmiento de Gomboa 1942: Pedro Sarmiento de Gamboa, *Historia de los Incas*, Original published in 1572. Atlas. Buenos Aires: Emecé, 1942.
Sarraga and Sarraga 1997: Marian Sarraga and Ramón E. Sarraga, "Some Episodes of Sephardic History as Reflected in Epitaphs of the Jewish Cemetery in Altona," in Michael Studemund-Halévy, ed., *Die Sefarden in Hamburg*, vol. 2. Hamburg: Buske, 1997, pp. 661–720.
Sass 1992: Louis S. Sass, *Madness and Modernism: Insanity in the Light of Modern Art, Literature and Thought*. Cambridge, MA: Harvard University Press, 1992.
Savoia 2018: Paolo Savoia, "Learned Hands: Skills, Experience, and Knowledge in Sixteenth-Century Surgery," in Gabriele Matino and Cynthia Klestinec, eds., *Art, Faith, and Medicine in Tintoretto's Venice*. Venice: Marsilio, 2018, pp. 94–103.

Savoia 2019: Paolo Savoia, "Skills, Knowledge, and Status: The Career of An Early Modern Italian Surgeon," *Bulletin of the History of Medicine* 93 (1) (2019), pp. 27–54.

Schaffner 1993: Kenneth Schaffner, *Discovery and Explanation in Biology and Medicine*. Chicago, IL: University of Chicago Press, 1993.

Schlick 1939: Moritz Schlick, *Problems of Ethics*. New York, NY: Prentice-Hall Inc, 1939.

Schmitz-Esser 2020: Romedio Schmitz-Esser, *The Corpse in the Middle Ages: Embalming, Cremating, and the Cultural Construction of the Dead Body*. Turnout: Brepols, 2020.

Schopenhauer 1966: Arthur Schopenhauer, *Die Welt als Wille und Vorstellung*. Berlin: DeGruyter, 2014. (Engl. transl. E.F J. Payne, *The World as Will and Idea*. New York, NY: Dover Publications, 1966).

Schön 1984: Donald A. Schön, *The Reflective Practitioner: How Professionals Think in Action*. New York, NY: Basic Books, 1984.

Schön 1987: Donald A. Schön, *Educating the Reflective Practitioner: Toward a New Design for Teaching and Learning in the Professions*. San Francisco: Jossey-Bass, 1987.

Schön 1991: Donald A. Schön, ed., *The Reflective Turn*. New York: Teachers College Press, 1991.

Schizophrenia Bulletin 2006: Robert Harland and Gareth Owen (eds), "Phenomenology and Psychiatry for the 21st century," *Schizophrenia Bulletin* 33 (1) (2006).

Schütz 1993: Alfred Schütz, *Der sinnhafte Aufbau der sozialen Welt*. Frankfurt: Suhrkamp, 1993.

Sechehaye 1951: Marguerite Sechehaye, *Autobiography of a Schizophrenic Girl*. New York, NY: Signet, 1951 (rpt. 1971).

Sen 1999: Amartya Sen, *Development as Freedom*. New York, NY: Oxford University Press, 1999.

Sen 2005: Amartya Sen, "Human rights and capabilities," *Journal of Human Development* 6.2 (2005), pp. 151–166.

Sen 2010: Amartya Sen, *The Idea of Justice*. London: Penguin, 2010.

Serroni 2007: Simonetta Serroni, "I resti di Grazia Deledda sono trasferiti in cimitero," *La Nuova Sassari* 2 (2007), p. 27.

Shankar 2001: Ravi Shankar, "Medical Humanities," *Journal of Medical and Allied Sciences* 2.1 (2011), pp. 26–28.

Shirley and Hoeninger 1985: John W. Shirley and David Hoeniger, eds, *Science and the Arts in the Renaissance*. Washington D.C.: Folger Shakespeare Library, 1985.

Shils 1972: Edward Shils, *The Intellectuals and the Powers, and Other Essays*. Chicago, IL: University of Chicago Press, 1972.

Sigerist 1946: Henry E. Sigerist, "Bedside Manners in the Middle Ages. The Treatise *De cautelis medicorum* Attributed to Arnald of Villanova," *Quarterly Bulletin of Northwestern University Medical School* 20 (1946), pp. 136–143.

Siraisi 1990: Nancy Siraisi, *Medieval and Early Renaissance Medicine: An Introduction to Knowledge and Practice*. Chicago, IL: University of Chicago Press, 1990.

Siraisi 2001: Nancy Siraisi, *Medicine and the Italian Universities, 1250–1650* (Education and society in the Middle Ages and Renaissance 12). Leiden: Brill, 2001.

Siraisi 2007: Nancy Siraisi, *History, Medicine, and the Traditions of Renaissance Learning*. Ann Arbor, MI: University of Michigan Press, 2007.

Smith 2004: Smith, Pamela, *The Body of the Artisan: Art and Experience in the Scientific Revolution*. Chicago, IL: University of Chicago Press, 2004.

Sobel 1999: Dava Sobel, *Galileo's Daughter: A Historical Memoir of Science, Faith and Love*. New York, NY: Walker and Co, 1999.

Sontag 2003: Susan Sontag, *Regarding the Pain of Others*. New York, NY: Farrar, Straus and Giroux, 2003.

Sorbelli 1929: Albano Sorbelli, *Storia della stampa in Bologna*. Bologna: Zanichelli, 1929.

Spila 2013: Cristiano Spila, "'Freddo cener ne l'urna, e fiamme al core': Pietro Della Valle e l'orazione funebre," *Studi Secenteschi* 54 (2013), pp. 147–185.

Spinks 2015: Jennifer Spinks, *Monstrous Births and Visual Culture in Sixteenth Century Germany*. London and New York, NY: Routledge, 2015.

Steer and Colin-Jones 1975: Howard Steer and Colin-Jones David, "Mucosal Changes in Gastric Ulceration and their Response to Carbenoxolone Sodium." *Gut* 16 (1975), pp. 590–597.

Stone 2008: Alison Stone, "Being, Knowledge, and Nature in Novalis," *Journal of the History of Philosophy* 46. 1 (2008), pp. 141–64.

Stroppiana 1956: Luigi Stroppiana, "Il *De adventu Medici ad aegrotum* di anonimo salernitano," *Gesnerus* 13 (1956), pp. 85–96.

Stephanson and Wagner 2015: Raymond Stephanson and Darren N. Wagner, eds., *The Secrets of Generation: Reproduction in the Long Eighteenth Century*. Toronto: University of Toronto Press, 2015.

StileArte.it: "Le relique corporee di Canova; il corpo tagliato in tre parti. Mano e cuore a Venezia," *StileArte.it* May 6, 2018. https://www.stilearte.it/le-reliquie-corporee-di-canova-il-corpo-tagliato-in-tre-parti/Accessed March 3, 2021.

Stoch and Smythe 1963: Mavis Belle Stoch and Patrick Monrose Smythe, "Does Undernutrition During Infancy Inhibit Brain Growth and Subsequent Intellectual Development?," *Archives of Diseases in Childhood* 38 (1963), PP. 546–552.

Struhal 2017: Eva Struhal, "Who Can Read the Book of Nature? Early Modern Artists and Scientists in Dialogue," *Nuncius*, 32 (3) (2017), pp. 501–513.

Sutherland 2005: Romy Sutherland, "Commanding Waves: The Films of Peter Weir," *Senses of Cinema*, Issue 34, (2005). https://www.sensesofcinema.com/2005/great-directors/weir/ Accessed January 15, 2022

Synapsis 2017: Synapsis: A Health Humanities Journal. https://medicalhealthhumanities.com/ Accessed March 5, 2022

Szasz, 1960: Thomas Szasz, "The Myth of Mental Illness," *American Psychologist*, 15 (1960), pp. 113: 18. http://psychclassics.yorku.ca/Szasz/myth.htm Accessed March 8, 2022.

Tanaron 1774: Pietro Paolo Tanaron, *Il Chirurgo-Raccoglitore Moderno, che assiste le Donne nei parti*. Bassano: Remondini di Venezia, 1774.

Taylor 1873: Alfred Swaine Taylor, *The Principles and Practice of Medical Jurisprudence*, vol. II. Philadelphia, PA: Henry C. Lea, 1873.

Tejada, Vizcarra and Perez 2003: Abelardo Tejada, Hugo Vizcarra, Jorge Perez, et al., "Estudio Clinico Epidemiologico de Barotonelosis Humana en el Valle del Monzon, Huanuco Huamalies," *Anales de la Facultad de Medicina*, Universidad Nacional Mayor de San Marcos, 64 (2003), pp. 211–217.

Temple 1949–1950: Ella Dunbar Temple, "Los testamentos ineditos de Paullu Inca, Don Carlos y Don Melchor Carlos Inca," *Documenta* 2 (1949–50), p. 643.

Thompson 1999: Anne Thompson, *Critical Reasoning in Ethics. A Practical Introduction*. London, New York, NY: Routledge, 1999.

Timberg and Halperin 2012: Craig Timberg and Daniel Halperin, *Tinderbox: How the West Sparked the Aids Epidemic and How the World can Finally Overcome It*. New York, NY: The Penguin Press, 2012.

Tomada 2014: Francesco Tomada, *Portarsi avanti con gli addii*. Rimini: Raffaelli Editore, 2014.

Tommasini 1836: Giacomo Tommasini, *Raccolta completa delle opere mediche del professore Giacomo Tommasini*. Bologna: Olmo e Tiocchi, 1836.

Tronto 1994: Joan C. Tronto, *Moral Boundaries: A Political Argument for an Ethic of Care*. London: Routledge, 1994.

Uhrová 2013: Petra Uhrová et al., "Stature Estimation from Various Foot Dimensions Among Slovak Population," *Journal of Forensic Sciences* 58 (2) (2013), pp. 448–451.
van der Ploeg 2001: Irma van der Ploeg, *Prosthetic Bodies: The Construction of the Fetus and the Couple as Patients in Reproductive Technologies*. Dordrecht and Boston, MA: Kluwer Academic, 2001.
Varchi 1858: Benedetto Varchi, *De la generazione de' mostri*, in *Opere di Benedetto Varchi*. Trieste: Sezione Letterario-Artistica del Lloyd Austriaco, 1858.
Vasari, G. 1832–1838: Le opere di Giorgio Vasari, pittore e architetto aretino, parte seconda. Firenze: David Passigli e Soci, 1832–1838.
Vattimo and Rovatti 2010: Gianni Vattimo and Pier Aldo Rovatti, eds, *Il pensiero debole*. Milan: Feltrinelli, 2010.
Veatch 1981: Robert M. Veatch, *A Theory of Medical Ethics*. New York, NY: Basic Books, 1981.
Visentin, 2007: Michele Visentin, "La coppia di fatto del '500," *La Tribuna di Treviso*, Feb. 11, 2007, p. 45.
Vistanet.it 2019: "Accadde oggi: 20 giugno 1959, la salma di Grazia Deledda da Roma arriva in Sardegna," https://www.vistanet.it/cagliari/2019/06/20/accadde-oggi-20-giugno-59-la-salma-di-grazia-deledda-da-roma-arriva-in-sardegna/ Accessed March 5, 2022.
Von Wessemberg 1845: Ignaz Heinrich Von Wessemberg, *Die grossen Kirchenversammlungen des 15ten und 16ten Jahrhunderts in Beziehung auf Kirchenverbesserung geschichtlich und kritisch dargestellt: mit einleitender Übersicht der frühern Kirchengeschichte*, vol 2. Constanz: Glükher, 1845.
Wack 1990: Mary Frances Wack, *Lovesickness in the Middle* Ages. *The "Viaticum" and its Commentaries*. Philadelphia, PA: University of Pennsylvania Press, 1990.
Wade and Finger 2001: Nicholas J. Wade and Stanley Finger, "The Eye as an Optical Instrument: From Camera Obscura to Helmholtz' Perspective," *Perception* 30 (10) (2001), pp. 1157–1177.
Walker 1997: Janet Walker, "The Traumatic Paradox: Documentary Films, Historical Fictions, and Cataclysmic Past Events," *Signs*, 22 (4) (1997), pp. 803–825.
Walker 2005: Janet Walker, *Trauma Cinema. Documenting Incest and the Holocaust*. Oakland, CA: University of California Press, 2005.
Walsham 1999: Alexandra Walsham, *Providence in Early Modern England*. Oxford: Oxford University Press, 1999.
Walter and Klein 2003: Jennifer K. Walter and Eran P. Klein, eds., *The Story of Bioethics. From Seminal Works to contemporary Exploration*. Washington D.C.: Georgetown University Press, 2003.
Warren 2009: Adam Warren, "An Operation for Evangelization: Friar Francisco González Laguna, the Cesarean Section, and Fetal Baptism in Late Colonial Peru," *Bulletin of the History of Medicine* 83.4 (2009), pp. 647–75.
Watson 2010: Katherine D. Watson, *Forensic Medicine in Western Society: A History*. New York, NY, and London: Routledge, 2010.
Watson 2019: Katherine D. Watson, *Medicine and Justice: Medico-Legal Practice in England and Wales, 1700–1914*. London and New York, NY: Routledge, 2019, pp. 143–198.
Watson and Breedlove 2016: Neil V. Watson and S. Marc Breedlove, *The Mind's Machine. Foundations of Brain Behavior*. Sunderland, MA: Sinauer Associates, Inc., 2016, pp. 361–ff.
Watzlawick, Beavin Bavelas, and Jackson 1967: Paul Watzlawick, Janet Beavin Bavelas, and Don D. Jackson, *Pragmatics of Human Communication: A Study of Interactional Patterns, Pathologies and Paradoxes*. New York, NY: Norton, 1967.
Wear, French, Lonie 1985: Andrew Wear, Roger K. French, and Iain M. Lonie, eds, *The Medical Renaissance in the Sixteenth Century*. Cambridge: Cambridge University Press, 1985.
Wehling 2008: Martin Wehling, "Translational Medicine: Science or Wishful Thinking?," *Journal of Translational Medicine* 6 (31) (2008), article 31. https://translational-medicine.biomedcentral.com/articles/10.1186/1479-5876-6-31 Accessed March 10 2022.

Weijin Wang 2019: Esmé Weijin Wang, *The Collected Schizophrenias*. New York, NY: Penguin, 2019.
Weir 1994: Peter Weir, *Fearless*. Los Angeles, CA: Warner Bros., 1994.
Wessling 1994: Mary Nagle Wessling, "Infanticide trials and forensic medicine: Württemberg 1757–93," in Michael Clark and Catherine Crawford, eds, *Legal Medicine in History*. Cambridge: Cambridge University Press, 1994, pp. 117–144.
Whitman 2004: Walt Whitman, *The Complete Poems*. London: Penguin Books, 2004.
Wilks and Bettany 1892: Samuel Wilks and Bettany George Thomas, *A Biographical History of Guy's Hospital*. London: Ward, Loock, Bowden & Co, 1892.
Wilson 1993: Dudley Wilson, *Signs and Portents: Monstrous Births from the Middle Ages to the Enlightenment*. London: Routledge, 1993.
Wind 1968: Edgar Wind, *Arte e anarchia*. Milan: Adelphi, 1972.
Wittgenstein 1953: Ludwig Wittgenstein, *Philosophische Untersuchungen*. Oxford: Blackwell, 1953.
Wittkower 1942: Rudolf Wittkower, "Marvels of the East: A Study in the History of Monsters," *Journal of the Warburg and Courtauld Institutes* 5 (1942), pp. 159–197.
Woods 2009: Robert Woods, *Death before Birth: Fetal Health and Mortality in Historical Perspective*. Oxford: Oxford University Press, 2009.
World Health Organization (WHO) 2019: World Health Organization, *Measles in Europe: record number of both sick and immunized*. Geneva: World Health Organization, 2019. http://www.euro.who.int/en/media-centre/sections/press-releases/2019/measles-in-europe-record-number-of-both-sick-and-immunized.Accessed October 15, 2018.
World Medical Association, *Declaration of Geneva* (1948). www.wma.net/what-we-do/medical-ethics/declaration-of-geneva Accessed March 10, 2022
Xerez 1891: Francisco de Xerez, *Verdadera relación de la conquista del Perú*. Madrid: J. C. García, 1891.
Zanatta, Zambieri, Scattolin and Bonati 2016: Alberto Zanatta, Fabio Zambieri, Giuliano Scattolin, and Maurizio Rippa Bonati, "Occupational Markers and Pathology of the Castrato Singer Gaspare Pacchierotti (1740–1821)," *Nature*, June 28, 2016, Article number: 28463. https://www.nature.com/articles/srep28463.pdf Accessed March 9, 2022.
Zanatta, Zambieri, Scattolin, and Thiene 2016: Alberto Zanatta, Fabio Zampieri, Giuliano Scattolin, and Gaetano Thiene, "Phrenology between Anthropology and Neurology in a Nineteenth-Century Collection of Skulls," *History of Psychiatry* 27 (4) (2016), pp. 482–492.
Zavoli 2011: Sergio Zavoli, *Il Ragazzo che io fui*. Milan: Mondadori, 2011.
Ziegler 2014: Joseph Ziegler, "The Medieval Foundations of Renaissance Medical Ethics: The Case of Gabriele Zerbi's *De cautelis medicorum*," in Mariacarla Gadebusch Bondio, ed., *Medical Ethics. Premodern Negotiations between Medicine and Philosophy* (Aurora. Schriften der Villa Vigoni 2). Stuttgart: Franz Steiner Verlag, 2014, pp. 117–129.
Ziemendorff 2018: Stephan Ziemendorff, "El hallazgo de las momias reales incaicas en el Cusco en 1559," *Historia y Cultura* 27 (2018), pp. 201–241.

Index of Names

https://doi.org/10.1515/9783110788501-023

www.ingramcontent.com/pod-product-compliance
Ingram Content Group UK Ltd.
Pitfield, Milton Keynes, MK11 3LW, UK
UKHW052008051026
15624UKWH00004B/625

9 783110 78800